Pitman Research Notes in Mathematics Series

Submission of proposals for consideration
Suggestions for publication, in the form of outlines and representative samples, are invited by the Editorial Board for assessment. Intending authors should approach one of the main editors or another member of the Editorial Board, citing the relevant AMS subject classifications. Alternatively, outlines may be sent directly to the publisher's offices. Refereeing is by members of the board and other mathematical authorities in the topic concerned, throughout the world.

Preparation of accepted manuscripts
On acceptance of a proposal, the publisher will supply full instructions for the preparation of manuscripts in a form suitable for direct photo-lithographic reproduction. Specially printed grid sheets are provided and a contribution is offered by the publisher towards the cost of typing. Word processor output, subject to the publisher's approval, is also acceptable.

Illustrations should be prepared by the authors, ready for direct reproduction without further improvement. The use of hand-drawn symbols should be avoided wherever possible, in order to maintain maximum clarity of the text.

The publisher will be pleased to give any guidance necessary during the preparation of a typescript, and will be happy to answer any queries.

Important note
In order to avoid later retyping, intending authors are strongly urged not to begin final preparation of a typescript before receiving the publisher's guidelines and special paper. In this way it is hoped to preserve the uniform appearance of the series.

Longman Scientific & Technical
Longman House
Burnt Mill
Harlow, Essex, UK
(tel (0279) 26721)

Titles in this series

1 Improperly posed boundary value problems
A Carasso and A P Stone
2 Lie algebras generated by finite dimensional ideals
I N Stewart
3 Bifurcation problems in nonlinear elasticity
R W Dickey
4 Partial differential equations in the complex domain
D L Colton
5 Quasilinear hyperbolic systems and waves
A Jeffrey
6 Solution of boundary value problems by the method of integral operators
D L Colton
7 Taylor expansions and catastrophes
T Poston and I N Stewart
8 Function theoretic methods in differential equations
R P Gilbert and R J Weinacht
9 Differential topology with a view to applications
D R J Chillingworth
10 Characteristic classes of foliations
H V Pittie
11 Stochastic integration and generalized martingales
A U Kussmaul
12 Zeta-functions: An introduction to algebraic geometry
A D Thomas
13 Explicit *a priori* inequalities with applications to boundary value problems
V G Sigillito
14 Nonlinear diffusion
W E Fitzgibbon III and H F Walker
15 Unsolved problems concerning lattice points
J Hammer
16 Edge-colourings of graphs
S Fiorini and R J Wilson
17 Nonlinear analysis and mechanics: Heriot-Watt Symposium Volume I
R J Knops
18 Actions of fine abelian groups
C Kosniowski
19 Closed graph theorems and webbed spaces
M De Wilde
20 Singular perturbation techniques applied to integro-differential equations
H Grabmüller
21 Retarded functional differential equations: A global point of view
S E A Mohammed
22 Multiparameter spectral theory in Hilbert space
B D Sleeman
24 Mathematical modelling techniques
R Aris
25 Singular points of smooth mappings
C G Gibson
26 Nonlinear evolution equations solvable by the spectral transform
F Calogero
27 Nonlinear analysis and mechanics: Heriot-Watt Symposium Volume II
R J Knops
28 Constructive functional analysis
D S Bridges
29 Elongational flows: Aspects of the behaviour of model elasticoviscous fluids
C J S Petrie
30 Nonlinear analysis and mechanics: Heriot-Watt Symposium Volume III
R J Knops
31 Fractional calculus and integral transforms of generalized functions
A C McBride
32 Complex manifold techniques in theoretical physics
D E Lerner and P D Sommers
33 Hilbert's third problem: scissors congruence
C-H Sah
34 Graph theory and combinatorics
R J Wilson
35 The Tricomi equation with applications to the theory of plane transonic flow
A R Manwell
36 Abstract differential equations
S D Zaidman
37 Advances in twistor theory
L P Hughston and R S Ward
38 Operator theory and functional analysis
I Erdelyi
39 Nonlinear analysis and mechanics: Heriot-Watt Symposium Volume IV
R J Knops
40 Singular systems of differential equations
S L Campbell
41 N-dimensional crystallography
R L E Schwarzenberger
42 Nonlinear partial differential equations in physical problems
D Graffi
43 Shifts and periodicity for right invertible operators
D Przeworska-Rolewicz
44 Rings with chain conditions
A W Chatters and C R Hajarnavis
45 Moduli, deformations and classifications of compact complex manifolds
D Sundararaman
46 Nonlinear problems of analysis in geometry and mechanics
M Atteia, D Bancel and I Gumowski
47 Algorithmic methods in optimal control
W A Gruver and E Sachs
48 Abstract Cauchy problems and functional differential equations
F Kappel and W Schappacher
49 Sequence spaces
W H Ruckle
50 Recent contributions to nonlinear partial differential equations
H Berestycki and H Brezis
51 Subnormal operators
J B Conway

52 Wave propagation in viscoelastic media
F Mainardi
53 Nonlinear partial differential equations and their applications: Collège de France Seminar. Volume I
H Brezis and J L Lions
54 Geometry of Coxeter groups
H Hiller
55 Cusps of Gauss mappings
T Banchoff, T Gaffney and C McCrory
56 An approach to algebraic K-theory
A J Berrick
57 Convex analysis and optimization
J-P Aubin and R B Vintner
58 Convex analysis with applications in the differentiation of convex functions
J R Giles
59 Weak and variational methods for moving boundary problems
C M Elliott and J R Ockendon
60 Nonlinear partial differential equations and their applications: Collège de France Seminar. Volume II
H Brezis and J L Lions
61 Singular systems of differential equations II
S L Campbell
62 Rates of convergence in the central limit theorem
Peter Hall
63 Solution of differential equations by means of one-parameter groups
J M Hill
64 Hankel operators on Hilbert space
S C Power
65 Schrödinger-type operators with continuous spectra
M S P Eastham and H Kalf
66 Recent applications of generalized inverses
S L Campbell
67 Riesz and Fredholm theory in Banach algebra
B A Barnes, G J Murphy, M R F Smyth and T T West
68 Evolution equations and their applications
F Kappel and W Schappacher
69 Generalized solutions of Hamilton-Jacobi equations
P L Lions
70 Nonlinear partial differential equations and their applications: Collège de France Seminar. Volume III
H Brezis and J L Lions
71 Spectral theory and wave operators for the Schrödinger equation
A M Berthier
72 Approximation of Hilbert space operators I
D A Herrero
73 Vector valued Nevanlinna Theory
H J W Ziegler
74 Instability, nonexistence and weighted energy methods in fluid dynamics and related theories
B Straughan
75 Local bifurcation and symmetry
A Vanderbauwhede
76 Clifford analysis
F Brackx, R Delanghe and F Sommen
77 Nonlinear equivalence, reduction of PDEs to ODEs and fast convergent numerical methods
E E Rosinger
78 Free boundary problems, theory and applications. Volume I
A Fasano and M Primicerio
79 Free boundary problems, theory and applications. Volume II
A Fasano and M Primicerio
80 Symplectic geometry
A Crumeyrolle and J Grifone
81 An algorithmic analysis of a communication model with retransmission of flawed messages
D M Lucantoni
82 Geometric games and their applications
W H Ruckle
83 Additive groups of rings
S Feigelstock
84 Nonlinear partial differential equations and their applications: Collège de France Seminar. Volume IV
H Brezis and J L Lions
85 Multiplicative functionals on topological algebras
T Husain
86 Hamilton-Jacobi equations in Hilbert spaces
V Barbu and G Da Prato
87 Harmonic maps with symmetry, harmonic morphisms and deformations of metrics
P Baird
88 Similarity solutions of nonlinear partial differential equations
L Dresner
89 Contributions to nonlinear partial differential equations
C Bardos, A Damlamian, J I Díaz and J Hernández
90 Banach and Hilbert spaces of vector-valued functions
J Burbea and P Masani
91 Control and observation of neutral systems
D Salamon
92 Banach bundles, Banach modules and automorphisms of C*-algebras
M J Dupré and R M Gillette
93 Nonlinear partial differential equations and their applications: Collège de France Seminar. Volume V
H Brezis and J L Lions
94 Computer algebra in applied mathematics: an introduction to MACSYMA
R H Rand
95 Advances in nonlinear waves. Volume I
L Debnath
96 FC-groups
M J Tomkinson
97 Topics in relaxation and ellipsoidal methods
M Akgül
98 Analogue of the group algebra for topological semigroups
H Dzinotyiweyi
99 Stochastic functional differential equations
S E A Mohammed

100 Optimal control of variational inequalities
V Barbu
101 Partial differential equations and dynamical systems
W E Fitzgibbon III
102 Approximation of Hilbert space operators. Volume II
C Apostol, L A Fialkow, D A Herrero and D Voiculescu
103 Nondiscrete induction and iterative processes
V Ptak and F-A Potra
104 Analytic functions – growth aspects
O P Juneja and G P Kapoor
105 Theory of Tikhonov regularization for Fredholm equations of the first kind
C W Groetsch
106 Nonlinear partial differential equations and free boundaries. Volume I
J I Díaz
107 Tight and taut immersions of manifolds
T E Cecil and P J Ryan
108 A layering method for viscous, incompressible L_p flows occupying R^n
A Douglis and E B Fabes
109 Nonlinear partial differential equations and their applications: Collège de France Seminar. Volume VI
H Brezis and J L Lions
110 Finite generalized quadrangles
S E Payne and J A Thas
111 Advances in nonlinear waves. Volume II
L Debnath
112 Topics in several complex variables
E Ramírez de Arellano and D Sundararaman
113 Differential equations, flow invariance and applications
N H Pavel
114 Geometrical combinatorics
F C Holroyd and R J Wilson
115 Generators of strongly continuous semigroups
J A van Casteren
116 Growth of algebras and Gelfand–Kirillov dimension
G R Krause and T H Lenagan
117 Theory of bases and cones
P K Kamthan and M Gupta
118 Linear groups and permutations
A R Camina and E A Whelan
119 General Wiener–Hopf factorization methods
F-O Speck
120 Free boundary problems: applications and theory, Volume III
A Bossavit, A Damlamian and M Fremond
121 Free boundary problems: applications and theory, Volume IV
A Bossavit, A Damlamian and M Fremond
122 Nonlinear partial differential equations and their applications: Collège de France Seminar. Volume VII
H Brezis and J L Lions
123 Geometric methods in operator algebras
H Araki and E G Effros
124 Infinite dimensional analysis–stochastic processes
S Albeverio
125 Ennio de Giorgi Colloquium
P Krée
126 Almost-periodic functions in abstract spaces
S Zaidman
127 Nonlinear variational problems
A Marino, L Modica, S Spagnolo and M Degiovanni
128 Second-order systems of partial differential equations in the plane
L K Hua, W Lin and C-Q Wu
129 Asymptotics of high-order ordinary differenti equations
R B Paris and A D Wood
130 Stochastic differential equations
R Wu
131 Differential geometry
L A Cordero
132 Nonlinear differential equations
J K Hale and P Martinez-Amores
133 Approximation theory and applications
S P Singh
134 Near-rings and their links with groups
J D P Meldrum
135 Estimating eigenvalues with *a posteriori/a pri* inequalities
J R Kuttler and V G Sigillito
136 Regular semigroups as extensions
F J Pastijn and M Petrich
137 Representations of rank one Lie groups
D H Collingwood
138 Fractional calculus
G F Roach and A C McBride
139 Hamilton's principle in continuum mechanics
A Bedford
140 Numerical analysis
D F Griffiths and G A Watson
141 Semigroups, theory and applications. Volume
H Brezis, M G Crandall and F Kappel
142 Distribution theorems of L-functions
D Joyner
143 Recent developments in structured continua
D De Kee and P Kaloni
144 Functional analysis and two-point differential operators
J Locker
145 Numerical methods for partial differential equations
S I Hariharan and T H Moulden
146 Completely bounded maps and dilations
V I Paulsen
147 Harmonic analysis on the Heisenberg nilpote Lie group
W Schempp
148 Contributions to modern calculus of variation
L Cesari
149 Nonlinear parabolic equations: qualitative properties of solutions
L Boccardo and A Tesei
150 From local times to global geometry, control physics
K D Elworthy

200 Stochastic analysis, path integration and dynamics
D Elworthy
201 Riemannian geometry and holonomy groups
S Salamon
202 Strong asymptotics for extremal errors and polynomials associated with Erdös type weights
D S Lubinsky
203 Optimal control of diffusion processes
V S Borkar
204 Rings, modules and radicals
B J Gardner
205 Numerical studies for nonlinear Schrödinger equations
B M Herbst and J A C Weideman
206 Distributions and analytic functions
R D Carmichael and D Mitrović
207 Semicontinuity, relaxation and integral representation in the calculus of variations
G Buttazzo
208 Recent advances in nonlinear elliptic and parabolic problems
P Bénilan, M Chipot, L Evans and M Pierre
209 Model completions, ring representations and the topology of the Pierce sheaf
A Carson
210 Retarded dynamical systems
G Stepan
211 Function spaces, differential operators and nonlinear analysis
L Paivarinta
212 Analytic function theory of one complex variable
C C Yang, Y Komatu and K Niino
213 Elements of stability of visco-elastic fluids
J Dunwoody
214 Jordan decompositions of generalised vector measures
K D Schmidt
215 A mathematical analysis of bending of plates with transverse shear deformation
C Constanda
216 Ordinary and partial differential equations Vol II
B D Sleeman and R J Jarvis
217 Hilbert modules over function algebras
R G Douglas and V I Paulsen
218 Graph colourings
R Wilson and R Nelson
219 Hardy-type inequalities
A Kufner and B Opic
220 Nonlinear partial differential equations and their applications. College de France Seminar Volume X
H Brezis and J L Lions
221 Workshop on dynamical systems
E Shiels and Z Coelho
222 Geometry and analysis in nonlinear dynamics
H W Broer and F Takens
223 Fluid dynamical aspects of combustion theory
M Onofri and A Tesei
224 Approximation of Hilbert space operators. Volume I. 2nd edition
D Herrero
225 Surveys of some recent results in operator theo Volume III
J B Conway and B B Morrel
226 Local cohomology and localization
J L Bueso Montero, B Torrecillas Jover and A Verschoren
227 Sobolev spaces of holomorphic functions
F Beatrous and J Burbea
228 Numerical analysis. Volume III
D F Griffiths and G A Watson
229 Recent developments in structured continua. Volume III
D De Kee and P Kaloni
230 Boolean methods in interpolation and approximation
F J Delvos and W Schempp
231 Further advances in twistor theory, Volume 1
L J Mason and L P Hughston
232 Further advances in twistor theory, Volume 2
L J Mason and L P Hughston
233 Geometry in the neighborhood of invariant manifolds of maps and flows and linearization
U Kirchgraber and K Palmer
234 Quantales and their applications
K I Rosenthal

D De Kee (Editor)

Université de Sherbrooke, Canada

and

P N Kaloni (Editor)

University of Windsor, Canada

Recent developments in structured continua VOLUME II

Longman Scientific & Technical

Copublished in the United States with
John Wiley & Sons, Inc., New York

Longman Scientific & Technical,
Longman Group UK Limited,
Longman House, Burnt Mill, Harlow
Essex CM20 2JE, England
and Associated Companies throughout the world.

Copublished in the United States with
John Wiley & Sons, Inc., 605 Third Avenue, New York, NY 10158

First published 1990

AMS Subject Classification: 76A 76D 76L 76T 76Z 70D 73B 73F

ISSN 0269-3674

British Library Cataloguing in Publication Data
Recent developments in structured continua – Vol. 2.
1. Fluids. Mechanics. Mathematics
I. De Kee, Daniel II. Kaloni, P.N. (Purna N)
523′.001′51

ISBN 0-582-05823-6

Library of Congress Cataloging-in-Publication Data
(Revised for vol. 2)
Recent developments in structured continua.
(Pitman research notes in mathematics series ; 143, 229)
Includes bibliographies and indexes.
1. Fluid dynamics 2. Transport theory.
I. De Kee, Daniel. II. Kaloni, P.N. III. Series.
QA911.R4 1986 532′.05 86-13264
ISBN 0-470-20364-1 (v. 1)
ISBN 0-470-21587-9 (v. 2)

Printed and bound in Great Britain
by Biddles Ltd, Guildford and King's Lynn

Contents

Preface

This collection of papers contains the full text of the invited lectures presented at the conference on "Recent Developments in Structured Continua II", held at the University of Sherbrooke, May 23-25, 1990. Recent rapid advances in the fields of transport phenomena, polymer mechanics, flow properties of biological fluids, suspension rheology, extended thermodynamics, diffusion, liquid crystals, mixture theories and the structured fluid theories clearly indicated that such a meeting was both desirable and timely. We hope that the present collection makes a significant contribution in the areas mentioned above. The technical content of the papers is also roughly divided into the broad categories stated above.

In chapter 1, Goldsmith et al. discuss some of their recent results in predicting physical factors governing pletelet aggregation. Aggregation measurement in rich plasma and in whole blood are discussed in detail. In chapter 2, Goddard and Bashir review and extend the century old Reynolds theory of dilatancy. Next, Larson obtains expressions for the birefringence and stress tensors, in nematic polymers via the computation of orientational distribution functions. In chapter 4, Stastna et al. deal with the problem of diffusion. Continuum mechanics ideas and sorption kinetics are presented.

In chapter 5, Hill and Soane discuss the flow properties of dispersions of rodlike particles in viscoelastic fluids. Rheological models for rigid and wormlike macromolecules are then treated by Carreau et al. Chapter 7 contains a brief discussion on the measurement of nonlinear viscoelastic properties of polymeric liquids.

Chapters 8 and 9 deal with the solutions of some problems in simple fluid theory. While Kaloni and Siddiqui treat the problem of generalized helical flow, Siginer embarks upon the solution of parallel and orthogonal superposition of pure and oscillatory shear flows. In chapter 10, Walters et al. discuss the influence of polymer conformation on the rheological properties of aqueous polymer solutions. In the next chapter, Schümmer et al. communicate rheological problems affecting jets.

In chapter 12, James reviews the state-of-the-art in extensional viscosity measurement. The next three chapters deal with the general theories of solid-fluid mixtures. Rajagopal et al. derive the appropriate constitutive equations, discuss boundary condition problems and provide solutions for homogeneous and inhomogeneous deformations of non-linear elastic solids. Passman and Drew on the other hand propose a simple multicomponent fluid theory with some physical bias. In the last chapter, Grmela presents an interesting interplay of thermodynamics and rheology.

Our thanks go to the invited speakers for submitting the material for this volume. We would like to acknowledge the financial support received from the Dean of the faculty of Applied Science of the University of Sherbrooke and of CERAM SNA. Finally, we wish to record our special thanks to Mrs. Louise Chapdelaine for excellent typing.

D. De Kee
P.N. Kaloni
January 15, 1990

List of Contributors

Dr. Y.M. Bashir
Department of Chemical Engineering
University of Southern California
Los Angeles, U.S.A.

Dr. C. Becker
Institut für Verfahrenstechnik
R.W.T.H. Aachen
D-5100 Aachen

Dr. D.N. Bell
McGill University
Montreal, Canada

Mr. A.Q. Bhatti
Department of Mathematics
University College of Wales
Aberystwyth, U.K.

Professor P.J. Carreau
Department of Chemical Engineering
Ecole Polytechnique
Montreal, Canada

Professor J.M. Dealy
Department of Chemical Engineering
McGill University
Montreal, Canada

Professor D. De Kee
Department of Chemical Engineering
University of Sherbrooke
Sherbrooke, Canada

Professor D.A. Drew
Rensselaer Polytechnic Institute
Troy, New York, U.S.A.

Mr. J.M. Ekmann
Pittsburgh Energy Technology Center
Pittsburgh, Pennsylvania, U.S.A.

Professor J.D. Goddard
R.J. Fluor Professor of Chemical Engineering
University of Southern California
Los Angeles, U.S.A.

Professor H.L. Goldsmith
Montreal General Hospital
McGill University
Montreal, Canada

Dr. M. Grmela
Department of Chemical Engineering
Ecole Polytechnique
Montreal, Canada

Dr. B. Harrison
Department of National Defence
D.R.E.O.
Ottawa, Canada

Dr. D.A. Hill
Center for Advanced Materials
Lawrence Berkeley Laboratory
Berkeley, California, U.S.A.

Professor D.F. James
Department of Mechanical Engineering
University of Toronto
Toronto, Canada

Professor P.N. Kaloni
Department of Mathematics
University of Windsor
Windsor, Canada

Dr. R. Larson
AT&T Bell Laboratories
Murray Hill, New Jersey, U.S.A.

Dr. M. Massoudi
Pittsburgh Energy Technology Center
Pittsburgh, Pennsylvania, U.S.A.

Dr. N. Mori
Department of Mechanical Engineering
Nara University
Nara, Japan

Dr. S.L. Passman
Pittsburgh Energy Technology Center
Pittsburgh, Pennsylvania, U.S.A.

Professor K.R. Rajagopal
Department of Mechanical Engineering
University of Pittsburgh
Pittsburgh, Pennsylvania, U.S.A.

Mr. A. Rollin
Department of Chemical Engineering
Ecole Polytechnique
Montreal, Canada

Professor P. Schümmer
Institut für Verfahrenstechnik
R.W.T.H. Aachen
D-5100 Aachen

Dr. A. Siddiqui
Department of Mathematics
Pennsylvania State University
York, Pennsylvania, U.S.A.

Professor A. Siginer
Department of Mechanical Engineering
Auburn University
Auburn, Alabama, U.S.A.

Professor D.S. Soane
Department of Chemical Engineering
University of California, Berkeley
Berkeley, California, U.S.A.

Mrs. S. Spain
McGill University
Montreal, Canada

Dr. J. Stastna
Department of Chemical Engineering
University of Sherbrooke
Sherbrooke, Canada

Dr. H.G. Thelen
Institut für Verfahrenstechnic
R.W.T.H. Aachen
D-5100 Aachen

Professor K. Walters
Department of Mathematics
University College of Wales
Aberystwyth, U.K.

Professor A.S. Wineman
Department of Mechanical Engineering
University of Michigan
Ann Arbor, Michigan, U.S.A.

H.L. GOLDSMITH, D.N. BELL AND S. SPAIN

Physical factors governing blood platelet aggregation in sheared suspensions

1. INTRODUCTION

Although outnumbered 10:1 by the red blood cell, and occupying only 0.3% of the particulate phase in blood, the platelet plays a very important role in haemostasis and thrombosis in the circulation. It does so by virtue of its ability to become activated both by substances in the plasma and by components of the blood vessel wall. Once activated, the cell changes shape and receptors on the membrane are exposed which allow circulating adhesive proteins to bind to them and form cross-bridges between activated cells and between cells and receptors on cells of the vessel wall. This process leads to the formation of platelet aggregates and the adhesion of such aggregates to the vessel wall, the latter being important in the plugging of holes in the vessel wall and the prevention of bleeding, as in haemostasis.

1.1 Haemostasis and Thrombosis. Haemostasis is a physiological defensive response to the injury of blood vessels whereby blood loss is prevented and the injured vessel is repaired. In contrast, thrombosis represents a pathological process which has been called "haemostasis in the wrong place" because the normal haemostatic plug is largely extravascular, while a thrombus, as illustrated in Fig. 1, is intravascular, making blood flow abnormal within the blood vessel, rather than preventing escape of blood out of the vessel. More than 100 years ago, Virchow [50] correctly enunciated the triad of factors involved in haemostasis and thrombosis: abnormalities of the vessel wall, alterations of blood flow, and changes in the composition of the blood. While haemostatic plug formation is always initiated by vessel damage, the initiating stimulus for thrombosis may involve intravascular activation of the cells and/or substances within the suspending plasma. A thrombus is an intravascular deposit consisting of a variable mixture of platelet aggregates, an intermeshing fibrin network and trapped white and red blood cells, depending on the conditions of formation.

During haemostasis, events occur which result in the formation of a platelet plug. When a blood vessel is damaged, the vessel constricts, blood is lost to the extravascular space, and platelets rapidly adhere to the subendothelial matrix, in particular to collagen fibres, and aggregate to form an unstable platelet plug. When fibrin consolidates the plug, trapping within it many white cells and red cells, the unstable platelet plug, a red thrombus, becomes stable. By contrast, in arterial thrombosis, a white or platelet

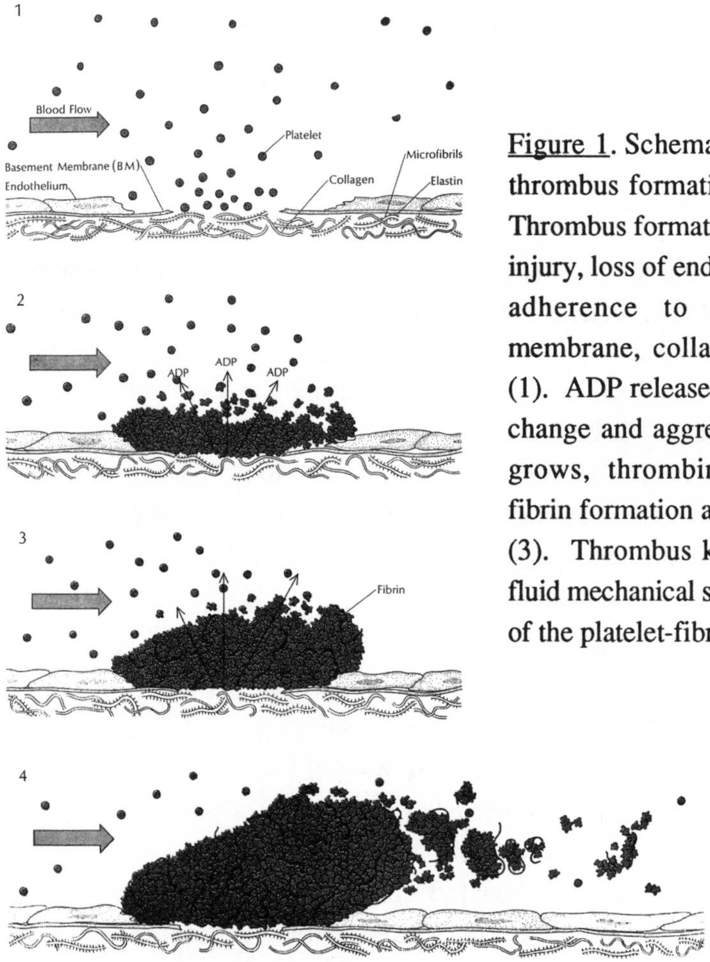

Figure 1. Schematic representation of thrombus formation in a blood vessel. Thrombus formation begins with vessel injury, loss of endothelium, and platelet adherence to exposed basement membrane, collagen and microfibrils (1). ADP release fosters platelet shape change and aggregation (2). As mass grows, thrombin generation causes fibrin formation and more ADP release (3). Thrombus keeps growing unless fluid mechanical stresses lead to breakup of the platelet-fibrin mass. (From [36]).

thrombus results, consisting essentially of aggregated platelets stabilized by a fibrin network (Fig. 1). When a thrombus is localized at a site, platelet aggregates (emboli) can break off and circulate intravascularly or deposit in the microcirculation.

1.2. Shape Change and Aggregation. There are several stimuli that can lead to platelet activation. In many cases, notably with adenosine diphosphate (ADP), platelet activation proceeds through an initial change of shape as the discoidal shape is lost and surface protrusions develop. As shown in Fig. 2, long thin pseudopods, 1-3 μm in length, appear at the periphery of discoidal platelets followed by short blunt pseudopods that cover the platelet surface as the overall shape of the main body (i. e. excluding pseudopods) varies from discoidal to spherical depending on the strength of the platelet activator and time of exposure to it.

Figure 2. Transmission electron micrographs of discoidal platelets (upper; 13,000 × magnification). Arrows indicate where channels of the open canalicular system communicate with the cell exterior. The lower photograph shows platelets fixed after exposure to ADP with loss of discoidal shape to form irregular spheres having long pseudopods (17,000 × mag.). (From [51]).

The change of shape is also associated with separate internal contractions as the formerly randomly dispersed organelles become centralized by a tight fitting web of microtubules and microfilaments. The microtubules are not contractile elements but serve to orientate the contractile mircofilaments toward the cell centre, and, by virtue of their inherent resistance to deformation, to facilitate shape change reversal in the event that the stimulus is insufficient to evoke the subsequent irreversible reactions [41]. Shape change functionally manifests itself through the exposure of fibrinogen binding sites on the platelet surface which is presumed to lead to cross-linking of the fibrinogen between adjacent cell surfaces and the formation of aggregates [37]. The platelet plasma membrane surface has receptors composed of glycoproteins anchored in the membrane which transmit information to the cell interior so that the platelet can respond to the stimuli produced by different activating agents. Among the receptors are the Glycoproteins IIb and IIIa, which upon stimulation by ADP form a complex which binds fibrinogen [39].

When low concentrations of ADP or several other chemical stimuli are added to a suspension of stirred platelets, shape change usually precedes aggregation, the latter being dependent on collisions between shape-changed platelets with exposed fibrinogen binding sites [37,40]. This primary aggregation is essentially reversible since within about two minutes of exposure to the aggregating agent, the platelet aggregates usually begin to break up as the cells exhibit refractoriness to the stimulus. However, if the concentration of the platelet agonist is high enough, secondary irreversible aggregation ensues due to the release of platelet granule contents.

Shape change usually precedes and apparently is a prerequisite for the platelet release reactions [41] which in turn are a result of internal contractions leading to the membrane fusion of granules with plasma membrane [51]. Moreover, it has been shown that platelets are capable of regaining their discoidal shape even after having extruded their granule contents [42].

Aggregation, perhaps the most important of the several platelet functions, is required for formation of an adequate white thrombus. The list of agents which will induce platelet aggregation includes thrombin, connective tissue particles or collagen, ADP, epinephrine, norepinephrine and serotonin [32]. Aggregation occurs promptly when any of these agents is added in appropriate concentration to a suspension of normal human platelets in the presence of Ca^{++} or Mg^{++}. When thrombin or collagen is the stimulating agent, aggregation is followed by organelle disintegration, plasma membrane disruption and generalized disintegration of platelet cytoplasm, resulting in the destruction of the platelet as an intact cell. However, when the stimulus is ADP, organelle disintegration is minimal if it occurs at all. The platelet plasma membrane remains intact, and the platelets of at least some subjects deaggregate following ADP-induced aggregation.

1.3. Fluid Mechanical Factors. It has long been suspected that blood flow, in particular the nature of the flow patterns and fluid stresses within the sheared blood, play an important role in thrombosis within the circulation as well as in extracorporeal devices [14,16]. Thus, the sites of atherosclerosis, the focal deposition of platelets resulting in thrombosis, and the formation of aneurysms and dilatation of the vessel wall are mainly localized in regions of geometrical irregularity where vessels branch, curve and change diameter and where blood is subjected to sudden changes in velocity and/or direction. In such regions, flow is disturbed and separation of streamlines from the wall and the formation of eddies are likely to occur. Our laboratory therefore undertook a number of studies using models of branching vessels [25,26,30] and stenoses [22-24] to elucidate the effects of recirculation zones, vortices and secondary flows on the flow behavior, interactions and wall adhesion of human red cells and platelets. More recently, these have been extended to studies of the flow patterns in sections of real arteries and veins, fixed and rendered transparent for viewing the motions of small suspended particles and blood cells [3,21,27,28,35]. However, basic to all considerations of thrombus formation is the fact that a velocity gradient is necessary for aggregation of platelets, the primary constituents of a thrombus, to occur. To study this aspect of the problem we have undertaken studies of the effect of shear rate and red blood cells on the aggregation of ADP-stimulated platelets in steady flow through circular cylindrical tubes [6-9]. Shear rate is the most important physical parameter governing platelet aggregation in flowing suspensions. It determines the platelet collision frequency, the shear and normal stresses which activate single cells and break up aggregates, and the interaction time of cell-cell and cell-surface collisions. The predilection of white platelet thrombi to form in regions of high wall shear rate in the arterial circulation emphasizes the need to focus on the effect of shear rate on platelet aggregation in well-defined flow. Time-averaged systemic arterial wall shear rate in humans is estimated to range from 50 - 2000 s^{-1} [20] based on a parabolic velocity profile for whole blood. Some values are given in Table I, where it can be seen that mean linear velocities over one cycle decrease from > 800 to < 1 mm s^{-1} and the mean tube Reynolds number, Re_t, from 6000 to < 10^{-3}, in going from the heart to the microcirculation where viscous effects dominate the flow. The corresponding mean wall shear rates, and hence shear stresses increase in going to the small vessels, reaching a maximum in the arterioles. It is evident, however, that wall shear rates at peak systole in some large arteries can be as large as those in small vessels.

2. <u>MEASUREMENT OF PLATELET AGGREGATION</u>

2.1 <u>Platelet Suspensions</u>. Venous blood was drawn from healthy volunteers into plastic syringes containing sodium citrate at a predetermined concentration. Platelet-rich plasma (PRP) containing approx. 3×10^5 cells was prepared by centrifuging the blood at 100g and adjusting the platelet concentration by adding plasma. In the case of whole blood, the citrate concentration was adjusted according to the donor hematocrit to give a final concentration of 0.62% in plasma. The platelet concentration per µl of plasma in the whole blood samples was 306,880 ± 60,060 (± S.D., n = 11), compared to 279,000 ± 29,100 (n = 144) per µl of plasma in the PRP samples.

2.2. <u>Flow System</u>. Platelet-rich plasma (or whole blood) and ADP were simultaneously infused into a small cylindrical mixing chamber using independent syringe pumps at a fixed volume ratio PRP(or whole blood) : ADP = 9:1. After rapid mixing, the suspension exited the chamber through lengths of 0.595 or 0.380 mm radius, R_o, polythene tubing corresponding to meant transit times $\bar{t} = X_3/\bar{U}$ between 1 and 86s, where X_3 (from 2 cm

TABLE I: PHYSIOLOGICAL PARAMETERS IN THE HUMAN ARTERIAL TREE

Vessel	Diameter mm	Mean Linear Velocity $\overline{U}$, mm s^{-1} min.	max.	mean	Mean Reynolds Number Re_t	Wall Shear Rate[1] G_w, s^{-1} min.	max.	mean
Ascending Aorta[2]	23.0–43.5	–	–	245–876	3210–6075	–	–	45–305
Femoral Artery[3]	5.0	–350	1175	188	283	-560	1885	302
Common Carotid[3]	5.9	99	388	187	332	134	526	253
Carotid Sinus[4]	5.2	85	325	156	244	130	500	240
External Carotid[4]	3.8	83	327	157	180	175	687	331
Small Arteries	0.3	–	–	50	2.3	–	–	1335
Arterioles	0.025	–	–	5	0.038	–	–	1600
Capillaries[5]	0.012	0.39	1.74	0.84	1.5 – 6.6 $\times 10^{-3}$	260	1290	560

1 Assuming Poiseuille Flow

2 Mean systolic values from MacDonald [34].

3 Values obtained *in vivo* by Anliker *et al.* [2]

4 Values calculated assuming 65% flow into the internal carotid. Measured wall shear rates from *ex vivo* studies in steady flow at $Re_t = 592$ [21] show that at the outer wall of the sinus there is a recirculating flow with G_w varying from –135 to +530 s^{-1} At the inner wall, $G_w > 2000$ s^{-1}. At the inner wall of the external carotid, $G_w > 1000$ s^{-1}.

5 Values from Bollinger *et al* [10] from red cell velocities in human nailfold capillaries of large diameter. Calculations of wall shear rate would correspond to that in plasma flow. In "Bolus Flow" of a train of red cells, much higher wall shear rates would exist in the plasma layer surrounding the red cells.

to 15.3 m) is the distance down the flow tube and $\overline{U}$ the mean linear velocity. Total volumetric flow rates, Q, were preset from 13 to 155 μl s^{-1} to generate volume flow averaged mean tube shear rates, $\overline{G}$, in Poiseuille flow, defined as:

$$\overline{G} = \frac{\int_0^{R_o} 2\pi U(R)G(R)R\,dR}{\int_0^{R_o} 2\pi U(R)R\,dR} \tag{1a}$$

$$= \frac{8}{15}G(R_o) = \frac{32Q}{15\pi R_o^3} \tag{1b}$$

from 41.9 to 1920 s^{-1}, where U(R) and G(R) are the respective fluid velocity and shear rate at a radial distance R from the tube axis, and $G(R_o)$ the shear rate at the tube wall. Tube Reynolds numbers ranged from 8 to 148. Control runs in which Tyrodes, a physiological salt solution, instead of ADP was infused, were also carried out.

The aggregation reaction was instantaneously and permanently arrested by collecting known volumes of the effluent into 20 × the volume of 0.5% isotonic glutaraldehyde. In the case of whole blood, 5 ml of the effluent fixed in glutaraldehyde were layered onto 40 ml of isotonic Percoll solution (density = 1.10 g ml^{-1}) and centrifuged at 4000g. The hardened red cells (density = 1.16 g ml^{-1}) formed a tight pellet at the bottom of the tube while hardened single platelets and aggregates (density = 1.04 g ml^{-1}) formed a thin layer at the glutaraldehyde-Percoll interface. The latter were carefully removed and analyzed as described below.

Measurement of the fraction of particles recovered from the chamber as a function of time after injection showed that only at flow rates ≤ 26 μl s^{-1} did the residence time exceed 4 s. Above 52 μl s^{-1}, 50% of the particles had exited by 0.7 s, and 90% by 1.3 s. At the highest flow rate tested, essentially all particles had exited the chamber within 1 s after injection, which was the shortest sampling time feasible with the current technique.

2.3 Particle concentration and size measurement. The number concentration and volume of single platelets and aggregates were measured using an electronic particle counter in conjunction with a logarithmic amplifier and a 100 channel pulse height analyzer to generate continuous 250 class log-volume histograms over the volume range 1 - 10^5 μm^3 [6]. The number concentration per histogram class is $N(x_i)$, particle volume $v(x_i)$, and volume fraction $\Phi(x_i) = N(x_i)v(x_i)$, where x_i is the mark of the ith class. Computer integration of the log-volume histograms yielded the number concentration, $N_{L,U}$, and volume fraction, $\Phi_{L,U}$ of particles between lower, $L = v(x_l)$ and upper, $U = v(x_u)$ volume limits. Individual histograms from multiple donors were averaged [7] yielding a histogram of the mean class volume fraction, $\overline{\Phi}(x_i)$, normalized to the maximum class content at $\bar{t} = 0$. The mean, modal and median single platelet volume were calculated from the mean and standard deviation of the log-volume distribution, assuming a normal distribution of the latter [4]. The assumption of log-normality of single platelet volume was tested using the Kolmogorov-Smirnov, K-S, one sample test. Unpaired, one-tailed t-tests were used to test the significance of differences between means.

3. AGGREGATION IN PLATELET-RICH PLASMA

3.1. Decrease in single platelet concentration. Figure 3 shows the single platelet number concentration after $\bar{t} = 43$ s exposure to 0.2 μM ADP, $N_{1,30}(\bar{t})$, normalized to a control at $\bar{t} = 0$ s, $N_{1,30}(0)$, as a function of the mean tube shear rate. The extent of aggregation of platelets from female donors was significantly greater than that of platelets from male donors ($p < 0.001$) over the range of shear rate $41.9 \leq \bar{G} \leq 1920$ s^{-1} [7,9]. The sex difference was greatest at $\bar{G} = 335$ s^{-1} where 76% ± 3.4 (S.E.M., n = 6) of single platelets from the female donors but only 49% ± 3.9 (n = 6) of those from male donors had aggregated. Significant changes ($p < 0.001$) in the single platelet concentration as a function of mean tube shear rate produced a similar pattern of aggregation for both groups of donors. Aggregation increased as the shear rate increased up to a maximum at $\bar{G} = 168$ and 335 s^{-1} for male and female donors, respectively. Thereafter, aggregation decreased linearly with increasing shear rate down to a minimum at $\bar{G} = 1000$ and 1335 s^{-1} for male and female donors, respectively. Further increases in shear rate, however, produced an increase in aggregation for both sexes. The two aggregation curves intersect at $\bar{G} = 1335$ s^{-1} where aggregation had begun to increase for the male donors while it was still decreasing for the female donors.

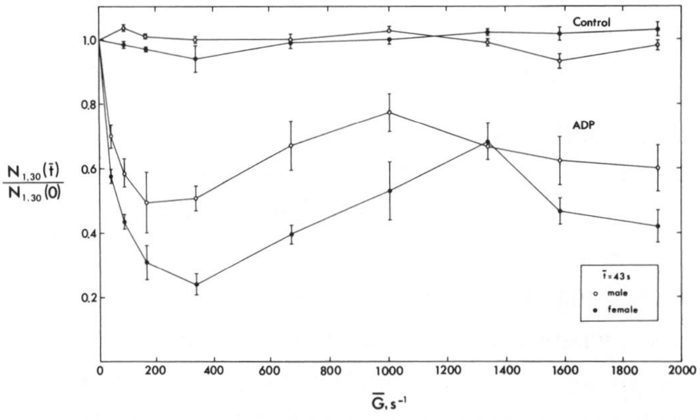

Figure 3. Normalized single platelet number concentration $N_{1,30}(\bar{t})/N_{1,30}(0)$ (±SEM) after $\bar{t} = 43$ s exposure to 0.2 μM ADP, as a function of mean tube shear rate for male and female donors. In the control runs, physiological salt solution was infused, instead of ADP. (From [7])

3.2 Collision efficiency. Theory applicable to dilute suspensions of rigid spheres was used to calculate the two-body collision efficiency in PRP in the initial stages of the reaction when the number of multiplets was small. The two-body collision frequency for monodisperse suspensions of rigid spheres in simple shear flow is given by [19,45]:

$$j = \frac{32}{3} G b^3 N, \tag{2}$$

where b is the sphere radius and N the number concentration. Collisions are defined as the rectilinear approach of sphere centres to within a distance of $\leq 2b$, and provided $b/R_0 << 1$, Eq. (2) has been shown to apply in Poiseuille flow [18]. Using the mean tube shear rate, the total number of two-body collisions per sec per unit volume of suspension is then:

$$J = \frac{1}{2} N j = \frac{4\Phi \bar{G} N}{\pi}, \tag{3}$$

where Φ is the volume fraction of suspended particles.

Inclusion of the orthokinetic collision capture efficiency, $\alpha_o = j_c / j$ in Eq. (2) where j_c is the collision capture frequency, accounts for the influence of both interaction and hydrodynamic forces on particle capture [49]. If $\alpha_o = 1$, then $j_c = j$ and every collision results in capture; however, in the absence of attractive forces permanent capture is impossible.

In the absence of aggregate break-up or the formation of higher order aggregates, the kinetics of aggregation are first order with respect to the total particle concentration, N_∞ [46]:

$$\frac{dN_\infty}{dt} = -\frac{4\Phi\alpha_0 \bar{G} N_\infty}{\pi}. \tag{4}$$

Integration of Eq. (4) yields:

$$\ln \frac{N_\infty(t)}{N_\infty(0)} = -\frac{4\Phi\alpha_0 \bar{G} t}{\pi}, \tag{5}$$

where $N_\infty(0)$ and $N_\infty(t)$ are the total particle concentrations at time 0 and t, respectively. Thus, the total particle concentration decays exponentially and a plot of $\ln N_\infty(t)$ vs t should give a straight line, the slope of which yields α_o.

In the present work the influence of Brownian motion on aggregation can be neglected due to the large value of the Péclet number, $Pe = \bar{G} b^2 / D_t > 1200$ where D_t is the translational diffusion coefficient for a single sphere calculated from the Stokes-Einstein equation. Thus, the measurement of the total particle concentration over the early stages of aggregation provides a value for α_o.

A plot of Eq. (5) using the data for female donors is shown in Fig. 4. Although the extent of aggregation was greatest at $\overline{G}$ =335 s^{-1}, the highest rate of aggregation occurred within the first $\bar{t}$ = 2 s at $\overline{G}$ = 41.9 s^{-1}, where the single platelet concentration decreased at a mean rate of 4.2% s^{-1}. At this shear rate, the rate of aggregation steadily decreased with increasing mean transit time. At higher $\overline{G}$, there was an initial lag phase followed by progressively increasing then decreasing rates of aggregation giving rise to the sigmoid curves in Fig. 4. The length of the lag phase increased with increasing mean tube shear rate reaching $\bar{t}$ ~ 11 s at $\overline{G}$ = 1335 s^{-1}. In addition, as the mean tube shear rate was raised, not only did the maximum rate of aggregation decrease, but it occurred at progressively increasing mean transit times. Male donors exhibited a pattern of aggregation similar to that of the female donors at the same shear rate but always with much longer lag phases and reduced rates of aggregation.

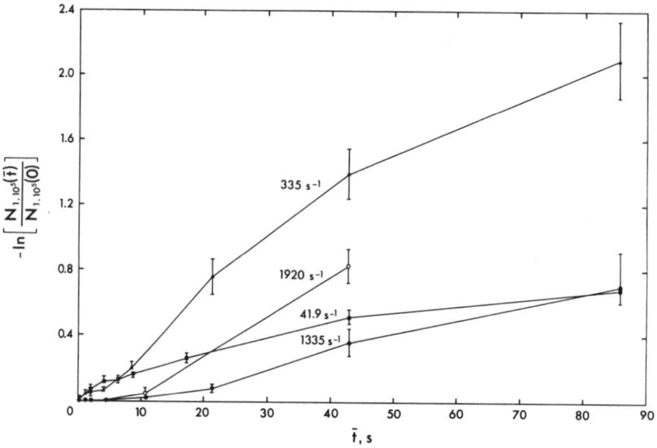

Figure 4. Decrease in total particle concentration with mean transit time at 0.2 μM ADP. Plot of Eq. (5) for the female donors at the mean shear rates shown on the graph. (From [7])

Figure 5 shows the collision efficiencies, averaged over three time intervals, plotted against log $\bar{t}\overline{G}$. In the time interval from $\bar{t}$ = 0 to 4.3 s, both the collision efficiency and its rate of decrease fell rapidly with increasing shear rate from a maximum of 0.26 at $\overline{G}$ = 41.9 s^{-1}. A similar decline in α_o in the interval from $\bar{t}$ = 4.3 to 8.6 s was interrupted between $\overline{G}$ = 168 and 335 s^{-1} before resuming at a higher rate of decrease beyond $\overline{G}$ = 335 s^{-1}. In contrast, during the time interval from $\bar{t}$ = 8.6 to 21 s, α_o decreased by only 12% up to $\overline{G}$ = 168 s^{-1}, but decreased sharply thereafter. Throughout all 3 time intervals, α_o either decreased or remained constant as mean tube shear rate increased up to $\overline{G}$ = 1,335 s^{-1} where $\alpha_o \leq 0.002$. At $\overline{G}$ = 1920 s^{-1}, there was a small but significant increase in α_o which, together with the high collision frequency was sufficient to support the high rate of aggregation shown in Fig. 4.

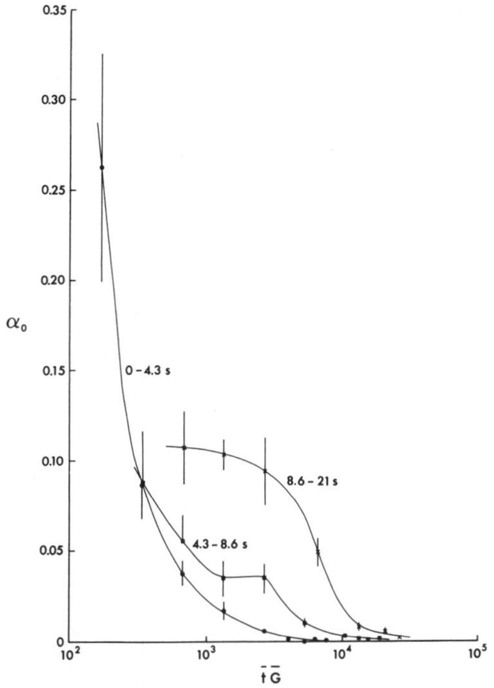

Figure 5. The collision efficiency, α_o ($\pm$ SEM.) plotted against log $\overline{tG}$ for the data shown in Fig. 4. The interruption in the decrease in α_o with increasing mean tube shear rate for mean transit times between 4.3 and 8.6 s is interpreted as indicating a transition from a weak to a strong platelet-platelet bond. (From [7])

3.3 Application of two-body collision theory to platelet aggregation. Theory, applicable to inert, charged colloidal-size particles, predicts that when net attractive forces operate between colliding particles, the two-body collision efficiency decreases with increasing shear stress [49]. This has been shown experimentally at shear stresses < 0.2 Nm^{-2} for the aggregation of model latex spheres in solutions of high ionic strength [12,46,52] and at shear stresses < 0.1 Nm^{-2} for human platelets exposed to 1 μM ADP [5]. In the present experiments the mean tube shear stress ranged from 0.08 to 3.5 N m^{-2} As the mean tube shear rate increases, the extent of aggregation is the result of a balance between an increased frequency of collision and an increased fluid shear stress. High collision rates support a high rate of aggregation in the absence of shear stresses sufficient to inhibit doublet formation. Beyond an optimum shear rate for aggregation, higher fluid shear stresses can prevent stable platelet-platelet bond formation and aggregation decreases. Short interaction times may limit stable doublet formation at high shear rates but, since this time is inversely proportional to the shear rate, it is impossible to separate the effects of short interaction time and high shear stress in limiting stable bond formation. The two-body collision efficiency, however, takes both effects into account by measuring the fraction of total platelet-platelet collisions that result in stable doublet formation.

The collision efficiencies were computed from changes in the total particle concentration using Eq. (5) which assumes that the decrease in total particle concentration with time is due only to the formation of doublets. This assumption can be tested by calculating the fraction of successful two-body collisions from the initial rate of single platelet decrease. The total number of two-body collisions per unit volume of suspension per second can be calculated from Eq. (3) using the initial number concentration and volume fraction of single platelets. If α_o computed from the total particle concentration is significantly greater than that computed using Eq. (3) then higher order aggregates are present in appreciable numbers. At 0.2 μm ADP and $\overline{G} = 41.9\ s^{-1}$ where the mean volume fraction of single platelets was 0.195% ± 0.023 (± S.D.) for male donors and 0.197% ± 0.023 for female donors, respective initial rates of single platelet decrease of 2.7 and 4.2% s^{-1} yielded collision efficiencies of 0.14 and 0.21. These values are remarkably similar to the collision efficiencies, 0.12 and 0.26, calculated according to Eq. (5) between $\bar{t} = 0$ and 4.3 s, well into the aggregation reaction. Actually, one would expect the collision efficiency calculated at long times to yield lower values due to the decrease in particle concentration with increasing aggregation. Thus, it is unlikely that the observed increase in collision efficiency with time in the present experiment is artifactual. It should be noted, however, that the presence of pseudopods on the activated platelets has been estimated to double the effective collision diameter of the cells [13]. This will result in an 8-fold increase in collision frequency and a corresponding decrease in α_o, and implies that the aggregation reaction is capable of propagation at very low collision efficiencies.

3.4. Interpretation of Results: Heterogeneity of Platelet-Platelet Bonds. The results show that the highest rate of aggregation and two-body collision efficiency (26%) occurred at the lowest shear rate but only for the first few seconds of the reaction. Although the fluid shear stress is low, the collision rate is not sufficient to sustain a high rate of aggregation. At higher shear rates there was generally a decrease in the fraction of efficient collisions but a high collision frequency accounts for the higher rates of aggregation. At the maximum shear rate in the present experiments, collision efficiencies between 0.6×10^{-3} and 2.3×10^{-3} were measured. The large decrease in collision efficiency at relatively low shear rates (Fig. 5) and the persistence of a nonzero collision efficiency at high shear rates suggest that more than one type of platelet-platelet bond mediates ADP-induced aggregation.

The increasing rates of aggregation with time as depicted by the sigmoid aggregation curves (Fig. 4) indicate that collision efficiency increases with time, even after a delay of up to 11 s in the onset of aggregation. Indeed, not only do calculations of collision efficiency confirm this but they also indicate that the heterogeneity among platelet bonds is time-dependent (Fig. 5). At early transit times the high rate of aggregation at low shear rates is sustained by a weak bond that is easily disrupted at higher shear rates resulting in a corresponding shear-dependent lag phase. Even at low shear rates the strength of this bond gradually diminishes with increasing transit time and, in conjunction with a low collision rate, produces a steadily decreasing rate of aggregation. At high shear rates, the increasing rate of aggregation at times beyond the lag phase reveals the emergence of a second stronger bond. Longer times are required before each bond is sufficiently strong, or is present in sufficient numbers, to support aggregation. The two types of bonds coexist at intermediate transit times where the weak bond is disrupted at low shear rates but higher shear rates are required to disrupt the stronger bonds. The interruption in the decrease in collision efficiency with increasing mean tube shear rate between 168 and 335 s^{-1} at mean transit times between 4.3 and 8.6 s (Fig. 5) points to a transition from weak to strong bond. At very long exposure times the strong bonds are maximally expressed through either strength or numbers, and only high shear rates are sufficient to disrupt them.

The cross-linking of bivalent fibrinogen molecules between activated glycoprotein IIb-IIIa (GPIIb-IIIa) complexes in the platelet membrane is the mechanism believed to underlie the ADP-induced aggregation of platelets [31]. The steadily increasing lag phase preceding aggregation with increasing shear rate in the present work points to a latency in the strength of the platelet-platelet cross-bridge. For any colloid whose aggregation is mediated by polymer cross-linking, unoccupied binding sites must be available on both surfaces for cross-linking to occur. In the case of platelets, the high concentration of fibrinogen in plasma (~ 2×10^7 molecules per platelet). would be expected to saturate all binding sites on platelets (~ 5×10^4 receptor sites per cell; [39]) before cross-linking could occur. In fact, cross-linking would require the simultaneous binding of opposite ends of the bivalent fibrinogen molecule to two platelets immediately after activation of GPIIb-IIIa complexes, and prior to saturation of these receptors with free fibrinogen. This scenario seems unlikely. It is more likely that all GPIIb-IIIa complexes would be saturated with free fibrinogen long before two platelets could simultaneously bind a single fibrinogen molecule. Instead, a model of aggregation requires either a low affinity for fibrinogen binding by activated platelets, and the subsequent continuous breaking and forming of new platelet-fibrinogen bonds, or the time-dependent exposure of new bonds that permits cross-linking during the interaction time of collision. There is evidence that both high and low affinity binding sites for fibrinogen exist on ADP stimulated platelets [40], and that binding increases with time [33,40]. In addition, fibrinogen itself appears to have a binding sequence in the carboxy terminus of the γ-chain that recognizes the GPIIb-IIIa complex, but is distinct from the arginine-glycine-aspartic acid- (RGD) containing sequence in the α chain [29,44]. Thus, the relatively slow kinetics of fibrinogen binding and the existence of heterogeneity in the affinity of fibrinogen provide a mechanism for platelet aggregation in the presence of high concentrations of the cross-linking ligand. Furthermore, the increase in the two-body collision efficiency with time and the formation of a high shear rate-resistant bond can be explained in terms of a time-dependent increase in fibrinogen binding.

There is at present, however, no conclusive proof that fibrinogen mediates aggregation by directly cross-linking activated platelets. An alternate mechanism proposes that GPIIb-IIIa receptor clustering is a prerequisite for fibrinogen binding and platelet aggregation [1,38]. It is possible that the role of fibrinogen is to stabilize such clusters permitting them to interact in some complementary manner between activated platelets. Thus, fibrinogen cross-linking between platelets *per se* would not be necessary for aggregation. If the platelets were initially aggregated by a mechanism independent of fibrinogen cross-linking but which maintained close contact, cross-linking could follow as fibrinogen binding sites were expressed. Coller [11] has proposed that both platelet binding of fibrinogen and the small radius of curvature of the pseudopods would be sufficient to lower the electrostatic repulsion between the similarly charged platelets and permit aggregation through van der Waals attraction. Since such aggregation is not mediated by specific fibrinogen cross-linking, it may not be resistant to high shear rates. It does, however, provide a mechanism for the relatively weak aggregation observed at short transit times in the present work.

3.5. Growth of Aggregates. The evolving pattern of aggregate growth for the female donors is shown in Fig. 6 where the normalized average class volume fraction $\overline{\Phi}(x_i)$ is plotted against particle volume at successive mean transit times. The decrease in single platelet concentration (1 - 30 μm^3) was accompanied by a sequential rise and fall of aggregates of successively increasing volume. The steadily decreasing rate of aggregation with time shown earlier (Fig. 4) at $\overline{G} = 41.9\ s^{-1}$ was associated with the formation of aggregates having a broad spectrum of size at $\bar{t} = 86$ s At $\overline{G} = 335\ s^{-1}$, no distinct aggregate peaks were present prior to $\bar{t} = 8.6$ s but by $\bar{t} = 21$ s aggregates of relatively discrete size had appeared. As aggregation continued, the upper limit of aggregate size increased. By $\bar{t} = 86$ s, most aggregates were present in one large group, a significant proportion of which exceeded $10^5\ \mu m^3$, the largest volume measured. Although there was considerable aggregation at $\overline{G} = 1920\ s^{-1}$ (not shown), the aggregates were smaller and occupied a size range considerably narrower than at the same mean transit time at

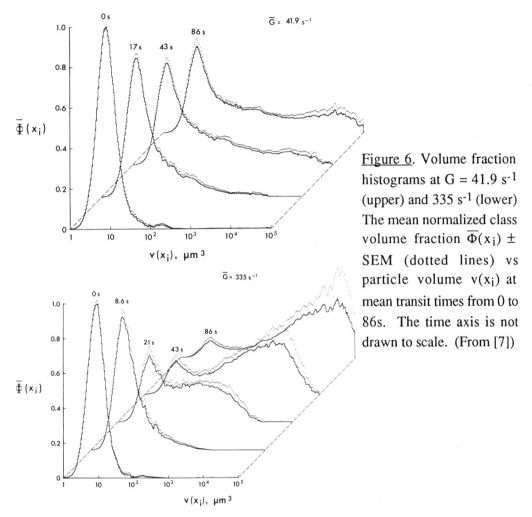

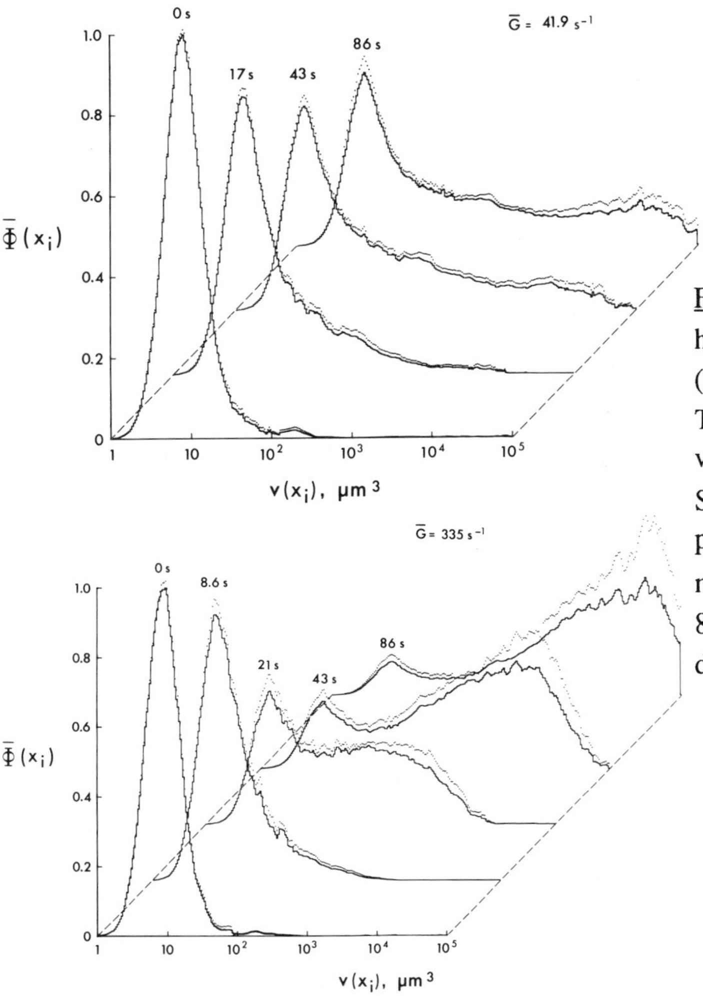

Figure 6. Volume fraction histograms at G = 41.9 s^{-1} (upper) and 335 s^{-1} (lower) The mean normalized class volume fraction $\overline{\Phi}(x_i)$ ± SEM (dotted lines) vs particle volume $v(x_i)$ at mean transit times from 0 to 86s. The time axis is not drawn to scale. (From [7])

$\overline{G}$ = 335 s^{-1}. A similar pattern of aggregate growth was exhibited by the male donors with aggregate size was much reduced compared to that of female donors at the same $\overline{t}$ and mean tube shear rate.

4. AGGREGATION IN WHOLE BLOOD (WB)

4.1. Decrease in Single Platelet Concentration. At all mean tube shear rates, and at both 0.2 and 1 µM ADP, the rate and extent of aggregation in whole blood were always much greater than those in PRP [8]. At 0.2 µM ADP, $\overline{G}$ = 41.9 s^{-1}, the rate of aggregation was highest over the first $\overline{t}$ = 1.7 s where 26% s^{-1} of single platelets aggregated, more than 7× the rate in PRP. By $\overline{t}$ = 43 s, only 13% of platelets remained unaggregated in WB as compared to 64% in PRP. At 1.0 µM ADP, the initial rate of aggregation in WB of 37% s^{-1} was 9× the mean value for the donors in PRP [7]. As shown in Fig. 7, the rate of aggregation at $\overline{G} = 335\ s^{-1}$ was also much greater than that in PRP at both 0.2 and 1.0 μM ADP. At 0.2 μM ADP, a low initial rate of aggregation led to the sigmoid aggregation curve characteristic of aggregation in PRP. At 1.0 μM ADP, however, the rate remained almost constant until $\bar{t} = 8.6$ s when 84% of singlets had aggregated. It then markedly decreased, but by $\bar{t} = 43$ s, 99% of the cells had aggregated. The pattern of aggregation at $\overline{G} = 1920\ s^{-1}$ in WB was almost identical to that at $\overline{G} = 335\ s^{-1}$.

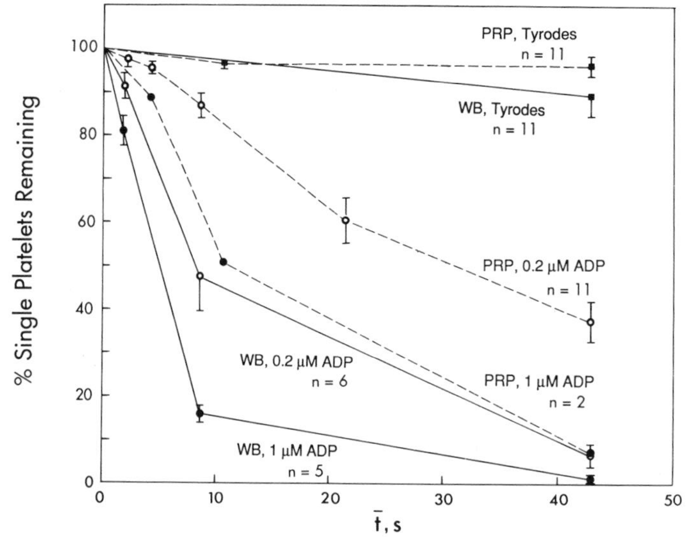

Figure 7. Mean values of the % single platelets remaining (± SEM, n = number of donors) at $\overline{G} = 335\ s^{-1}$ plotted against mean transit time $\bar{t}$ at 0.2 and 1.0 μM ADP. The control runs were carried out with Tyrodes infused instead of ADP. (From [8]).

4.2. Aggregate Growth. In whole blood, the rapid formation and size of the aggregates was most remarkable. At 1 μM ADP and $\overline{G} = 41.9\ s^{-1}$, the visible flake-like aggregates were red in colour due to red cells trapped within the interior. At $\overline{G} > 335\ s^{-1}$, aggregate growth was initially slower, with successive formation of distinct populations of increasing size, an effect even more pronounced at $\overline{G} = 1920\ s^{-1}$.

4.3. Aggregation in Control Samples. There was significant aggregation in the controls of whole blood at $\overline{G} = 41.9$ and $335\ s^{-1}$ (Fig. 7), not seen in PRP. This was not due to release of ADP from platelets or from haemolysed red cells [8]. Nevertheless, recent experiments have shown that an enzyme which converts ADP to ATP inhibits aggregation in WB control samples. This suggests that red cells release small quantities of ADP which lead to the formation of aggregates, stable at $\overline{G} = 41.9\ s^{-1}$ but not at $\overline{G} > 335\ s^{-1}$.

4.4. Mechanism of Red Cell-Enhanced Platelet Aggregation. It is known that red cells markedly increase the diffusivity of platelets in the plasma of blood undergoing tube flow The effective translational diffusion coefficient, D_t, increases by at least two orders of magnitude from 10^{-9} to > 10^{-7} cm^2 s^{-1} [48]. The effect is likely due to an increase in the lateral dispersion of platelets caused by the erratic motions of the continuously colliding and deforming red cells, as has been demonstrated in optically transparent suspensions of red cell ghosts flowing through small tubes [15,17]. The two-body collision frequency due to translational Brownian motion of rigid spheres, $j_D = 16\pi bND_t$ [45], and the ratio of shear-induced [Eq. (2)] to Brownian motion-induced collision frequency is then:

$$\frac{j}{j_D} = \frac{2\bar{G}b^2}{3\pi D_t} . \tag{6}$$

Assuming $D_t = 2 \times 10^{-7}$ cm^2 s^{-1} in WB, and an equivalent sphere radius b = 1.2 μm based on a mean platelet volume = 7.5 μm^3, $j/j_D = 0.64$ and 5.2 at $\bar{G} = 41.9$ and 335 s^{-1}, respectively, corresponding to increases of 156 and 19% in the collision frequency. However, results from the present work show that collision frequencies in WB would be 7 and 15 times greater than in PRP at $\bar{G} = 335$ and 41.9 s^{-1}, respectively, if the collision efficiency remained unchanged. An explanation for the higher rates of aggregation in WB may thus be sought in terms of a much greater collision efficiency resulting from an increased velocity of approach and time of interaction during collision due to the presence of the red cells. We are testing this hypothesis using 40% suspensions of red cell ghosts containing 3×10^5 /μl plasma of 2 μm latex spheres serving as models of platelets. The spheres are coated with polyethylene oxide to eliminate interaction forces between particles. The suspension flows through a 100 μm diameter tube and cine films of two-body collisions between spheres are analyzed to obtain the doublet lifetime, τ_{meas}, during which the particles are in apparent contact. Measured τ are compared with τ_{calc} predicted by theory assuming the spheres rotate together as a spheroid of axis ratio = 1.98 [19]:

$$\tau_{calc} = \frac{5}{G(R)} \tan^{-1}\left(\frac{1}{2}\tan\phi_1^0\right) \tag{7}$$

Here, ϕ_1^0 is the azimuthal angle of orientation of the doublet axis when the spheres first make contact (apparent angle of collision), referred to X_1 as the polar, and vorticity axis. In the absence of three-body interactions and electrostatic or van der Waals forces between sphere surfaces, the particles rotate as a rigid dumbbell until at an angle $\phi_1^{0'}$, the reflection of the apparent angle of collision, they separate. Analysis of 170 collisions at G(R) from 5 - 40 s^{-1}, has shown that the mean measured τ is 2.0 × greater than predicted from theory. A histogram of the distribution in τ_{meas}/τ_{calc} is shown in Fig. 8.

It should also be noted that, the collision frequency itself would be higher in tubular vessels if , at the periphery, where the shear rate is highest, the platelet concentration is two-fold greater than at the centre of the vessel, as observed in small arteries [47]. It

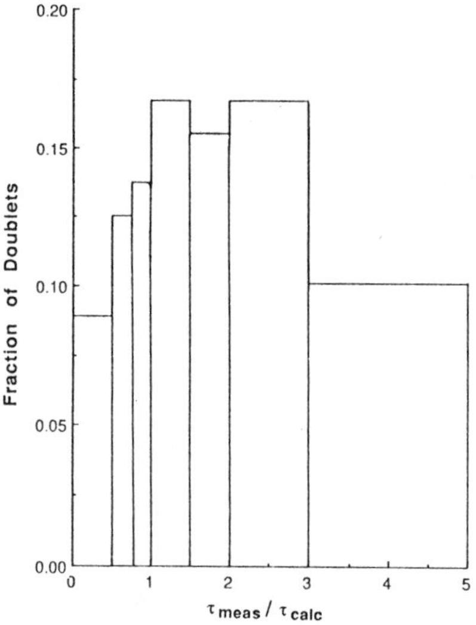

Figure 8. Histogram showing the distribution in the ratio of measured to calculated lifetimes of doublets formed by two-body collisions between 2 μm latex spheres in a 40% suspension of red cell ghosts flowing through a 100 μm tube. The values of τ_{calc} were obtained from Eq. (7) using the measured apparent collision angle ϕ_1^o and the shear rate G(R) at the doublet centre of rotation.

should be emphasized, however, that an increase in collision efficiency due to locally released ADP or other metabolites from red cells cannot be entirely excluded [43].

5. NOMENCLATURE

b	Sphere radius
D_t	Translational diffusion coefficient
G(R), G_w; $\overline{G}$	Shear rate at radial distance R, value at wall; volume flow-averaged value in Poiseuille flow
i	Histogram class number
j; j_D	Two-body collision frequency; due to Brownian motion
j_c	Two-body capture frequency
J	Total two-body collision frequency per unit volume of suspension
l, u	Lower, upper histogram classes
L, U	Lower, upper volume of classes *l* and *u*
N; $N_{L,U}$; $N(x_i)$	Number concentration of equal-sized spheres; between L and U; per histogram class;
$N_\infty(t)$	Total particle concentration at time t
Pe	Péclet number
Q	Volume flow rate
R; R_0	Radial distance from tube axis; tube radius
Re_t	Tube Reynolds number
t; $\bar{t}$	Time; mean transit time

$U(R); \overline{U}$	Linear velocity in the axial direction at a radial distance R in Poiseuille flow; mean value
$v(x_i)$	Volume of class i
x_i	Mark of class i of the log-volume histogram
X_3	Distance down flow tube from exit of chamber
α_o	Orthokinetic collision capture efficiency
$\Phi; \Phi(x_i), \Phi_{L,U}$	Particle volume fraction; of histogram class i, between L and U
$\phi_1^o, \phi_1^{o'}$	Respective apparent angles of collision and separation of spheres
τ_{meas}, τ_{calc}	Respective measured and theoretical doublet lifetimes

6. REFERENCES

1. AASCH, A.S., LEUNG, L.K., POLLEY, M.J. and NACHMAN, R.L. 1985. Platelet membrane topography: colocalization of thrombospondin and fibrinogen with glycoprotein IIb-IIIa complex. *Blood* **66**: 926-934.
2. ANLIKER, M., CASTY, M., FRIEDLI, P., KUBLI, R. and KELLER, H. 1977. Non-invasive measurement of blood flow. *In*: "Cardiovascular Flow Dynamics and Measurements" (Hwang, N.H.C. and Normann, N.A., eds.) University Park Press, Baltimore. pp. 43-88.
3. ASAKURA, T. and KARINO, T. 1990. Flow patterns and spatial distribution of atherosclerotic lesions in human coronary arteries. *Circ. Res.* In Press.
4. BELL, D.N. 1988. Physical factors governing the aggregation of human platelets in flow through tubes. Ph.D. Thesis, McGill University, Montreal, Canada.
5. BELL, D.N. and GOLDSMITH, H.L. 1984. Platelet aggregation in Poiseuille flow. II. Effects of shear rate. *Microvasc. Res.* **27**: 316-330.
6. BELL, D.N., SPAIN, S. and GOLDSMITH, H.L. 1989. The ADP-induced aggregation of human platelets in flow through tubes. I. Measurement of the concentration and size of single platelets and aggregates. *Biophys. J.* **56**: 817-828.
7. BELL, D.N., SPAIN, S. and GOLDSMITH, H.L. 1989. The ADP-induced aggregation of human platelets in flow through tubes. II. Effect of shear rate, donor sex and ADP concentration. *Biophys J.* **56**: 829-843.
8. BELL, D.N., SPAIN, S. and GOLDSMITH, H.L. 1989. The effect of red blood cells on the ADP-induced aggregation of human platelets in flow through tubes. *Thromb. Haemost.* **63**: February 1990.
9. BELL, D.N., SPAIN, S. and GOLDSMITH, H.L. 1989. Extracellular Ca^{2+} accounts for the sex difference in the aggregation of human platelets in citrated platelet-rich plasma. *Thromb. Res.*, submitted.
10. BOLLINGER, A., BUTTI, P., BARRAS, P., TRACHLER, H. and SIEGENTHALER, N. 1974. Red blood cell velocity in nailfold capillaries of man, measured by a television microscopy technique. *Microvasc. Res.* **6**: 61-72.
11. COLLER, B.S. 1983. Biochemical and electrostatic considerations in primary platelet aggregation. *Ann. N.Y. Acad. Sci.* **416**: 693-708.

12. CURTIS, A.S.G. and HOCKING, L.M. 1970. Collision efiiciency of equal spherical particles in shear flow. *Trans. Faraday Soc.* **66**: 1381-1390.
13. FROJMOVIC, M.M., LONGMIRE, K.A. and VAN DE VEN. T.G.M. 1990. Platelet pseudopod number and length in mediating platelet aggregation. Theory and observations. Submitted for publication.
14. GLAGOV, S. 1972. Hemodynamic Risk Factors: Mechanical stress, mural architecture, medical nutrition and the vulnerability of arteries to atherosclerosis. *In*: "The Pathogenesis of Atherosclerosis" (Wissler, R.W. and Geer, J.C., eds.) Williams and Wilkins, Baltimore. Chapter 6.
15. GOLDSMITH, H.L. 1971. Red cell motions and wall interactions in tube flow. *Fed. Proc.* **30**: 1578-1588.
16. GOLDSMITH, H.L. and KARINO, T. 1978. Mechanically induced thromboemboli. *In*: "Quantitative Cardiovascular Studies: Clinical and Research Applications" (Hwang, N.H.C., Gross, D.R. and Patel, D.J., eds.) University Park Press, Baltimore. pp. 289-351.
17. GOLDSMITH, H.L. and MARLOW, J. 1979. Flow behavior of erythrocytes. II. Particle motions in sheared suspensions of ghost cells. *J. Colloid Interface Sci.* **71**: 383-407.
18. GOLDSMITH, H.L. and MASON, S.G. 1964. The flow of suspensions through tubes. III. Collisions of small uniform spheres. *Proc. Roy. Soc. (London)* **A282**: 569-591.
19. GOLDSMITH, H.L. and MASON, S.G. 1967. The microrheology of dispersions. *In*: "Rheology: Theory and Applications" Vol. 4, (Eirich, F.R., ed.) Academic Press, New York. pp. 85-250.
20. GOLDSMITH, H.L. and TURITTO, V.T. 1986. Rheological aspects of thrombosis and haemostasis. Basic principles and applications. *Thrombos. Haemost.* **55**: 415-435.
21. KARINO, T. 1986. Microscopic structure of disturbed flows in the arterial and venous systems, and its implication in the localization of vascular diseases. *Intern. Angiology* **4**: 297-325.
22. KARINO, T. and GOLDSMITH, H.L. 1977. Flow behaviour of blood cells and rigid spheres in an annular vortex. *Phil. Trans. Roy. Soc. (London)* **B279**: 413-445.
23. KARINO, T. and GOLDSMITH, H.L. 1979. Aggregation of platelets in an annular vortex distal to a tubular expansion. *Microvasc. Res.* **17**: 217-237.
24. KARINO, T. and GOLDSMITH, H.L. 1979. Adhesion of human platelets to collagen on the walls distal to a tubular expansion. *Microvasc. Res.* **17**: 238-262.
25. KARINO, T. and GOLDSMITH, H.L. 1985. Particle flow behavior in models of branching vessels. II. Effect of branching angle and diameter ratio on flow patterns. *Biorheology* **22**: 87-104.
26. KARINO, T., KWONG, H.H.M. and GOLDSMITH, H.L. 1979. Particle flow behavior in models of branching vessels: I. Vortices in 90° T-junctions. *Biorheology* **16**: 231-247.
27. KARINO, T. and MOTOMIYA, M. 1984. Flow through a venous valve and its implication in thrombus formation. *Thromb. Res.* **36**: 245-257.
28. KARINO, T., MOTOMIYA, M. and GOLDSMITH, H.L. 1989. Flow patterns at the major T-junctions of the dog descending aorta. *J. Biomechanics* In Press.

29. KLOCZEWIAK, M., TIMMONS, S, LUKAS, T.J. and HAWIGER. J. 1984. Platelet receptor recognition site on human fibrinogen. Synthesis and structure-function relationship of peptides corresponding to the carboxy-terminal segment of the γ chain. *Biochem.* **23**: 1767-1774.
30. LEVINE, R. and GOLDSMITH, H.L. 1977. Particle behavior in flow through small bifurcations. *Microvasc. Res.* **14**: 319-344.
31. LEUNG, L. and NACHMAN, R. 1986. Molecular mechanisms of platelet aggregation. *Ann. Rev. Med.* **37**: 179-186.
32. MARCUS, A.J. 1982. Platelet aggregation. *In*: "Hemostatis and Thrombosis" (Colman, R.W., Hirsh, J., Marder, V.J. and Salzman, E.W., eds.) J.B. Lipincott, Philadelphia. pp. 380-389.
33. MARGERIE, G.A., EDGINGTON, G.S. and PLOW, E.F. 1980. Interaction of fibrinogen with its receptor as a part of a multistep reaction in ADP-induced platelet aggregation. *J. Biol. Chem.***255**: 154-161.
34. MacDONALD, D.A. 1974. Blood Flow in Arteries, 2nd ed., Edward Arnold, Baltimore, 496 pp.
35. MOTOMIYA, M. and KARINO, T. 1984. Particle flow behavior in the human carotid artery bifurcation. *Stroke* **15**: 50-56.
36. MUSTARD, J.F. 1974. Platelets and thrombosis in acute myocardial infarction. *In*: "The Myocardium: Failure and Infarction" (Braunwald, E., ed.) HP Publishing Co., Inc., New York. pp. 177-190.
37. MUSTARD, J.F., PACKHAM, M.A., KINLOUGH-RATHBONE, R.L., PERRY, D.W. and REGOECZI, E. 1978. Fibrinogen and ADP-induced platelet aggregation. *Blood* **52**: 453-466.
38. NEWMAN, P.J., McEVER, R.P., DOERS. M.P. and KUNICKI, T.S. 1987. Synergistic action of two murine monoclonal antibodies that inhibit ADP-induced platelet aggregation without blocking fibrinogen binding. *Blood* **69**: 668-676.
39. NURDEN, A.T. 1987. Platelet membrane glycoproteins and their clinical aspects. In: "Thrombosis and Haemostasis" (Verstraete, M., Vermylen, J., Lijnen, R and Arnout, J., eds.) Leuven Univ. Press, Leuven, Belgium. pp. 93-125.
40. PEERSCHKE, E.I., ZUCKER, M.B., GRANT, R.A., EGAN, J.J. and JOHNSON, M.M. 1980. Correlation between fibrinogen binding for human platelet and platelet aggregability. *Blood* **55**: 841-847.
41. PHILLIPS, D.R. and POH AGIN, P. 1977. Platelet plasma membrane glycoproteins. *J. Biol. Chem.* **252**: 2121-2128.
42. REIMERS, H.J., PACKHAM, M.A., KINLOUGH-RATHBONE, R.L. and MUSTARD, J.F. 1977 Adenine nucleotides in thrombin-degranulated platelets: effect of prolonged circulation in vivo. *J. Lab. Clin.Med.* **90:** 490-501.
43. REIMERS, H.J., SUTERA, S.P. and JOIST, J.H. 1984. Potentiation by red cells of shear-induced platelet aggregation: relative importance of chemical and physical mechanisms. *Blood* **64**: 1200-1206.
44. SANTORO, S.A. and LAWING, W.J. 1987. Competition for related but nonidentical binding sites on the glycoprotein IIb-IIIa complex by peptides derived from platelet adhesive proteins. *Cell* **48**: 867-873.
45. SMOLUCHOWSKI, M. VON. 1917. Versuch einer Mathematischen Theorie der Koagulationskinetik kolloider Lösungen. *Z. Physik. Chem.* **92**:129-168.
46. SWIFT, D.L. and FRIEDLANDER, S.K. 1964. The coagulation of hydrosols by Brownian motion and laminar shear flow. *J. Colloid Sci.* **19**: 621-647.

47. TANGELDER, G.J., TEIRLINCK, H.C. , SLAAF, D.W. and RENEMAN, R.S. 1985. Distribution of blood platelets in flowing arterioles. *Am. J. Physiol. (Heart Circ. Physiol.)*: **248**: H318-H323.
48. TURITTO, V.T., BENIS, A.M. and LEONARD, E.F. 1972. Platellet diffusion in flowing blood. *Ind . Eng. Chem. Fundam.* **11**: 216-223.
49. VAN DE VEN, T.G.M. and MASON, S.G. 1977. The microrheology opf colloidal dispersions. VII. Orthokinetic doublet formation of psheres. *Colloid Polymer Sci.* **255**: 468-479.
50. VIRCHOW, R. VON. 1862. Phlogose und Thrombose im Gefässystem. Gessamelte Abhandlungen zur Wissenschaftlichen Medizin. Max Hirsch, Berlin.
51. WHITE, J.G. 1987. Platelet Ultrastructure. *In*: "Haemostasis and Thrombosis" (Bloom, A.L. and Thomas, D.P., eds.) Churchill Livingstone, Edinburgh. pp. 20-46.
52. ZEICHNER, G.R. and SCHOWALTER, W.R. 1977. Use of trajectory analysis to study stability of colloidal dispersions in flow fields. *Am. Inst. Chem. Eng. J.* **23**: 243-254.

From the McGill University Medical Clinic, The Montreal General Hospital, 1650 Cedar Avenue, Montreal, Quebec, H3G 1A4, Canada.

ACKNOWLEDGMENTS

The work was supported by grant MT-1835 from the Medical Research Council of Canada and by the Quebec Heart Foundation. The authors gratefully acknowledge helpful discussions with Drs. M.M. Frojmovic and T.G.M. van de Ven.

J.G. GODDARD AND Y.M. BASHIR

On Reynolds dilatancy

"Without attempting anything like a complete dynamical theory ⋯ I would point out the existence of a singular fundamental property of granular media ⋯ I have called ⋯ 'dilatancy', because the property consists in a definite change of bulk, consequent on a definite change of shape or distortional strain ⋯" O. Reynolds (1885) [18].

1 Introduction

In the veritable gem of Victorian science quoted above, Reynolds first revealed, conceptually and experimentally, the remarkable phenomenon of *volume-coupled shape change*, a revelation which has far-reaching implications for the mechanics of granular media and soils.

Much importance has been attached to Reynolds' concept in the literature on granular media and soil mechanics [16, 19, 20, 21, 22] and various attempts have been made to incorporate it into constitutive models [8, 10, 11, 12]. However, with few exceptions [9, 19, 20][1], it appears that past workers have either failed to recognize Reynolds dilatancy as a *kinematical constraint* or otherwise have not pursued its full mathematical consequences.

Unfortunately, the term "dilatancy" is also applied nowadays to the rheological phenomenon of "shear thickening", apparently because of the erroneous association by Freundlich and coworkers [5] of two phenomena later revealed to be independent [13]. (See [2] for a review of the latter.)

The purpose of the present article is to recast Reynolds' concept in modern continuum mechanical terms, which will not only serve to distinguish it from related ideas, but will also clarify further its fundamental significance for the mechanics of granular materials. In particular, we assert at the outset that, in order to achieve its full mathematical utility, Reynolds dilatancy must be interpreted as an internal kinematical constraint.

In the absence of any such internal constraint, the volume or density of a compressible material is of course independent of its shape; that is, volumetric and shear strains represent independent kinematic quantities. Of course, certain materials may exhibit a tendency towards shape- (or shear-) dependent expansion or contraction, owing to the constitutive dependence of isotropic stress or pressure on shear strain, as in the static deformation of

[1] Rowe [19, 20] mentions the "dilatancy constraint" but does not explicitly identify the stress indeterminacy. Kanatani [9] extends Reynolds' constraint on infinitesimal strains and appears to identify the stress indeterminacy associated with a standard plasticity model.

linear-anisotropic or nonlinear elastic bodies. Similar effects can also occur in the flow of fluids and history-dependent materials. For example, in an isothermal steady simple shear or *viscometric motion* of a compressible fluid-like material, the pressure (isotropic part of stress) p should depend in general on the density and shear rate. Such an effect, anticipated in the *dilatant fluid* of Reiner [17], corresponds to Bagnold's [1] *dispersive pressure* in high-speed granular flows, as well as the shear-rate dependent pressures observed in recent "molecular-dynamics" computer simulations [4] of hard-sphere fluids. In such materials, it is evident that, at fixed ambient pressure p, the density will depend on the shear rate.

This tendency toward shape-dependent dilation represented by *Reiner-Bagnold dilatancy* is to be clearly distinguished from the absolute constraint represented by Reynolds dilatancy, and reflected by his careful appellation "definite" in the above quotation. While Reynolds dilatancy might be viewed mathematically as a limiting form of Reiner-Bagnold dilatancy, their physical origins in granular materials appear to be quite different, the latter being the result of granular kinetic energy or inertia, and the former reflecting steric or geometric effects which are operative in the quasi-static motion of nearly rigid granules.

We present next the outlines of a continuum theory of Reynolds dilatancy, regarded as an *evolutionary* or *history-dependent* kinematical constraint on deformation. For the purposes of this discussion, we assume that the continuum in question can be treated as a *simple material*, in the sense of Noll [23], but we suppress the standard labelling of material particles and notation for (strain) history dependence, *etc.*. Of course, since the most plausible realm of application and the examples considered here lie within the field of granular materials[2] , we assume that these can in some appropriate sense be represented in the above continuum framework. We note that the assumption of nearly rigid grains, with perhaps slight contact elasticity, constitutes an excellent approximation for many real granular materials.

2 Review and Extension of the Reynolds Theory

Proceeding from microstructural ideas based on assemblages of equal rigid spheres, Reynolds [18] asserts that, in a state of maximum density, any contraction in one direction of a granular material is accompanied by equal extensions in mutually perpendicular directions. In a more modern tensor notation and in terms of the principal values $\dot{\epsilon}_i$ of the strain-rate tensor (or stretching) $\mathbf{D}$, the Reynolds hypothesis becomes:

$$\dot{\epsilon}_2 = \dot{\epsilon}_3 = -\dot{\epsilon}_1 \tag{1}$$

[2]despite Reynolds own disclaimer and apparent predilections to the contrary (*pp.* 476 *ff.* of [18]).

Since the rate of volumetric strain $\dot{\epsilon}_v$ is

$$\dot{\epsilon}_v = tr(\mathbf{D}) = \dot{\epsilon}_1 + \dot{\epsilon}_2 + \dot{\epsilon}_3 = -\dot{\epsilon}_1 \tag{2}$$

it follows that

$$\dot{\epsilon}_v = |\dot{\epsilon}_1| \tag{3}$$

at the state of maximum density.

In order to extend Reynolds' elementary arguments to arbitrary states of deformation we adopt here, without elaboration, a generalization of the theory of internally constrained continua (*pp.* 69-73 of [23]) in which allowance is made for *non-holonomic, evolutionary* constraints, that is, constaints on kinematics which are not generally derivable from constraints on static configurations (a matter well known in the classical mechanics of systems with finite degrees of freedom [14]) and which, moreover, may otherwise depend on configuration history.

As the simplest generalization of the Reynolds infinitesimal-strain theory, and in an invariant form applicable for any deformation, we write:

$$\dot{\epsilon}_v = \alpha\,|\mathbf{D}| \tag{4}$$

where

$$|\mathbf{D}| = [tr(\mathbf{D}^2)]^{1/2} \tag{5}$$

or

$$\dot{\epsilon}_v = \alpha(\dot{\epsilon}_1^2 + \dot{\epsilon}_2^2 + \dot{\epsilon}_3^2)^{1/2} \tag{6}$$

Eqs. (1), (3) and (6) then yield

$$\alpha = 1/\sqrt{3} \tag{7}$$

at the state of maximum density. In any other state, the scalar quantity α, which we call the *coefficient of dilatancy*, may in general depend on the entire history of deformation. The classical strain-based (*holonomic*) theory of Reynolds produces a simple expression for α with a periodic dependence on strain and allows for compaction, $\alpha < 0$, as well as dilation, $\alpha > 0$, but it is doubtful that his elementary arguments would hold far from the state of maximum density. Also, while Reynolds appeals to the notion of an *amorphous* or *isotropic* structure, the state of maximum density for mono-sized spheres (or discs in two dimensions) corresponds in fact to hexagonal close-packed arrays which are mechanically *anisotropic* or *crystalline*. As a consequence, there are evident inconsistencies between microstructural and continuum models. More is said below about the functional form of α in (4).

To illustrate some of the microstructural considerations involved, let us repeat Reynolds' argument in a form suitable for two-dimensional arrays of equal circular discs or cylinders. Assuming a similar deformation mechanism, one can see from Fig. 1 that the ratio of the area A of the deformed triangle $a'b'c'$ to the area A_0 of the undeformed triangle abc, representing the two-dimensional volumetric strain ϵ_v, is given by

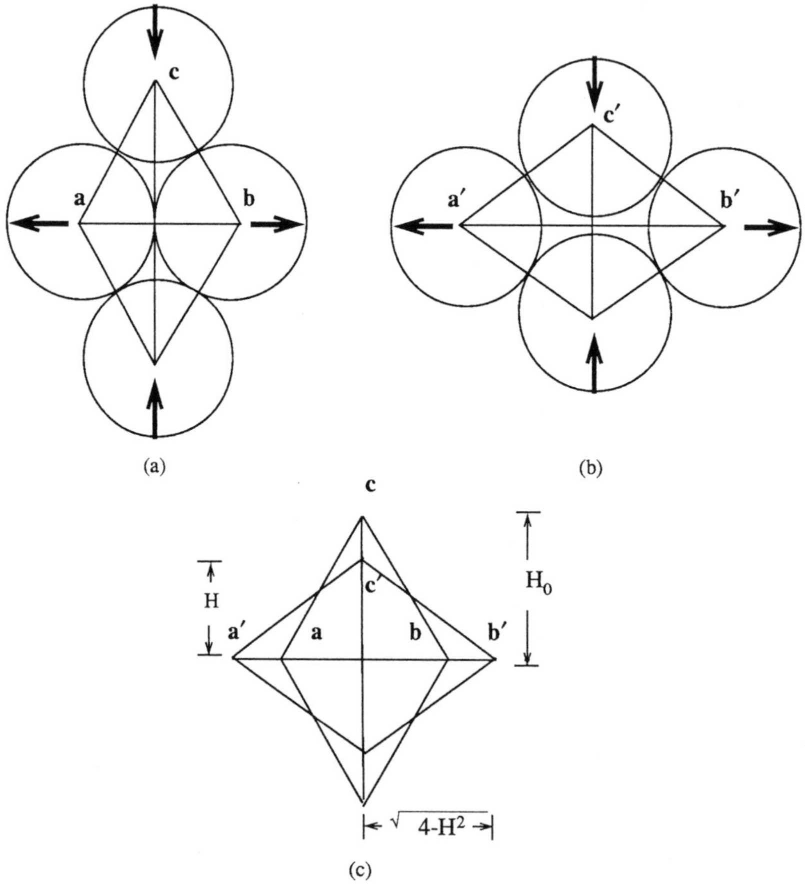

Figure 1: Dilatancy mechanism for 2-D discs.

$$\epsilon_v = \ln \frac{A}{A_0} = \ln \frac{H\sqrt{(2R)^2 - H^2}}{H_0\sqrt{(2R)^2 - H_0^2}} = \ln \lambda_v \tag{8}$$

where

$$\lambda_v = \lambda_1\lambda_2, \quad \lambda_1 = \frac{H}{H_0}, \quad \lambda_2 = \frac{\sqrt{(2R)^2 - H^2}}{\sqrt{(2R)^2 - H_0^2}}$$

For close-packed discs with unit radius, $R = 1$, $H_0 = \sqrt{3}$ and $H = \sqrt{3}\lambda_1$. Therefore

$$\lambda_2 = \sqrt{4 - 3\lambda_1^2} \tag{9}$$

and

$$\epsilon_v = \ln\left(\lambda_1\sqrt{4 - 3\lambda_1^2}\right) \tag{10}$$

Taking time derivatives in (10), with $\dot{\epsilon}_i = \dot{\lambda}_i/\lambda_i$ we, obtain

$$\dot{\epsilon}_v = \frac{-3{\lambda_1}^3\dot{\epsilon}_1}{(4 - 3\lambda_1^2)} + \lambda_1\dot{\epsilon}_1 \tag{11}$$

and

$$\dot{\epsilon}_2 = \frac{-3\lambda_1\dot{\epsilon}_1}{\sqrt{4 - 3\lambda_1^2}} \tag{12}$$

At the maximum density, $\lambda_1 = 1$, and Eqs. (11) and (12) become

$$\dot{\epsilon}_v\big|_{\lambda_1=1} = -3\dot{\epsilon}_1 + \dot{\epsilon}_1 = -2\dot{\epsilon}_1 = 2\,|\dot{\epsilon}_1| \tag{13}$$

and

$$\dot{\epsilon}_2\big|_{\lambda_1=1} = -3\dot{\epsilon}_1 \tag{14}$$

But since

$$|\mathbf{D}| = \left[\dot{\epsilon}_1^2 + \dot{\epsilon}_2^2\right]^{1/2} = \left[\dot{\epsilon}_1^2 + (3\dot{\epsilon}_1)^2\right]^{1/2} = \sqrt{10}\,|\dot{\epsilon}_1| \tag{15}$$

one obtains from equations (13), (14) and (15) for the two-dimensional case that

$$\dot{\epsilon}_v = \sqrt{\frac{2}{5}}\,|\mathbf{D}| \tag{16}$$

or $\alpha = \sqrt{2/5}$. The above reasoning is similar to that employed by Rowe [19], whose experiments on hexagonally close-packed spheres and rods give close agreement with the Reynolds maximum-density values.

To illustrate the application to a dilatant simple shear, involving a mandatory simultaneous planar isotropic expansion, we note that the x, y, z (1,2,3) components of $\mathbf{D}$ are given by

$$\mathbf{D} = \frac{1}{2}\begin{pmatrix} \dot{\epsilon}_v & \dot{\gamma} & 0 \\ \dot{\gamma} & \dot{\epsilon}_v & 0 \\ 0 & 0 & 0 \end{pmatrix} \tag{17}$$

so that Eq. (4) gives for the *shearing dilation*

$$s := \frac{d\epsilon_v}{d\gamma} = \frac{\dot{\epsilon}_v}{\dot{\gamma}} = \frac{\alpha}{(2-\alpha^2)^{1/2}} \tag{18}$$

Thus, it is seen by (16) that $s = 1/2$ for discs in 2-D, whereas for equisized spheres in 3-D one obtains from (4) $s = 1/\sqrt{5} = 0.447...$ for the dilatant simple shear (17), and $s = \sqrt{3}/4 = 0.433...$ for simple shear with isotropic 3-D expansion, while $s = 1/(2\sqrt{2}) = 0.353...$ for a uniaxial compression with γ denoting the usual definition of shear strain.

In a computer simulation of random 3-D systems of spheres, Zhang and Cundall [24] find $s \doteq 0.36$ compared to the Reynolds-theory value 0.353... at maximum density. Given that their systems were not the monosized spheres assumed by Reynolds [18], the agreement is remarkable, which fact appears to have been overlooked in [24]. It is also noteworthy that the slope of Zhang and Cundall's curve of volumetric strain versus shear strain is almost constant throughout entire deformation range (up to shears of greater than 0.35) and not limited to the state of maximum density.[3] On the other hand, in our own numerical simulations of 2-D systems of monosized discs [3], slightly expanded and randomized about triangular-close packing, we have found $s \doteq 0.86$ *vs.* the value of 0.5 given above, a decidedly poorer agreement which may be due to anisotropy. At any rate, in many of the numerical simulations to date it appears that s or α are almost independent of intergranular (*Coulombic*) friction.

Although one can conceive of much more general expressions we shall, for the purposes of the present discussion, largely adhere to the elementary form (4) which, while isotropic in appearance (involving simple proportionality between isotropic scalar invariants of $\mathbf{D}$), can exhibit anisotropy through dependence of α on strain, strain history or strain-rate $\mathbf{D}$. The form of this dependence is of course severely restricted if (4) is to have the standard rate-independent character associated with the plasticity of granular media; *in particular, α cannot depend on* $|\mathbf{D}|$.

3 Normality Constraint and Stress Indeterminacy

In contrast to the usual holonomic formulation [23], we postulate here that an internal material constraint can be represented by means of a second-rank symmetric tensor $\mathbf{N}$, in the form:

$$\mathbf{N} \cdot \mathbf{D} \equiv tr(\mathbf{N}\mathbf{D}) \equiv N_{ij} D_{ij} = 0 \tag{19}$$

[3] We recall that, in initially compacted states, real granular materials such as sand exhibit numerically comparable dilatancy, which appears eventually to saturate at a "critical-state" $s = 0$ for large strains [15].

where, as in the following, the centered dot represents the corresponding scalar product. In the six-dimensional vector space of strains, $\mathbf{N}$ may be viewed as the vector normal to a hyperplane (linear space) to which the strain-rate $\mathbf{D}$ is constrained. In general, each point in strain space through which a material point moves has a different hyperplane attached to it.

In the usual holonomic theories [23, 9] where $\mathbf{N}$ is assumed derivable from a scalar constraint on strain, the hyperplane in question represents the local *tangent space* or *linear support* for the associated manifold. As standard examples we have incompressible materials, where $\mathbf{N}$ is proportional to the unit tensor $\mathbf{1}$ ($N_{ij} = const. \times \delta_{ij}$), or materials inextensible along certain material elements or "fibers", in which case $\mathbf{N}$ is proportional to the dyadic or tensor product $\mathbf{ee} := \mathbf{e} \otimes \mathbf{e}$ ($N_{ij} = const. \times e_i e_j$), where $\mathbf{e}$ is a physical-space vector lying along the fiber direction.

For the dilatant granular mass defined by (4), we might take

$$\mathbf{N} = const. \times (\alpha\hat{\mathbf{D}} - \mathbf{1}) \tag{20}$$

in (19) where, as below, we write

$$\hat{\mathbf{B}} = \frac{\mathbf{B}}{|\mathbf{B}|}, \quad \text{with} \quad |\mathbf{B}| = \left(tr\mathbf{B}^2\right)^{1/2} \tag{21}$$

for the *versor* or *director* of the symmetric second-rank tensor $\mathbf{B}$ [6, 7], in terms of which (4) reads

$$tr\hat{\mathbf{D}} = \hat{\mathbf{D}} \cdot \mathbf{1} = \sqrt{3}\hat{\mathbf{D}} \cdot \hat{\mathbf{1}} = \alpha \tag{22}$$

We note here that (20) is a special case of a more general form satisfying (4) and (19):

$$\mathbf{N} = \alpha\mathbf{A} - (\mathbf{A} \cdot \hat{\mathbf{D}})\mathbf{1} \tag{23}$$

where $\mathbf{A}$ is a history-dependent, symmetric second-rank tensor such that, for arbitrary strain histories,

$$\mathbf{A} \cdot \hat{\mathbf{D}} \neq 0 \tag{24}$$

This in general requires that $\mathbf{A}$ be given by a history-dependent linear form in $\hat{\mathbf{D}}$ (defined in turn by a *sign-definite* fourth-rank tensor [6, 7]). Equation (20) arises from the choice $\mathbf{A} = const. \times \hat{\mathbf{D}}$, which appears to be the simplest isotropic form in $\hat{\mathbf{D}}$ compatible with (4).

Whenever the quantity $\mathbf{N}$ in forms such as (19)-(23) is not derivable from a functional restriction on strain relative to a fixed state but, rather, depends on the entire strain-path or strain-history of a material point, we refer to (19) as *non-holonomic.* Because of the particulate microstructure and the general tendency towards hysteretic behavior, we expect that dilatancy in the quasi-static motion of granular media should be treated as non-holonomic. Thus, in essence the constraint itself becomes a rheological variable.

Whether holonomic or non-holonomic, the existence of internal constraints must lead to *rheologically indeterminate stresses* [23], a fact which does not appear to be fully appreciated in much of the literature on soil mechanics and granular media. In particular, as a direct extension of an existing principle (*pp.* 69-73 of [23]), we assert that, subject to the internal constraint (19), the stress tensor $\hat{\mathbf{T}}$ is rheologically determinate only up to an additive stress $\mathbf{T}^\circ$, say, which *does no work* in any deformation compatible with the constraint [23] . It follows that $\mathbf{T}^\circ$ is given by

$$\mathbf{T}^\circ = \lambda \mathbf{N} \tag{25}$$

where the constant of proportionality λ is a generalized *Lagrange multiplier* representing reaction to the constraint. As with pressure p in an incompressible material ($\mathbf{N} = \mathbf{1}$, $\lambda = -p$) or the "fiber tension" σ in an inextensible material ($\mathbf{N} = \mathbf{ee}$, $\lambda = \sigma$), the stress $\mathbf{T}^\circ$ in (25) represents a dynamical field variable which, not being determined by material deformation, is governed instead by the balance of momentum and imposed forces. Thus, the quantity λ resulting from the substitution (20) into (25) has to be determined from the solution to a specific mechanical boundary-value problem.

To illustrate the above principle for the representation (20) and the dilatant planar shear of (17), we recall that:

$$\mathbf{D} = \frac{\dot{\gamma}}{2}\begin{pmatrix} s & 1 \\ 1 & s \end{pmatrix}, \quad \mathbf{T} = \begin{pmatrix} -p_{11} & \tau \\ \tau & -p_{22} \end{pmatrix} \tag{26}$$

with

$$p = -(p_{11} + p_{22})/2 \tag{27}$$

where s is given by (18), while p denotes pressure and τ shear stress. Whence, it follows by (20) and (25) that

$$\mathbf{T}^\circ = const. \times \begin{pmatrix} -1 & s \\ s & -1 \end{pmatrix} \tag{28}$$

That is, the indeterminate stress represents a state in which

$$\tau = \ sp, \quad p_{11} = p_{22} = p \tag{29}$$

This becomes entirely transparent on recalling that the stress-work rate $\dot{W} = \mathbf{T} \cdot \mathbf{D}$ is given by (26) as

$$\dot{W} = \tau\dot{\gamma} - p\dot{\epsilon}_v = (\tau - sp)\dot{\gamma} \tag{30}$$

with the $\tau\dot{\gamma}$ representing the shear ("shape") work and $p\dot{\epsilon}_v$ the volume work. While independent for unconstrained materials these are connected by the relation $\dot{\epsilon}_v = s\dot{\gamma}$ for the dilatant material. Thus, that part of the stress which does the work of deformation is given by $\tau - sp$, so that any stress state of the form (29) is obviously work free.

As appreciated by Reynolds [18] and also by Rowe [19, 20], the above arguments have interesting consequences for the quasi-static motion of granular materials composed of rigid frictional particles, for which kinetic energy and elastic strain energy are strictly negligible. In this case, it can be seen that all the energy input (30) must be entirely accounted for by dissipation associated with (Coulombic) sliding friction. Therefore, it follows for the limiting case of *ideal frictionless* particles where $\mathbf{T} = \mathbf{T}^{\circ}$, that $\dot{W}$ in (30) must vanish identically and the stress state for simple shear takes on the form (29). That is, *all* stresses represent work-free reactions. It is noteworthy that one obtains in this limit a kind of hybrid of the inviscid fluid and the rigid solid.

In general, the above ideas appear crucial to the formulation of continuum rheological equations, for if the dilatancy and the associated form of the tensor $\mathbf{N}$ in (19) are known, then for any stress state $\mathbf{T}$ the rheologically determinate (frictional) stress $\mathbf{T}'$, say, is given generally by the *projection* normal to $\mathbf{N}$:

$$\mathbf{T}' = \mathbf{T} - (\mathbf{T} \cdot \hat{\mathbf{N}})\hat{\mathbf{N}} \equiv \mathbf{T} - tr(\mathbf{T}\hat{\mathbf{N}})\hat{\mathbf{N}} \tag{31}$$

(the analog of deviatoric stress for incompressible materials), where $\hat{\mathbf{N}}$ is the versor of $\mathbf{N}$. Viewed in these terms, the micromechanical energy-dissipation arguments and granular-plasticity model of Rowe [19] take on a much more transparent meaning, as indicated by the following discussion.

4 Granular Plasticity and the Stress-Dilatancy Model of Rowe

While long since superceded in the soil mechanics literature by much more elaborate models, the ideas of Rowe [19, 20] serves to illustrate nicely the role of the dilatancy constraint in the rate-independent plasticity and yield of granular materials. For that reason, it is worth reformulating in a modern form. To that end, we note that if α is independent of $|\mathbf{D}|$ then the relations (4) and (22) define a *cone* with director $\hat{\mathbf{1}}$ in strain-rate space (*i.e.*, a set $\mathcal{C}$ such that $\mathbf{D} \in \mathcal{C} \Rightarrow const. \times \mathbf{D} \in \mathcal{C}$) with included semi-angle $\theta = cos^{-1}(\alpha/\sqrt{3})$, as illustrated schematically in Figure 2. If α and, hence, θ are allowed to depend on $\hat{\mathbf{D}}$, *e.g.*, on $tr\hat{\mathbf{D}}^3$ for isotropic materials, then the cone is non-circular. As apparently appreciated by Reynolds [18] and Rowe [19, 20], the kinematical constraint manifold, represented by the hypersurface $\mathcal{C}$ in Figure 2, generates a *yield locus* in stress space for the frictionless material. In particular, it is easy to show that the reaction stress implied by (29) and (22) lies on the *orthogonal cone* $\mathcal{C}'$, defined by

$$\hat{\mathbf{T}} \cdot \hat{\mathbf{1}} = -\cos\theta' \tag{32}$$

where $\theta' = \pi/2 - \theta$. This situation is depicted in Figure 2 where the (six-dimensional) space of symmetric tensors is partitioned into stress and strain-rate spaces by the hyperplane passing through the origin $\perp$ to $\hat{\mathbf{1}}$. Without further assumptions whatsoever, we thus obtain the standard "normality" rule for plastic flow in this frictionless limit, even when $\mathcal{C}$ and $\mathcal{C}'$ are non-circular cones. Of course $\mathcal{C}'$ suffers an indeterminacy whenever $\mathcal{C}$ exhibits

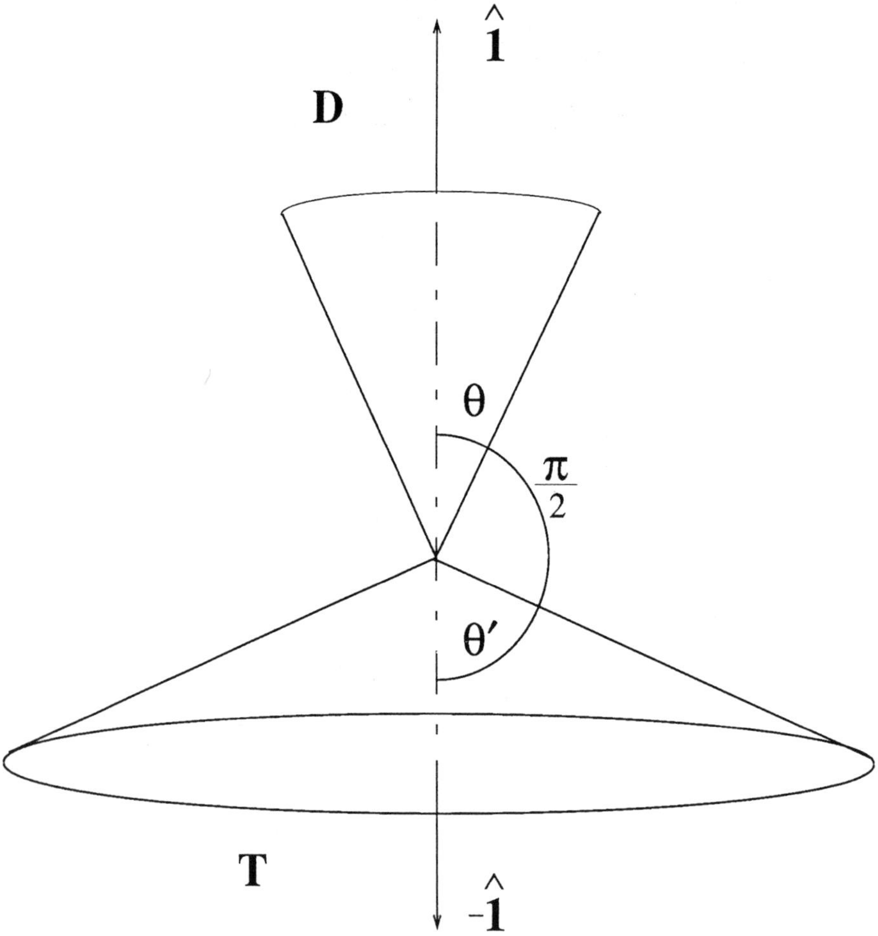

Figure 2: Constraint and yield-stress cones for a frictionless granular material.

sharp corners, the obverse of the usual corner singularity in stress-space models of plasticity. With the usual allowance for path-dependence or "work-hardening", the constraint and yield cones $\mathcal{C}$ and $\mathcal{C}'$ depend generally on the history of plastic deformation, so that Figure 2 represents a "snapshot" at the current time corresponding to a given plastic-strain history.

In the simplest interpretation of the "stress-dilatancy" model of Rowe [19, 20] the yield behavior for a frictional material is determined from that of the frictionless material and the coefficient of Coulombic intergranular friction. Thus, for the frictional material, one merely replaces $\cos\theta'$ in (32) by $\cos\phi$, where $\phi > \theta'$, corresponding to an exterior angle $< \pi/2$ between the cones of Figure 2 and ensuring positive dissipation:

$$(\mathbf{T} - \mathbf{T}^{\circ}) \cdot \mathbf{D} \equiv \mathbf{T} \cdot \mathbf{D} > 0 \tag{33}$$

However, it can be seen that this assumption of a dissipative conical yield surface in stress space vitiates the standard normality rule for plastic flow.

Many of the above issues, including the uniqueness of the reaction stress $\mathbf{T}^{\circ}$ and a more general treatment of the dilatancy constraint, appear worthy of further study which we hope this article will serve to motivate.

Aknowledgement

This work was supported in part by NSF Grant CTS-8615160, AFOSR Grant 87-0284 and by a computational grant from the San Diego Supercomputer Center. One of us (J.D.G.) benifitted from participation in the University of Minnesota IMA Minisymposium on Plasticity in Febuary 1989, where parts of this article took shape.

J. D. Goddard and Y. M. Bashir, Department of Chemical Engineering, University of Southern California, Los Angeles, CA 90089-1211.

References

[1] Bagnold, R. A., 1954, *Proc. Roy. Soc. A*, **225**, 49.

[2] Barnes, H. A., 1989, *J. of Rheology*, **33(2)**, 329.

[3] Bashir, Y. M., 1990, Ph. D. Dissertation, in progress, University of Southern California.

[4] Evans, D. J. , 1983, *J. Chem. Phys.*, **78**, 2297.

[5] Freundlich, H. and Roder, H. L., 1938, *Trans. Faraday Soc.*, **34**, 308.

[6] Goddard, J. D. , 1984, *J. of Non-Newtonian Fluid Mech.*, **14**, 141.

[7] Goddard, J. D. , 1986, *Acta Mechanica*, **63**, 3.

[8] Harris, D. , 1985, *J. Mech. Phys. Solids*, **33**, 51.

[9] Kanatani, K.-I., 1982, in *Deformation and Failure of Granular Materials*, (Eds., P. A. Vermeer and H. J. Luger), A. A. Balkema.

[10] Mehrabadi, M. M. and Cowin, S. C., 1978, *J. Mech. Phys. Solids*, **26**, 269.

[11] Mehrabadi, M. M. and Cowin, S. C., 1982, in *Deformation and Failure of Granular Materials* (Eds., P. A. Vermeer and H. J. Luger), A. A. Balkema.

[12] Mehrabadi, M. M. and Nemat-Nasser, S. , 1983, *Mechanics of Materials*, **2**, 155.

[13] Metzner, A. B. and Whitlock, M., 1958, *Trans. Soc. Rheology*, **2**, 239.

[14] Neimark, Ju. I. and Fufaev, N. A. , 1972, *Dynamics of Non-holonomic Systems* (Translation from the Russian by J. R. Barbour), American Mathematical Society.

[15] Newland, R. L. and Allely, B. L. , 1957, *Geotechnique*, **7**, 17.

[16] Oda, M. , Konishi, J. and Nemat-Nasser, S. , 1983, in *Mechanics of Granular Materials: New Models and Constitutive Relations*, J. T. Jenkins and M. Satake (Eds.), Elsevier.

[17] Reiner, M., 1945, *American Journal of Mathematics*, **67**, 350.

[18] Reynolds, O., 1885, *Phil. Mag.* , **20**, 469.

[19] Rowe, P. W. , 1962, *Proc. Roy. Soc. Lond.*, **A264**, 500.

[20] Rowe, P. W. , 1963, *JSMF-ASCE*, **89**, 37.

[21] Seed, H. B. and Lee, K. L., 1966, *SMFD-ASCE*, **92**, 105.

[22] Seed, H. B. , 1987, *Journal of Geotech. Eng.*, *13*, 827.

[23] Truesdell, C. and Noll, W. , 1965, Non-Linear Field Theories of Mechanics, in *Encyclopedia of Physics* (ed. S. Flügge), Vol. 3, Springer-Verlag.

[24] Zhang, Y. and Cundall, P. A. , 1986, *Proc. Tenth U. S. Congress Appl. Mech.*, Austin, Texas.

R.G. LARSON

A method for calculating orientational distribution functions in flows of nematic polymers

I. INTRODUCTION

When rod-like polymers are concentrated sufficiently, their statics and dynamics are significantly influenced by the effects of orientation-dependent excluded volume. When strong enough, these excluded-volume effects can lead to the formation of an equilibrium anisotropic orientation distribution of the molecules — i.e., a liquid crystalline phase. Even when not quite this strong, such effects can influence the flow-induced stress and birefringence in solutions that are isotropic at equilibrium[7]. If the rod-like molecules are rigid and long and thin enough, excluded-volume effects become strong even at rather low volume fractions of molecules, low enough that the effects can accurately be described by the second order term of a virial expansion in concentration. Onsager[10] calculated this second-order virial term by considering pairwise interactions of particles. From Onsager's results, one finds that the repulsive potential V_O that a rod with orientation described by a unit vector $\underset{\sim}{u}$ feels because of the presence of identical neighboring particles is

$$V_O(\underset{\sim}{u}) = 2cdL^2kT \int \psi(\underset{\sim}{u}')\sin(\underset{\sim}{u}', \underset{\sim}{u})\, du'^2 \qquad (1)$$

Here d is the rod diameter, L its length, c is the number of molecules per unit volume, $\psi(\underset{\sim}{u}')$ is the probability that a rod has an orientation given by the unit vector $\underset{\sim}{u}'$, and $\sin(\underset{\sim}{u}', \underset{\sim}{u})$ is the sine of the angle between the unit vectors $\underset{\sim}{u}'$ and $\underset{\sim}{u}$ describing the orientations of two rods. The integral is over the surface of a unit sphere in orientation space. Note that the excluded volume potential is most repulsive when the rods are perpendicular to each other.

Doi[1] has incorporated excluded-volume effects of this kind into a dynamic theory of orientation and stress in flows of solutions of rod-like particles. The Doi equation is

$$\frac{\partial \psi}{\partial t} + \frac{\partial}{\partial \underset{\sim}{u}} \cdot [(\underset{\sim}{u} \cdot \underset{\sim}{\nabla} \underset{\sim}{v} - \underset{\sim}{u}\,\underset{\sim}{u}\,\underset{\sim}{u} : \underset{\sim}{\nabla} \underset{\sim}{v})\psi]$$

$$- \bar{D}_r \frac{\partial}{\partial \underset{\sim}{u}} \cdot [\frac{\partial \psi}{\partial \underset{\sim}{u}} + \psi \frac{\partial}{\partial \underset{\sim}{u}}(\frac{V_{ev}}{kT})] = 0 \tag{2}$$

where $\underset{\sim}{\nabla} \underset{\sim}{v}$ is the velocity gradient, $\bar{D}_r$ is an effective orientation-independent rotary diffusivity, and V_{ev} is an effective excluded-volume potential, such as that of Onsager in Eqn. (1). $\frac{\partial}{\partial \underset{\sim}{u}}$ is the gradient operator on the unit sphere.

Because of the complexity of this theory, calculations of flow-induced stress and orientation have thus far been carried out only with the aid of two important approximations. The first of these is to replace the Onsager excluded-volume potential V_O by the phenomenological Maier-Saupe form, $V_{MS} = UkT \underset{\sim}{u}\,\underset{\sim}{u} : \underset{\approx}{S}$, where $\underset{\approx}{S}$ is the tensor order parameter,

$$\underset{\approx}{S} \equiv \langle \underset{\sim}{u}\,\underset{\sim}{u} \rangle - \frac{1}{3}\underset{\approx}{\delta} \tag{3}$$

U is the strength of the excluded volume potential, and the brackets denote an average over ψ,

$$<\cdots> \equiv \int \psi(\underset{\sim}{u}) \cdots du^2 \tag{4}$$

The Maier-Saupe potential can be derived from the Onsager from by truncating the latter at low order in $\underset{\approx}{S}$, but the result differs considerably in its quantitative predictions from the Onsager potential when $\underset{\approx}{S}$ takes on values typical of the liquid crystalline phase. Furthermore, unless the Maier-Saupe potential given above is augmented by adding additional terms, it is incapable of predicting the range of concentrations over which equilibrium isotropic and liquid crystalline phases can coexist. Thus with the Maier-Saupe potential one cannot make clear-cut predictions of the conditions under which shearing might induce the discontinuous transformation of an equilibrium isotropic phase into a liquid crystalline one[7].

A second, even more serious, approximation is to replace fourth moments of the orientation distribution function by products of the second moment in the evolution equation for the second moment and in the equation for the stress tensor. The equation for the second moment is obtained from Eqn. (2) by multiplying it by $\underset{\sim}{u}\,\underset{\sim}{u}$ and integrating over the unit sphere du^2. By replacing the fourth moment $<\underset{\sim}{u}\,\underset{\sim}{u}\,\underset{\sim}{u}\,\underset{\sim}{u}>$ that results from this procedure by a product of second moments $<\underset{\sim}{u}\,\underset{\sim}{u}><\underset{\sim}{u}\,\underset{\sim}{u}>$, a closure is effected on the equation for the second moment, decoupling it from equations for the higher moments. By a similar approximation, the stress tensor is obtained from an equation involving the second moment alone. Hence in this way the stresses can be calculated without needing to obtain the distribution of rod orientations. Although this closure procedure vastly simplifies the problem, an important property of shearing flows of liquid crystalline phases is thereby lost, namely the tendency of rods to *tumble* collectively at low shear rates. The tumbling phenomenon was predicted by Kuzuu and Doi[4], who solved Doi's original dynamic equation (2) with the Onsager potential and without any closure approximations by a perturbation technique valid only at shear rates so low that the shear stress is linear in the shear rate. At higher shear rates, Marrucci and Maffettone[6] were able to solve a two-dimensional analog of the Doi equation without any closure approximations; this solution shows that the tumbling mechanism can lead directly or indirectly to a prediction of *negative first normal stress differences*, an experimentally observed[3,9] phenomenon unique to liquid crystalline polymers. The predictions of tumbling and of negative first normal stress differences disappear when one uses the closure approximation described above.

From the above remarks, one gathers that it is important to calculate the three dimensional orientational distribution function ψ in flows of rod-like polymers with as few approximations as possible. In what follows, we present a scheme for calculating ψ and the stresses with the full Onsager potential and without artificial closures. The only approximation to the original Doi equation that we retain is the use of an averaged rather than an orientation-dependent rotary Brownian diffusivity.

II. EXPANSION IN SPHERICAL HARMONIC FUNCTIONS

Our approach is to expand the distribution function ψ in terms of the spherical harmonic functions Y_l^m. Recursion formulas for the Y_l^m and the lowest order Y_l^m are given in Messiah[8]. An expansion of this kind was carried out by Doi and Edwards[2] for Eqn. (2), but without the excluded volume potential V_{ev}. Doi and Edwards were seeking to describe the flow dynamics of less concentrated solutions of rod-like polymers for which V_{ev} can be neglected. Here we extend this earlier work by expanding V_{ev} in terms of the Y_l^m, derive the contribution of V_{ev} to the stress tensor in terms of the Y_l^m, and we also correct an error we found in the convection term of the earlier expansion in ref. [2].

Following Doi and Edwards[2], we define x as the flow direction of the simple shearing flow, z as the direction of the gradient of the flow, and y as the vorticity direction. Using the symmetries of the shear flow, Doi and Edwards expressed ψ as

$$\psi(\underset{\sim}{u},t) = \sum_{\substack{\ell=0 \\ \ell\,\text{even}}}^{\infty} \sum_{m=0}^{\ell} b_{\ell m}\,|\ell m) \tag{5}$$

where

$$|\ell m) \equiv \begin{cases} Y_l^m(\underset{\sim}{u}) & \text{for } m=0 \\ \dfrac{1}{\sqrt{2}}(Y_l^m(\underset{\sim}{u}) + (-1)^m Y_\ell^{-m}(\underset{\sim}{u})) & \text{for } m\neq 0 \end{cases} \tag{6}$$

We now substitute this expression for ψ into Eqn. (2):

$$\frac{\partial}{\partial t}|\psi) = -|\frac{\partial}{\partial \underset{\sim}{u}}\cdot\{[\underset{\sim}{u}\cdot\underset{\sim}{\nabla}\underset{\sim}{v} - \underset{\sim}{u}\underset{\sim}{u}\underset{\sim}{u}\underset{\sim}{\nabla}\underset{\sim}{v}]\psi\})$$

$$+ \bar{D}_r|\frac{\partial}{\partial \underset{\sim}{u}}\cdot\frac{\partial}{\partial \underset{\sim}{u}}\psi) + \frac{\bar{D}_r}{kT}|\frac{\partial}{\partial \underset{\sim}{u}}\cdot[\psi\frac{\partial}{\partial \underset{\sim}{u}}V_{ev}]) \qquad (7)$$

In the above, we use the convention that an expression enclosed within the symbols "|)", should be replaced by its expansion in terms of spherical harmonics. For example,

$$|\psi) \equiv \sum_{\substack{\ell'=0 \\ \ell' \text{ even}}}^{\infty} \sum_{m'=0}^{\ell'} b_{\ell' m'}\,|\ell', m') \qquad (8)$$

Now the inner product $(\ell m|X)$, where X is some expression, is formed by multiplying $|X)$ by $|\ell\ m)$ and then integrating over the unit sphere. That is,

$$(\ell m|X) \equiv \int |\ell m)X\,du^2 \qquad (9)$$

We note that $|\ell m)$ is its own Hermitian conjugate — i.e., $(\ell m| = |\ell m)$ — so that $(\ell m|X)$ is an inner product. By the orthonormality of the spherical harmonic functions,

$$(\ell' m'|\ell m) = \delta_{mm'}\delta_{\ell\ell'} \qquad (10)$$

where δ_{jk} is unity if j=k; zero otherwise.

We now multiply Eqn. (7) by $|\ell m)$ and integrate over the unit sphere to give

$$\frac{\partial}{\partial t}(\ell m\,|\,\psi) = -\,(\ell m\,|\,\frac{\partial}{\partial \underset{\sim}{u}}\cdot\{[\underset{\sim}{u}\cdot\nabla \underset{\sim}{v} - \underset{\sim}{u}\,\underset{\sim}{u}\,\underset{\sim}{u}:\nabla \underset{\sim}{v}]\psi\})$$

$$+\ \bar{D}_r(\ell m\,|\,\frac{\partial}{\partial \underset{\sim}{u}}\cdot\frac{\partial}{\partial \underset{\sim}{u}}\psi) + \frac{\bar{D}_r}{kT}(\ell m\,|\,\frac{\partial}{\partial \underset{\sim}{u}}\cdot[\psi\frac{\partial}{\partial \underset{\sim}{u}}V_{ev}]) \tag{11}$$

Thus for each pair of values $\{\ell,m\}$, there is an inner-product equation (11) and there is an unknown coefficient $b_{\ell m}$. Hence (11) is a well-posed system of ordinary differential equations for the coefficients $b_{\ell m}$.

It remains to evaluate $\bar{D}_r$ and the inner products in (11) in terms of the $b_{\ell m}$'s. Expressions for $\bar{D}_r$ and two of these inner products were given correctly by Doi and Edwards[2]:

$$\bar{D}_r = D_r\left(1 - 8\pi \sum_{\substack{\ell=2\\ \ell\ \text{even}}}^{\infty} \sum_{m=0}^{\ell} \left(\frac{\ell-1}{\ell+2}\right)\left[\frac{(\ell-3)!!}{\ell!!}\right]^2 b_{\ell m}^2\right)^{-2} \tag{12}$$

$$(\ell,m\,|\,\psi) = b_{\ell m} \tag{13}$$

$$(\ell m\,|\,\frac{\partial}{\partial \underset{\sim}{u}}\cdot\frac{\partial}{\partial \underset{\sim}{u}}\psi) = \ell(\ell+1)b_{\ell m} \tag{14}$$

Here we use the notation

$$n!! \equiv \begin{cases} n(n-2)(n-4).....(1)\,, & \text{for n odd} \\ n(n-2)(n-4).....(2)\,, & \text{for n even} \\ 1\,, & \text{for } n \leq 1 \end{cases} \tag{15}$$

Doi and Edwards give an apparently erroneous expression for the convection term, so we here provide one that is correct, and give an expression for the term involving the Onsager potential.

A. Convection Term

With Doi and Edwards, we find that $\frac{\partial}{\partial \underset{\sim}{u}} \cdot \{[\underset{\sim}{u} \cdot \nabla \underset{\sim}{v} - \underset{\sim}{u}\underset{\sim}{u}\underset{\sim}{u} : \nabla \underset{\sim}{v}]\psi\}$ can be written as $\dot{\gamma}\Gamma\psi$ where $\dot{\gamma}$ is the shear rate and

$$\Gamma = \cos^2\theta \cos\phi \frac{\partial}{\partial \theta} - \cot\theta \sin\phi \frac{\partial}{\partial \phi} - 3 \sin\theta \cos\theta \cos\phi \tag{16}$$

Here the polar and azimuthal angles θ and ϕ are defined in such a way that

$$x = \sin\theta \cos\phi \ ; \quad y = \sin\theta \sin\phi \ ; \quad z = \cos\theta \tag{17}$$

Γ can be written in terms of spherical harmonic functions first by representing the products of trigonometric functions by their corresponding spherical harmonic functions, and second by representing the angular derivatives by angular momentum operators. For the first step we use

$$\begin{aligned} Y_2^0 &= \sqrt{5/16\pi}\,(3\cos^2\theta - 1) \\ Y_2^1 &= -\sqrt{15/8\pi}\,\sin\theta \cos\theta\, e^{i\phi} \\ Y_2^{-1} &= -(Y_2^1)^* \end{aligned} \tag{18}$$

where the asterisk denotes the complex conjugate, and i is the pure imaginary unit. For the second step we use the angular momentum operators,

$$L_\alpha \equiv -i(\underset{\sim}{u} \times \frac{\partial}{\partial \underset{\sim}{u}})_\alpha \ ; \quad \alpha = x, y, z \tag{19}$$

where $\times$ represents the vector cross product. Hence

$$L_z = -i\frac{\partial}{\partial\phi} ;$$

$$L_x = i\sin\phi\frac{\partial}{\partial\theta} + i\cot\theta\cos\phi\frac{\partial}{\partial\phi}$$

$$L_y = -i\cos\phi\frac{\partial}{\partial\theta} + i\cot\theta\sin\phi\frac{\partial}{\partial\phi} \tag{20}$$

With Eqns. (18) and (20), we are able to write Eqn. (16) as

$$\Gamma = \sqrt{16\pi/45}\,Y_2^0 iL_y + \frac{i}{3}L_y + \sqrt{2\pi/15}\,(Y_2^1 + Y_2^{-1})L_z$$

$$- 3\sqrt{2\pi/15}(-Y_2^1 + Y_2^{-1}) \tag{21}$$

This expression corrects the sign errors in Eqn. (B.1) of Doi and Edwards.

To assist in our evaluation of the inner products, we use the definitions

$$L_+ \equiv L_x + iL_y ; \qquad L_- \equiv L_x - iL_y \tag{22}$$

and define the shorthand

$$[\ell,m,p,q,\ell']^- \equiv \sum_{m'=0}^{\ell'} b_{\ell'm'}(\ell m\,|\,Y_p^q(L_+ - L_-)\,|\,\ell'm')$$

$$[\ell,m,p,q,\ell']^+ \equiv \sum_{m'=0}^{\ell'} b_{\ell'm'}(\ell m\,|\,Y_p^q(L_+ + L_-)\,|\,\ell'm')$$

$$[\ell,m,p,q,\ell']^z \equiv \sum_{m'=0}^{\ell'} b_{\ell'm'}(\ell m\,|\,Y_p^q L_z\,|\,\ell'm')$$

$$[\ell,m,p,q,\ell']^0 \equiv \sum_{m'=0}^{\ell'} b_{\ell'm'}(\ell m\,|\,Y_p^q\,|\,\ell'm') \tag{23}$$

In terms of the above, we find that the convection inner product is

$$(\ell m \,|\, \frac{\partial}{\partial \underset{\sim}{u}} \cdot \{[\underset{\sim}{u} \cdot \nabla \underset{\sim}{v} - \underset{\sim}{u}\,\underset{\sim}{u}\,\underset{\sim}{u} : \nabla \underset{\sim}{v}]\psi\}) = \dot{\gamma}(\ell m \,|\, \Gamma \psi)$$

$$= \frac{1}{2}\sqrt{16\pi/45}\,[\ell,m,2,0,\ell']^{-} + \frac{1}{6}\sqrt{4\pi}\,[\ell,m,0,0,\ell']^{-}$$

$$+ \sqrt{2\pi/15}\,[\ell,m,2,1,\ell']^{z} + \sqrt{2\pi/15}\,[\ell,m,2,-1,\ell']^{z}$$

$$+ 3\sqrt{2\pi/15}\,[\ell,m,2,1,\ell']^{0} - 3\sqrt{2\pi/15}\,[\ell,m,2,-1,\ell']^{0} \qquad (24)$$

We now must evaluate the shorthand expressions defined by (23). First we note that the operators L_z, L_+, and L_- can be eliminated from the above with the aid of the following identities from Messiah[8]:

$$\begin{aligned} L_z Y_l^m &= \ell(\ell+1) Y_l^m \\ L_+ Y_l^m &= [\ell(\ell+1) - m(m+1)]^{1/2}\, Y_\ell^{m+1} \\ L_- Y_l^m &= [\ell(\ell+1) - m(m-1)]^{1/2}\, Y_\ell^{m-1} \end{aligned} \qquad (25)$$

This leaves in Eqn. (23) only the inner products of the form $(\ell m \,|\, Y_p^q \,|\, \ell' m')$. To obtain these, we must evaluate integrals over products of three spherical harmonic functions, such as $\int Y_\ell^{-m} Y_p^q Y_{\ell'}^{m'} du^2$, which we represent by the shorthand

$$\int Y_\ell^{-m} Y_p^q Y_{\ell'}^{m'} du^2 \equiv \langle \ell m \,|\, Y_p^q \,|\, \ell' m' \rangle \qquad (26)$$

Again from Messiah[8], these can be evaluated using

$$\langle \ell m \,|\, Y_p^q \,|\, \ell' m' \rangle = (-1)^{m'} \left[\frac{(2\ell+1)(2\ell'+1)(2p+1)}{4\pi} \right]^{1/2}$$

$$\times \begin{pmatrix} \ell & p & \ell' \\ -m & q & m' \end{pmatrix} \begin{pmatrix} \ell & p & \ell' \\ 0 & 0 & 0 \end{pmatrix} \qquad (27)$$

Here the "6j" symbols $\begin{pmatrix} a & b & c \\ \alpha & \beta & \gamma \end{pmatrix}$ have been introduced; these can be obtained from the Racah formula given as Eqns. (C.21) and (C.22) in Messiah[8].

B. The Onsager Term

We next evaluate the inner product in Eqn. (11) that involves the excluded volume potential:

$$(\ell m\,|\,\frac{\partial}{\partial \underset{\sim}{u}}\cdot[\psi\frac{\partial}{\partial \underset{\sim}{u}}V_{ev}]) = \int |\,\ell m)\frac{\partial}{\partial \underset{\sim}{u}}\cdot[\psi\frac{\partial}{\partial \underset{\sim}{u}}V_{ev}]du^2 \tag{28}$$

where $|\,\ell m)$ is given by Eqn. (6). Integrating by parts gives

$$(\ell m\,|\,\frac{\partial}{\partial \underset{\sim}{u}}\cdot[\psi\frac{\partial}{\partial \underset{\sim}{u}}V_{ev}]) = -\int \frac{\partial}{\partial \underset{\sim}{u}}[\,|\,\ell m)]\cdot\psi\frac{\partial}{\partial \underset{\sim}{u}}V_{ev}\;du^2 \tag{29}$$

Next we invoke the identity

$$(\underset{\sim}{u}\times\frac{\partial}{\partial \underset{\sim}{u}}f)\cdot(\underset{\sim}{u}\times\frac{\partial}{\partial \underset{\sim}{u}}g) = (\frac{\partial}{\partial \underset{\sim}{u}}f)\cdot(\frac{\partial}{\partial \underset{\sim}{u}}g) \tag{30}$$

where f and g are arbitrary functions of $\underset{\sim}{u}$. Making use of this identity, Eqn. (29) can be rewritten as

$$(\ell m\,|\,\frac{\partial}{\partial \underset{\sim}{u}}\cdot[\psi\frac{\partial}{\partial \underset{\sim}{u}}V_{ev}]) = -\int \underset{\sim}{u}\times\frac{\partial}{\partial \underset{\sim}{u}}[\,|\,\ell m)]\cdot\psi\;\underset{\sim}{u}\times\frac{\partial}{\partial \underset{\sim}{u}}V_{ev}\;du^2$$

$$= \int L_x[\,|\,\ell m)]\,\psi L_x[V_{ev}]\;du^2 + \int L_y[\,|\,\ell m)]\,\psi L_y[V_{ev}]\;du^2$$

$$+ \int L_z[\,|\,\ell m)]\,\psi L_z[V_{ev}]\;du^2 \tag{31}$$

In the above, we have made use of Eqn. (19).

We now choose the Onsager potential, Eqn. (1), as our excluded volume potential V_{ev}:

$$V_{ex} = V_O = U \int \sin(\underset{\sim}{u}, \underset{\sim}{u}') \psi(\underset{\sim}{u}') \, du'^2 \tag{32}$$

where $U \equiv 2ckTdL^2$. We next express $\sin(\underset{\sim}{u}, \underset{\sim}{u}')$ in terms of the spherical harmonic functions:

$$\sin(\underset{\sim}{u}, \underset{\sim}{u}') = -2\pi^2 \sum_{\substack{\ell'=0 \\ \ell' \text{ even}}}^{\infty} \sum_{m'=-\ell'}^{\ell'} \left(\frac{\ell'-1}{\ell'+2} \right) \left[\frac{(\ell'-3)!!}{\ell'!!} \right]^2 Y_{\ell'}^{m'}(\underset{\sim}{u}) \, Y_{\ell'}^{m'}(\underset{\sim}{u}') \tag{33}$$

Inserting (33) and expression (5) for $\psi(\underset{\sim}{u}')$ in terms of the spherical harmonic functions into Eqn. (32), and using the orthonormal properties of the spherical harmonic functions gives

$$V_{ev} = -2\pi^2 U \sum_{\substack{\ell''=0 \\ \ell'' \text{ even}}}^{\infty} \sum_{m''=0}^{\ell''} \left(\frac{\ell''-1}{\ell''+2} \right) \left[\frac{(\ell''-3)!!}{\ell''!!} \right]^2 b_{\ell'' m''} \, | \ell'' m'') \tag{34}$$

When this is inserted into (31), with ψ in that expression expanded in terms of spherical harmonic functions, and when the operators L_x, L_y, and L_z are eliminated using (22) and (25), the result is a series of integrals each over a product of three spherical harmonic functions. These are evaluated using the Racah formula given in Messiah[8].

C. Expressions for the Birefringence and Stress Tensors

Having outlined our method for evaluating the inner products, the time dependence of the coefficients $b_{\ell m}$ after start-up of shearing can be obtained by Runge-Kutta time integration of Eqn. (11). Note in Eqn. (11) that $\partial(\ell m \,|\, \psi)/\partial t$

$= \partial b_{\ell m}/\partial t$. From the values of the $b_{\ell m}$'s, we wish to calculate the optical and mechanical properties of the fluid; i.e., the birefringence and the stress tensors. To within a constant of proportionality, the non-zero components of the birefringence tensor are given by Doi and Edwards,

$$
\begin{aligned}
<u_x u_z> &= -\sqrt{4\pi/15}\, b_{21} \\
<u_x^2 - u_z^2> &= \sqrt{4\pi/15}\,(b_{22} - \sqrt{3}\, b_{20}) \\
<u_z^2 - u_y^2> &= \sqrt{4\pi/15}\,(b_{22} + \sqrt{3}\, b_{20})
\end{aligned}
\tag{35}
$$

The stress tensor is given by the sum of two terms,

$$
\underset{\approx}{\sigma} = \underset{\approx}{\sigma}^{b} + \underset{\approx}{\sigma}^{ev} \tag{36}
$$

The first term is contributed by Brownian forces and is given by

$$
\underset{\approx}{\sigma}^{b} = 3ckT \int \psi(\underset{\sim}{u}\,\underset{\sim}{u} - \frac{1}{3}\underset{\approx}{\delta})\, du^2 = 3ckT < \underset{\sim}{u}\,\underset{\sim}{u} - \frac{1}{3}\underset{\approx}{\delta} > \tag{37}
$$

The nonzero components of $\underset{\approx}{\sigma}^{b}$ can be obtained using Eqn. (35).

The second term comes from the excluded volume potential, and is given by

$$
\underset{\approx}{\sigma}^{ev} = c \int \psi(\frac{\partial}{\partial \underset{\sim}{u}} V_{ev})\, \underset{\sim}{u}\, du^2 \tag{38}
$$

The components of the operator $\frac{\partial}{\partial \underset{\sim}{u}})_i u_j$ that we shall need are

$$\frac{\partial}{\partial \underset{\sim}{u}})_x u_x = \sin\theta\cos\phi\,[\frac{1}{2}\cos\theta\,(L_+ - L_-) - i\sin\theta\sin\phi\,L_z]$$

$$\frac{\partial}{\partial \underset{\sim}{u}})_y u_y = \sin\theta\sin\phi\,[\frac{1}{2i}\cos\theta\,(L_+ + L_-) + i\sin\theta\cos\phi\,L_z]$$

$$\frac{\partial}{\partial \underset{\sim}{u}})_z u_z = -\cos\theta\,[\frac{1}{2}\sin\theta\cos\phi\,(L_+ - L_-) + \frac{1}{2i}\sin\theta\sin\phi\,(L_+ + L_-)]$$

$$\frac{\partial}{\partial \underset{\sim}{u}})_z u_x = -\frac{1}{2}\sin^2\theta\cos^2\phi\,(L_+ - L_-) - \frac{1}{2i}\sin^2\theta\sin\phi\cos\phi\,(L_+ + L_-) \quad (39)$$

To eliminate the sine and cosine functions in favor of spherical harmonic functions, we use the following easily derived identities

$$\sin\theta\cos\theta\sin\phi = i\sqrt{2\pi/15}\,(Y_2^1 + Y_2^{-1})$$

$$\sin\theta\cos\theta\cos\phi = -\sqrt{2\pi/15}\,(Y_2^1 - Y_2^{-1})$$

$$\sin^2\theta\cos^2\phi = \sqrt{2\pi/15}\,(Y_2^2 + Y_2^{-2}) - \frac{1}{3}\sqrt{4\pi/5)}\,Y_2^0 + \frac{1}{3}\sqrt{4\pi}\,Y_0^0$$

$$\sin^2\theta\cos\phi\sin\phi = -i\sqrt{2\pi/15}\,(Y_2^2 - Y_2^{-2})$$

$$\cos^2\theta = \frac{1}{3}\sqrt{16\pi/5}\,Y_2^0 + \frac{1}{3}\sqrt{4\pi}\,Y_0^0 \quad (40)$$

Substituting expressions (39) and (40) into (38), expanding ψ in terms of spherical harmonic functions using (5), and inserting expression (34) for V_{ev}, the results can be written as

$$\sigma_{xx}^{ev} - \sigma_{zz}^{ev} = -2\pi^2 U \sum_{\substack{\ell=0 \\ \ell\,\text{even}}}^{\infty} \sum_{m=0}^{\ell} \sum_{\substack{\ell'=0 \\ \ell'\,\text{even}}}^{\infty} \left(\frac{\ell'-1}{\ell'+2}\right)\left[\frac{(\ell'-3)!!}{\ell'!!}\right]^2 b_{\ell m}(2\pi/15)^{1/2}$$

$$\times \left\{ -[\ell,m,2,1,\ell']^- + [\ell,m,2,-1,\ell']^- + \frac{1}{2}[\ell,m,2,1,\ell']^+ \right.$$

$$\left. + \frac{1}{2}[\ell,m,2,-1,\ell']^+ - [\ell,m,2,2,\ell']^z + [\ell,m,2,-2,\ell']^z \right\} \quad (41)$$

$$\sigma_{zz}^{ev} - \sigma_{yy}^{ev} = -2\pi^2 U \sum_{\substack{\ell=0 \\ \ell\,\text{even}}}^{\infty} \sum_{m=0}^{\ell} \sum_{\substack{\ell'=0 \\ \ell'\,\text{even}}}^{\infty} \left(\frac{\ell'-1}{\ell'+2}\right)\left[\frac{(\ell'-3)!!}{\ell'!!}\right]^2 b_{\ell m}(2\pi/15)^{1/2}$$

$$\times \left\{ -[\ell,m,2,1,\ell']^+ - [\ell,m,2,-1,\ell']^+ + \frac{1}{2}[\ell,m,2,1,\ell']^- \right.$$

$$\left. - \frac{1}{2}[\ell,m,2,-1,\ell']^- - [\ell,m,2,2,\ell']^z + [\ell,m,2,-2,\ell']^z \right\} \quad (42)$$

$$\sigma_{xz}^{ev} = -2\pi^2 U \sum_{\substack{\ell=0 \\ \ell\,\text{even}}}^{\infty} \sum_{m=0}^{\ell} \sum_{\substack{\ell'=0 \\ \ell'\,\text{even}}}^{\infty} \left(\frac{\ell'-1}{\ell'+2}\right)\left[\frac{(\ell'-3)!!}{\ell'!!}\right]^2 b_{\ell m}(2\pi/15)^{1/2}$$

$$\times \left\{ -\frac{1}{2}[\ell,m,2,2,\ell']^- - \frac{1}{2}[\ell,m,2,-2,\ell']^- + \frac{1}{\sqrt{6}}[\ell,m,2,0,\ell']^- \right.$$

$$\left. - \sqrt{5/6}\,[\ell,m,0,0,\ell']^- + \frac{1}{2}[\ell,m,2,2,\ell']^+ - \frac{1}{2}[\ell,m,2,-2,\ell']^+ \right\} \quad (43)$$

III. COMPUTER ALGORITHM

The time evolution of the $b_{\ell m}$'s was obtained by a fourth order Runge-Kutta integration of Eqn. (11). The infinite sums appearing in the inner-product expressions were truncated at $\ell = \ell_{max}$. Values of 4—16 for ℓ_{max} were considered; sufficient accuracy was obtained with $\ell_{max} = 12$.

The computer algorithm contains three major subroutines. The first computes the convection term, the second the Onsager term, and the third calculates the stress tensor from the $b_{\ell m}$'s. Each of these subroutines was debugged by comparing computational results from each subroutine to exact results known for various limits.

The convection subroutine was checked by setting the diffusion and excluded-volume terms to zero, leaving only the convection term to influence the evolution of the $b_{\ell m}$'s. Starting from an isotropic initial distribution, i.e., $b_{\ell m} = 0$ for $\ell \neq 0$ and $b_{00} = 1/\sqrt{4\pi}$, the $b_{\ell m}$'s are computed as a functions of shear strain after initiation of steady shearing. Because in this limit only the convection term controls the evolution of the $b_{\ell m}$'s, the ratio $<u_x^2 - u_z^2>/<u_x u_z>$ at any instant during shear must equal γ, the cumulative shear strain imposed. This is an analog of the Lodge-Meissner relationship[5]. We find that for a fixed ℓ_{max}, the calculated ratio $<u_x^2 - u_z^2>/<u_x u_z>$ remains within 1% of the correct value (namely γ) up to a critical strain γ_c and thereafter deviations become larger. The critical strain γ_c increases as ℓ_{max} increases; that is, the calculation is accurate to higher and higher strains as more and more spherical harmonic functions are used. The following table summarizes the results.

ℓ_{max}	γ_c
4	1.
6	2.
12	3.

The expression for the convection term given by Doi and Edwards does not converge; i.e., for values of $\gamma < 1$, $<u_x^2 - u_z^2>/<u_x u_z>$ deviates significantly from γ even when ℓ_{max} is as large as 24.

The Onsager subroutine was checked by dropping the convection term and calculating the equilibrium orientation distribution with only the Onsager and Brownian terms present. In this limit, other numerical work has shown that at a value of U = 10.67, equilibrium anisotropy occurs and the equilibrium scaler order parameter for the Onsager potential is $S^{eq} = 0.7922$. The equilibrium scaler order parameter S^{eq} is defined by the following equation:

$$<\underset{\sim}{u}\,\underset{\sim}{u} - \frac{1}{3}\underset{\approx}{\delta}> \; = \; S^{eq}(\underset{\sim}{n}\,\underset{\sim}{n} - \frac{1}{3}\underset{\approx}{\delta}) \tag{44}$$

where the unit vector $\underset{\sim}{n}$ is the *director* — i.e., the direction of average orientation. S^{eq} must lie between the values of zero for an isotropic distribution of orientations and unity for perfect anisotropy. The numerical calculations give the values tabulated below.

ℓ_{max}	S^{eq}
4	0.6059
8	0.7867
12	0.7920

Thus convergence towards the exact value, $S^{eq} = 0.7922$, occurs as the number of spherical harmonic functions is increased.

The algorithm for computing the excluded volume contribution $\underset{\approx}{\sigma}^{ev}$ to the stress tensor was checked by computing the total stress $\underset{\approx}{\sigma} = \underset{\approx}{\sigma}^{b} + \underset{\approx}{\sigma}^{ev}$ in the absence of flow or under very low rates of shear. $\underset{\approx}{\sigma}$ under these conditions must be zero or very small, which implies that the two contributions $\underset{\approx}{\sigma}^{b}$ and $\underset{\approx}{\sigma}^{ev}$ cancel or nearly cancel each other. This was in fact found to occur, verifying the algorithm for computing $\underset{\approx}{\sigma}^{ev}$.

IV. SUMMARY

An algorithm is presented for computing the three-dimensional orientation distribution function, stresses, and birefringences in a solution of rod-like polymers that is concentrated enough that excluded-volume effects are strong. The algorithm is developed by expanding the diffusive, convective, and excluded volume terms of the Doi theory in a series of products of spherical harmonic functions, and taking inner products of the equation of motion with the spherical harmonic functions. The time evolution of the coefficients of the terms of the series after start-up of steady shearing is determined by Runge-Kutta integration. The algorithm avoids closure approximations, such as replacing fourth moments of the distribution function by products of second moments. It uses the Onsager potential and so can be used for precise calculations of the conditions under which weak shearing might induce an abrupt transition to a highly anisotropic state in a fluid that is isotropic at equilibrium. It can also be used to determine the conditions under which negative first normal stress differences occur during shearing of liquid crystalline materials.

References

1. Doi, M., J. Polym. Sci., Polym. Phys. Ed., *19*, 229 (1981).
2. Doi, M. and Edwards, S.F., J. Chem. Soc., Faraday Trans. II, *74*, 918 (1978).
3. Kiss, G. and Porter, R.S., J. Polym. Sci., Polym. Symp., *65*, 193 (1978).
4. Kuzuu, N. and Doi, M., J. Phys. Soc. Japan, *53*, 1031 (1984).
5. Lodge,A.S. and Meissner, J., Rheol. Acta, *11*, 351 (1972).
6. Marrucci, G. and Maffettone, P.L., Macromolecules, *22*, 4076 (1989).
7. Mead, D.W. and Larson, R.G., Macromolecules, in press 1990.
8. Messiah, A., Quantum Mechanics, Vols. 1 & 2, North Holland, Amsterdam (1972).
9. Navard, P., J. Polym. Sci., Poly. Phys. Ed., *24*, 435 (1986).

10. Onsager, L., Ann. N.Y. Acad. Sci., *51* 627 (1949).

NOMENCLATURE

D_r rotary diffusivity of isotropic solution of rod-like molecules
$\overline{D}_r$ effective rotary Brownian diffusivity
L length of rod-like molecule
L_α angular momentum operator defined in Eqn. (19)
L_+, L_- defined in Eqn. (22)
$\underset{\approx}{S}$ order parameter tensor, defined in Eqn. (3)
T absolute temperature
V_{ev} excluded volume potential
V_O Onsager potential given in Eqn. (1)
Y_ℓ^m spherical harmonic function
$b_{\ell m}$ coefficient of spherical harmonic expansion in Eqn. (5)
c concentration of rod-like molecules, number per unit volume
k Boltzmann constant
$\underset{\sim}{u}$ unit vector describing the orientation of a rod
$\underset{\sim}{v}$ flow velocity
$\underset{\approx}{\delta}$ unit tensor
i pure imaginary unit
Γ convection operator defined in Eqn. (21)
$\dot{\gamma}$ shear rate
θ polar angle; see Eqn. (17)
$\underset{\approx}{S}$ stress tensor
ψ distribution function of rod orientations
ϕ azimuthal angle; see Eqn. (17)
$\dfrac{\partial}{\partial \underset{\sim}{u}}$ gradient operator on the unit sphere
$|\ell m)$ defined in Eqn. (6)
$[\,]^-$, $[\,]^+$, $[\,]^z$, $[\,]^0$ defined in Eqn. (23)

Ronald Larson
Room 6E-320, AT&T Bell Laboratories
600 Mountain Ave.
Murray Hill, N.J. 07974-2070

J. STASTNA, D. DE KEE AND B. HARRISON

Diffusion in macromolecular continua

1. INTRODUCTION

Traditionally, diffusion processes have been studied in thermodynamics [11]. One has to be aware that non-equilibrium thermodynamics asks many fundamental and difficult questions which have a long history of often controversial answers, such as for example: the fundamental and unresolved question of ergodicity. In everyday life, these questions are rarely important. We know (believe), that many systems do come to equilibrium if given sufficient time. After a perturbation of the state of equilibrium, relaxation processes bring a system into a new equilibrium state. During these processes various quantities such as density, concentration, temperature, etc. are equilibrated; that is: macroscopic fluxes of mass, energy, electric charge, etc. are observed within the system. The equilibrium state is then characterized by zero macroscopic fluxes. At every moment of a slow irreversible process, there exist nonzero gradients of concentration, temperature, electrical potential, ... among parts of the system. If the gradients of individual macroscopic quantities are small, it is possible to express macroscopic fluxes as linear functions of these gradients. As an example one can consider a system of two non-reacting components. Let the concentration of the studied component be c. The concentration of the second component is then 1-c. During the relaxation processes there exists a mass flux $\underset{\sim}{j}$, leading to the equilibriation of concentration in different parts of the system. Consider also the existence of the heat flux $\underset{\sim}{q}$. Let n_1 and n_2 be the numbers of particles of the individual components in a unit of mass of the mixture. Let m_1 and m_2 represent the masses of these particles. Then: $n_1m_1+n_2m_2=1$. The thermodynamic equation per unit mass is then given by:

$$d\bar{u} = Tds - pdv + \mu_1 dn_1 + \mu_2 dn_2 \qquad (1)$$

where $\bar{u}$ is the internal energy per unit mass, s is the entropy per unit of mass, T is the temperature, p is the pressure, v is the specific volume (density $\rho = v^{-1}$) and μ_1 and μ_2 are chemical potentials. Substituting $c = n_1 m_1$ and $\mu = \mu_1 m_1^{-1} - \mu_2 m_2^{-1}$ one can write:

$$d\bar{u} = Tds + \frac{p}{\rho^2} d\rho + \mu dc \tag{2}$$

Gibbs potential, g, per unit mass of the mixture is given by:

$$g = \bar{u} - Ts + \frac{p}{\rho} = \mu c \tag{3}$$

Using equation (2) one can write the following equation for the production of entropy $\dot{S}$;

$$\dot{S} = \frac{\partial}{\partial t} \int \rho s dV = - \int \frac{1}{T^2} (\underset{\sim}{q} - \mu \underset{\sim}{j}) \cdot \underset{\sim}{\nabla} T dV - \int \frac{\underset{\sim}{j}}{T} \cdot \underset{\sim}{\nabla} \mu dV - \\ - \oint \frac{1}{T} (\underset{\sim}{q} - \mu \underset{\sim}{j}) \cdot d\sigma \tag{4}$$

If on the surface of volume V, $\underset{\sim}{q} = \mu \underset{\sim}{j}$, (the energy flux is given by vector $\underset{\sim}{q} - \mu \underset{\sim}{j}$) the entropy production given by eq. (4) has the form required by the theory of irreversible processes [25]

$$\dot{S} = \int \sum_i J_i X_i dV$$

Here J_i is the i-th flux and X_i is the associated thermodynamic force.

In our example the forces are components of vectors $-\underset{\sim}{\nabla} T/T^2$, $-\underset{\sim}{\nabla}\mu/T$, and fluxes are components of vectors $\underset{\sim}{q} - \mu \underset{\sim}{j}$ and $\underset{\sim}{j}$. In a linear approximation, the fluxes J_i are given by a linear combination of thermodynamic forces:

$$\underset{\sim}{j} = -\alpha T \left(\frac{\underset{\sim}{\nabla}\mu}{T} \right) - \beta T^2 \left(\frac{\underset{\sim}{\nabla} T}{T^2} \right) \tag{5}$$

$$\underset{\sim}{q} - \mu \underset{\sim}{j} = -\varepsilon T \left(\frac{\underset{\sim}{\nabla}\mu}{T} \right) - \wp T^2 \left(\frac{\underset{\sim}{\nabla} T}{T^2} \right) \tag{6}$$

In this notation Onsager's coefficients are given as follows:

$$L_{11} = \alpha T, L_{12} = \beta T^2, L_{21} = \varepsilon T, L_{22} = \wp T^2$$

from the symmetry of L_{ik} we have

$$\varepsilon = \beta T$$

Eliminating $\underset{\sim}{\nabla}\mu$ from equ. (6) one can write:

$$\underset{\sim}{j} = -\alpha\underset{\sim}{\nabla}\mu - \beta\underset{\sim}{\nabla}T \tag{7}$$

$$\underset{\sim}{q} = \left(\mu + \frac{\beta T}{\alpha}\right)\underset{\sim}{j} - \kappa\underset{\sim}{\nabla}T \tag{8}$$

where $\kappa = \wp - \beta^2 T/\alpha$.

Vector $\underset{\sim}{j}$ represents the total flux of mass (diffusion flux). For $\underset{\sim}{j} = \underset{\sim}{O}$, the heat flux becomes $\underset{\sim}{q} = -\kappa\underset{\sim}{\nabla}T$, where κ represents the thermal conductivity coefficient. For the gradient of chemical potential $\mu(c,p,T)$ one can write:

$$\underset{\sim}{\nabla}\mu = \left(\frac{\partial\mu}{\partial c}\right)_{p,T}\underset{\sim}{\nabla}c + \left(\frac{\partial\mu}{\partial p}\right)_{c,T}\underset{\sim}{\nabla}p + \left(\frac{\partial\mu}{\partial T}\right)_{c,p}\underset{\sim}{\nabla}T \tag{9}$$

It follows from equ. (3), that ($v=\rho^{-1}$):

$$\left(\frac{\partial\mu}{\partial p}\right)_{c,T} = \left(\frac{\partial v}{\partial c}\right)_{p,T}$$

Using the following notation:

$$D = \frac{\alpha}{\rho}\left(\frac{\partial\mu}{\partial c}\right)_{p,T}, \quad \frac{\rho D\upsilon}{T} = \alpha\left(\frac{\partial\mu}{\partial T}\right)_{c,p} + \beta$$

and $\dfrac{\rho Db}{p} = \alpha\left(\dfrac{\partial\mu}{\partial p}\right)_{c,T}$

one can write the fluxes as

$$\underset{\sim}{j} = -\rho D\left((\underset{\sim}{\nabla}c + \frac{\upsilon}{T}\underset{\sim}{\nabla}T + \frac{b}{p}\underset{\sim}{\nabla}p\right) \tag{10}$$

$$\underset{\sim}{q} = \left[\upsilon\left(\frac{\partial\mu}{\partial c}\right)_{p,T} - T\left(\frac{\partial\mu}{\partial T}\right)_{c,p} + \mu\right]\underset{\sim}{j} - \kappa\underset{\sim}{\nabla}T \tag{11}$$

One can see that the flux of mass $\underset{\sim}{j}$ consists of three terms:

$\underset{\sim}{j}_c = -\rho D\underset{\sim}{\nabla}c$ regular diffusion; $\underset{\sim}{j}_T = -\frac{1}{T}\rho D\upsilon\underset{\sim}{\nabla}T$ thermodiffusion and

$\underset{\sim}{j}_p = -\frac{1}{p}\rho Db\underset{\sim}{\nabla}p$ barodiffusion. The coefficient D determines the mass flux generated by the gradient of concentration, i.e. D is the diffusion coefficient $D\upsilon$ is the coefficient of thermodiffusion and Db is the barodiffusion coefficient. Assuming that the influence of $\underset{\sim}{\nabla} p$ is negligible and that $\underset{\sim}{\nabla} T$ and $\underset{\sim}{\nabla} c$ are small; i.e.: D and υ are constants in this approximation, we obtain from the conservation of mass $\left(\frac{\partial(\rho c)}{\partial t} = -\operatorname{div} \underset{\sim}{j}\right)$ and equation (2) the equations for the distribution of concentration and temperature. For pure diffusion and pure heat conduction, we thus obtain the known parabolic equations

$$\frac{\partial c}{\partial t} = D\Delta c \quad \text{and} \quad \frac{\partial T}{\partial t} = \chi \Delta T, \quad \chi = \kappa/\rho c_p \tag{12}$$

1.1 Continuum Mechanics

The method outlined in the introduction represents the classical treatment of diffusion processes. A different approach has been followed by modern continuum mechanics which is based on the study of constitutive equations [40,5,24]. Consider again our example of a two component system where for example the first component (index 1) represents a gas (penetrant) and the second component (index 2) is a polymer. We shall use no index for quantities referring to a mixture of those two components. For the momentum balance of the gas we can write:

$$\rho_1 \frac{{}'d \underset{\sim}{v}_1}{dt} = \operatorname{div} \underset{\sim}{T}_1 + \rho_1 \underset{\sim}{b}_1 + \underset{\sim}{k}_1 \tag{13}$$

where $\frac{{}'dA}{dt} = \frac{\partial A}{\partial t} + \underset{\sim}{v}_1 . \underset{\sim}{\nabla} A$, $\underset{\sim}{T}_1$ is the partial stress tensor, $\underset{\sim}{b}_1$ is the external body force acting on component 1, $\underset{\sim}{v}_1$ is the velocity of the gas and $\underset{\sim}{k}_1$ is the interaction body force from the polymer acting on the gas. Now we can assume that our system is described by constitutive equations for free energy $\left(\bar{f}_i\right)$, entropy (s_i), stress $\underset{\sim}{T}_i$, interaction force $\left(\underset{\sim}{k}_i\right)$, heat flux $(\underset{\sim}{q})$ and (for simplicity only one) internal parameter $(\bar{\beta})$.

Constitutive equations for internal parameters take the form of evolution equations (differential equations $\dot{\bar{\beta}} = B(\bullet,\bullet,\ldots,\bullet)$ [6]. Through such internal parameters, one can introduce memory effects into the system. For a nonreacting mixture of non-simple components one can assume that the set of dependent variables for the constitutive equations consists of

$\left\{\rho_1, \underset{\sim}{F}_2, \underset{\sim}{\nabla}\rho_1, \underset{\sim}{\nabla}\underset{\sim}{F}_2, \underset{\sim}{w}, T, \underset{\sim}{\nabla}T, \bar{\beta}\right\}$, where the diffusion velocity $\underset{\sim}{w} = \underset{\sim}{v}_1 - \underset{\sim}{v}_2$,

and $\underset{\sim}{F}_2$ is the deformation gradient of the polymer. Let us note that

$\underset{\sim}{\nabla}\underset{\sim}{F}_2$ and $\underset{\sim}{\nabla}\rho_1$ are necessary for the description of material interaction between the components. Applying the constitutive principles, including the important entropy principle of Coleman and Noll [7] one can eliminate some variables from some constitutive equations. For example Samohyl et al. [34,35] have discussed the situation considered in our example and obtained a constitutive equation for the chemical potential of the gas (identical with the partial Gibbs potential g_1) in the form

$$g_1 = g_1\left(\rho_1, \underset{\sim}{F}_2, T, \bar{\beta}\right) \tag{14}$$

and a constitutive equation for the interaction force $\underset{\sim}{k}_1$ in the form:

$$\underset{\sim}{k}_1 = \underset{\sim}{K}_0 - \underset{\sim}{K}_1 . \underset{\sim}{w} - \underset{\sim}{K}_1 . \underset{\sim}{\nabla} T + \rho_2 \frac{\partial \bar{f}_2}{\partial \rho_1} . \underset{\sim}{\nabla}\rho_1 - \rho_1 \frac{\partial \bar{f}_1}{\partial \underset{\sim}{F}_2} . \underset{\sim}{\nabla}\underset{\sim}{F}_2 . \underset{\sim}{F}_2^{-1} \tag{15}$$

In combination with equation (13) and using the definition of the Gibbs potential $g_1 = \bar{f}_1 + p_1/\rho_1$ one can obtain the equation for the diffusion velocity $\underset{\sim}{w}$ [35]

$$-\underset{\sim}{K}_1 \underset{\sim}{w} = -\underset{\sim}{K}_o + \rho_1 \left(\underset{\sim}{\nabla} g_1\right)_T + \rho_1 \frac{{}'d\underset{\sim}{v}_1}{dt} - \rho_1 \underset{\sim}{b}_1 + \left(\frac{\partial p_1}{\partial T} \underset{\sim}{I} + \underset{\sim}{K}_2\right) . \underset{\sim}{\nabla} T \tag{16}$$

where $\left(\underset{\sim}{\nabla} g_1\right)_T = \underset{\sim}{\nabla} g_1 - \frac{\partial \bar{g}_1}{\partial T} \underset{\sim}{\nabla} T$

Now it is possible to express the diffusion flux $\underset{\sim}{j} = \rho_1 \underset{\sim}{w}$ as

$$-\underset{\sim}{j} = \underset{\sim}{L}_o + \underset{\sim}{L} \cdot \underset{\sim}{X} + \underset{\sim}{L}_1 \cdot \underset{\sim}{\nabla} T \tag{17}$$

The thermodynamic force $\underset{\sim}{X}$ is defined as:

$$\underset{\sim}{X} = (\underset{\sim}{\nabla} g_1)_T - \underset{\sim}{b}_1 + \frac{'d\underset{\sim}{v}_1}{dt} \tag{18}$$

and the phenomenological coefficients are:

vector $\underset{\sim}{L}_o = -\rho_1 \underset{\sim}{K}_1^{-1} \cdot \underset{\sim}{K}_o$ (19)

and tensors

$$\underset{\sim}{L} = \rho_1^2 \underset{\sim}{K}_1^{-1}, \quad \underset{\sim}{L}_1 = \rho_1 \underset{\sim}{K}_1^{-1} \cdot \left(\frac{\partial p_1}{\partial T} \underset{\sim}{I} + \underset{\sim}{K}_2 \right) \tag{20}$$

The kinetic coefficients are functions of $\rho_1, \underset{\sim}{F}_2$, T and $\bar{\beta}$. In the case of isothermal diffusion $(\underset{\sim}{\nabla} T = \underset{\sim}{0})$ and assuming that $\underset{\sim}{b}_1 = \underset{\sim}{0}$ and $\frac{'d\underset{\sim}{v}_1}{dt} = \underset{\sim}{0}$ we have

$$-\underset{\sim}{j} = \underset{\sim}{L}_o + \underset{\sim}{L} \cdot \left(\frac{\partial \bar{g}}{\partial \rho_1} \underset{\sim}{\nabla} \rho_1 + \frac{\partial \bar{g}}{\partial \underset{\sim}{F}_2} \cdot \underset{\sim}{\nabla} \underset{\sim}{F}_2 \cdot \underset{\sim}{F}^{-1} + \frac{\partial \bar{g}}{\partial \bar{\beta}} \underset{\sim}{\nabla} \bar{\beta} \right) \tag{21}$$

where vector $\underset{\sim}{L}_o$ and the second order tensor $\underset{\sim}{L}$ are both functions of $\rho_1, \underset{\sim}{F}_2$, T and $\bar{\beta}$.

Vector $\underset{\sim}{L}_o$ is zero for isotropic polymers and also if the relaxation processes related to $\bar{\beta}$ are either very fast or negligible [35]. The relation to the classical case is apparent from equation $\rho_1 = c\rho \; (\rho_2 = (1-c)\rho)$. It is clear from eq. (21) that deformation of the polymer and the gradient of the internal parameters will lead to the cases of non-classical (non-Fickian) diffusion in the considered model.

Thus the methods of modern continuum mechanics seem to be able to describe classical as well as anomalous diffusion processes. However one has to note that the mentioned method should be accompanied by specification of interface and boundary processes. In the chemical and in the engineering literature, the classical method is usually combined with various boundary condition models, see for example the monograph by Crank [8] and the review by Frisch [13].

2. Experimental

The classical Fick's first law provides the simplest technique for diffusion experiments. One can again consider the diffusion of a gas in a membrane. One side of the membrane, at thickness l, is brought in contact with the diffusant at constant activity, i.e. eventually a constant surface concentration c_o is established at the surface of the membrane. The other side of the membrane is maintained at zero concentration, and the amount Q, of emerging diffusant is measured as a function of time. According to Fick's first law the flux, j, of diffusant, at steady state, is given by:

$$j = (1/A_r)dQ/dt = Dc_o/l \tag{22}$$

Then, knowing the membrane area, A_r, the thickness l, and the surface concentration c_o, it is possible to calculate the diffusion coefficient D. Solving the diffusion equation for an initially gas-free membrane with one surface at zero concentration of penetrant and the other surface at concentration c=c' at time t=0, one can obtain at large values of t the relation [10]

$$Q/A_r = (Dc'/l)(t - l^2/6D) \tag{23}$$

From the plot Q/A_r versus t one can obtain the time lag t' (t' is the time at which the linear asymptote intersects the t-axis)

$$t' = l^2/6D \tag{24}$$

This relation enables one to obtain D directly from permeation experiments. Another solution of the diffusion equation [8] yields:

$$j = 2c'(D\pi t)^{1/2} \sum_i \exp(-(2n+1)^2 l^2/4Dt) \tag{25}$$

At low values of t one can write

$$\ln(jt^{1/2}) = \ln(2c'(D\pi)^{1/2}) - l^2/4Dt \tag{26}$$

and D can be obtained from the slope of plot $jt^{1/2}$ versus $1/t$ [30]. A discussion of various boundary value problems related to experimental techniques can be found in [8].

The dependence on temperature is basically described by the Arrhenius relation

$$D = D_{\infty} \exp(-\Delta E_a/RT) \qquad (27)$$

where the Arrhenius activation energy is ΔE_a. This equation is satisfactory for many gases diffusing in polymers. A useful empiricism has been proposed by Barrer and Chio [4]

$$\ln(D_0) = -18.9 + 0.626\ \Delta E_a/RT,\ D_0[=]\ m^2s^{-1}$$

This relation is used not only for simple gases, but also for the diffusion of more complicated molecules. In combination with equ. (27) the last relation allows one to estimate the activation energies even if D is only known at a single temperature. Even for simple gases diffusing through a polymeric membrane one can, in some cases observe a discontinuity in the Arrhenius plot if the temperature passes through a critical glass transition temperature Tg. The situation is even more complicated for the diffusion of organic molecules in glassy polymers. Here the diffusion is subjected to time effects which produce various complications, as discussed elsewhere [13,28,16].

Basically the methods mentioned for diffusion experiments with simple gases can be used also for experiments with organic penetrants (vapours). However, larger diffusant molecules have very low diffusion coefficients, which lead to low permeation rates and long lag times. This problem can be overcome by using thin membranes and by immersing them in the organic penetrants at constant pressure. It is assumed that the constant activity of the penetrant leads to a constant surface concentration at both faces of the membrane. Then solving the diffusion equation for an initially diffusant-free membrane (or for a membrane containing a constant concentration of diffusant) one can obtain the following two approximations [8]:

$$M_t/M_{\infty} = \left(16\,Dt/\pi l^2\right)^{1/2} \quad \ldots \text{for } M_t/M_{\infty} < 1/2 \text{ (short times)} \qquad (28)$$

and

$$\ln\left(M_{\infty} - M_t\right) = \ln\left(8M_{\infty}/\pi^2\right) - \pi^2 Dt/l^2 \quad \ldots \text{for } M_t/M_{\infty} > 1/2 \qquad (29)$$

Here M_t is the mass of penetrant absorbed by a membrane of thickness l, at time t, and M_∞ is the mass absorbed by the membrane at equilibrium. Equation (28) enables one to obtain the diffusion coefficient from plots of M_t/M_∞ versus $t^{1/2}$ at the initial stage of sorption while at the end of the sorption process a plot of $\ln(M_\infty - M_t)$ versus time also yields the diffusion coefficient.

The difference between the sorption of organic vapours by polymeric membranes and the sorption of simple gases is the much larger concentration of diffusant in the first case. Usually one has to deal with problems caused by swelling [8].

Approximation (28) is valid for times, much longer than those suggested by the exact solution. Also, approximation (29) covers a much shorter period than the one predicted by the exact solution. These differences are caused by the dependence of D on the concentration in the membrane. Crank [8] showed that for sorption from zero concentration of penetrant to an equilibrium concentration, c* in the membrane, approximation (28) gives an average diffusion coefficient, $\bar{D}$:

$$\bar{D} = \frac{1}{c^*}\int_0^{c^*} D\,dc \qquad (30)$$

The concentration dependence varies from polymer to polymer. In many cases, there seems to be a correlation between D_0 (the diffusion coefficient at zero concentration) and the dependence of D on concentration, c.

There is ample experimental evidence, suggesting that the diffusion coefficient depends on the shape and size of the diffusant molecules. The variation of D with the geometry of diffusants is usually discussed in terms of the diffusion coefficient at zero concentration, D_0. Prager et al. [29] were the first to recognize that D_0 decreases with increasing penetrant size. Park and Saleem [26] have shown that the flexible n-hexadecane molecule diffusing through plasticized polyvinylchloride has a diffusion coefficient almost 30 times the diffusion coefficient of the more rigid DDT molecule, eventhough the molar volume of DDT is less than that of n-hexadecane. In a more recent study, Saleem et al. [32] have studied the diffusion of several hydrocarbons through LDPE films. They have found a profound effect of the shape and size of the penetrants on D_0, as shown in Figure 1. One can expect an increase in D_0, with increasing flexibility of the diffusant molecule, since these molecules can move through the network of polymer chains in coordinated steps, disturbing the network less than would be the case for large and less flexible molecules. Furthermore, the diffusion coefficients for organic penetrants, as well as those for simple gases, show a strong increase with increasing temperature [32], see Figure 2.

For small temperature variations, reasonable Arrhenius plots can be obtained and one can interpret the temperature dependence in terms of activation energy. It has been suggested [31] that the activation energy of diffusion increases with the penetrant size until a value associated with the activation energy of viscous flow is reached. No evidence for this effect has been found in a more recent study [33] of diffusion of saturated hydrocarbons in LDPE. This may be due to the partial crystallinity of the used LDPE.

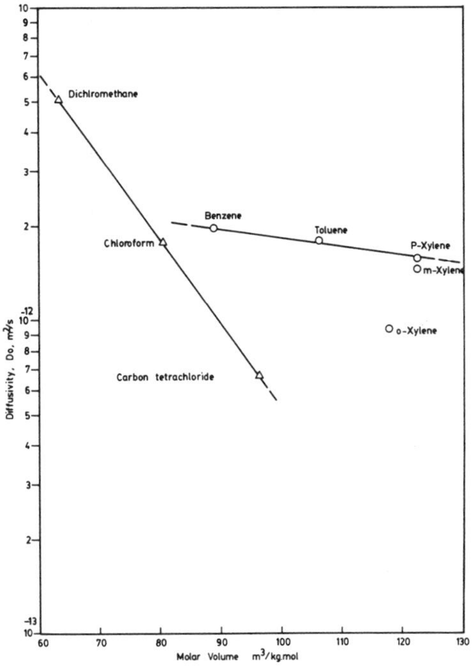

Figure 1: Variation of D_0 with molar volume, taken from Saleem et al. [32]

The effect of the nature of the polymer on the diffusion is, in experimental studies, usually described by a correlation of D_0 with Tg. The increase of D_0 with decreasing values of Tg is accompanied by a weaker concentration dependence of the diffusion coefficient.

The variations of the diffusion coefficient with concentration, temperature, glass transition temperature and penetrant size are reasonably well described by the free volume theory of diffusion [14,41]. This theory is based on the assumption that a diffusing molecule can only jump from one place to another if the local specific volume, i.e. the local amount of empty space, called free volume, accessible to the diffusing molecule, exceeds a certain critical value. For the thermodynamic diffusion coefficient D_T, the free volume theory yields:

$$D_T = RTA_p \exp(-B_v/f_v) \tag{31}$$

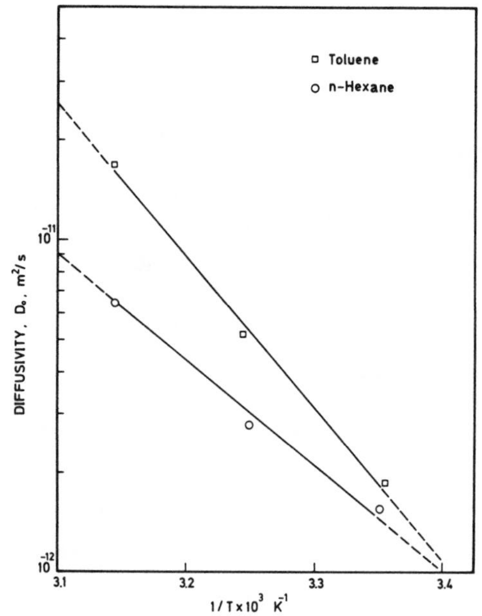

Figure 2: Dependence of D_0 on temperature of n-hexane (O) and toluene (❑) in LDPE, taken from Saleem et al. [32]

where A_p is a proportionality factor relating the probability of finding sufficient local free space to the diffusion coefficient; B_v is the amount of local free volume needed and f_v is the volume fraction of empty space. Even if not completely satisfactory, the free volume theory is a useful tool.

Finally, we have to mention various anomalies of the sorption in polymers. It is believed that various types of non-Fickian diffusion are characteristic for the glassy state. Studying the vapour sorption in polystyrene and other glassy polymers, Crank and Park [9] found sorption and desorption curves of a shape different than the one predicted by equation (28). In Fickian sorption the time to obtain the value M_t/M_∞ should be proportional to l^{-2}. However, in anomalous systems, this time $\sim l^{\alpha}$, where $\alpha \in (-2,0)$ [12]. Recently, Williams et al. [42] tried to explain the problem with the help of the random walk model and an apparent local fractal dimension. It is also known that the rate of advancing of the boundary of penetrant concentration in polymer membranes is not proportional to $t^{1/2}$ as predicted by the Fickian theory. In the early stages of sorption, the flux through the membrane surface of a glassy polymer is often greater for thin membranes than for thick ones. This effect, which can be explained by the differential stress mechanism, has been reported on by Park [27]. Alfrey et al. [1] have introduced a classification of diffusion processes in polymers, based on the motion of the advancing front of penetrant in the polymer. Processes in which this advance is proportional to $t^{1/2}$ are case I processes, and the other processes for which the front is advancing at a rate proportional to t, belong to case II. In case I, the time needed to attain $1/2\ M_t/M_\infty$ is proportional to l^{-2}; in case II it is proportional to l^{-1}. According to Lewis [20] in case III, this time is independent of l. Lewis found this type of behavior for cellulose acetate films plasticized with nitroglycerine. In some glassy polymers it is possible to observe a process for which $M_t/M_\infty \sim t^m$, $m > 1$ [17]. In many polymers, a two-stage sorption has been observed. This was first reported in [3]. On increasing the vapour pressure, the surface concentration immediately increases which leads to Fickian sorption with a relatively high diffusion coefficient. At the end of this stage, the surface concentration starts to increase due to the relaxation of internal stresses in the membrane. The change in surface concentration leads to a new rapid diffusion. This second stage of sorption, controlled by relaxation processes, does not depend on membrane thickness. The second stage process is very dependent on penetrant concentration and on the type of polymer. Generally it is believed that the concentration dependence is much stronger for non-crystalline glassy polymers than for those with some sort of order (crystalline or dipolar) [15]. A characteristic feature of case II diffusion is the strong discontinuity in penetrant concentration that forms the advancing front of the penetrant . Thomas and Windle [39] formulated a theory in which the sharp concentration front is explained by the concentration dependence of viscous flow.

Finally, since sometimes the shape of the nonlinear uptake curves is similar to the ones for a linear model, one has to be cautious in calculating the kinetic parameters from those curves [38].

3. Sorption Kinetics

From the point of view of penetrant molecules a polymer is a disordered system with a dispersion in the separation distances between nearest neighbor localized sites available for jumping molecules and a dispersion in the potential barriers between these sites. Both of these variables have a strong effect on the jumping time, the time for a molecule to arrive on successive sites. Hence, the distribution of jumping times, $\psi(t)$, will generally have a long tail. Adopting this hypothesis, one can use the model of continuous-time random walk (CTRW) on a lattice and study the transport properties of our system (diffusion in a polymer). Montroll and coworkers have extensively studied the transport of carriers in amorphous media, by CTRW, [23,18].

Before we try to apply CTRW, we shall mention some characteristic features of the desorption curves.

As noted in the previous section, a numerical analysis of desorption curves is a common method for determining the diffusion coefficient. A simple gravimetric desorption experiment, described in [32], has been performed with low density polyethylene (LDPE) film, at 25°C. The amount of penetrant $\left(Q(t) \equiv M_t - M_\infty\right)$ has been measured using a Cahn

2000 electrorecording balance. A typical desorption curve ÷ log $(M_t - M_\infty)$ versus t, for the system LDPE-octane is shown in Figure 3.

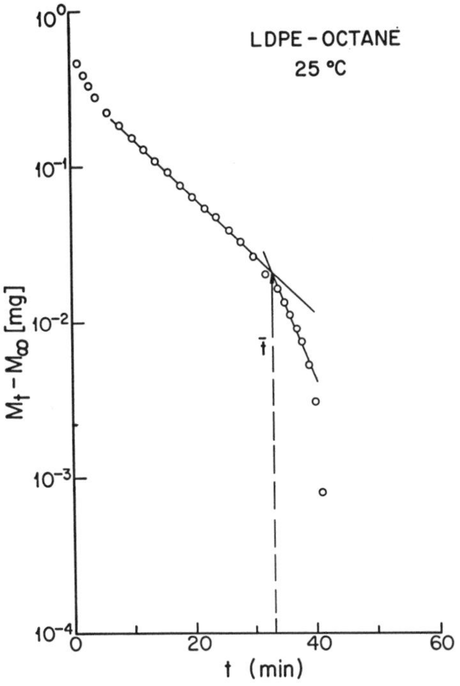

Figure 3: Desorption curve for the system LDPE-octane at 25°C

The desorption kinetics have been studied for various systems (different types of LDPE, rubbers and various organic penetrants) [2]. Analyzing these experiments one can notice a similar character in all of the desorption curves. These curves have three distinct regions. Immediately after the onset of desorption the amount of penetrant is quickly decreasing; this is followed by a slower decrease or transition region, followed finally by a tail region. Representing the onset time of the tail region is the transit time $\bar{t}$, normalizing Q(t) to $Q(\bar{t})$ and measuring the time in relative units of $\bar{t}$ one obtains a plot as shown in Figure 4. The similarity of desorption curves (for different penetrants and the same LDPE film) is even more pronounced on a log-log plot of $Q(t)/Q(\bar{t})$ versus $t/\bar{t}$. Such a plot is shown in Figure 5.

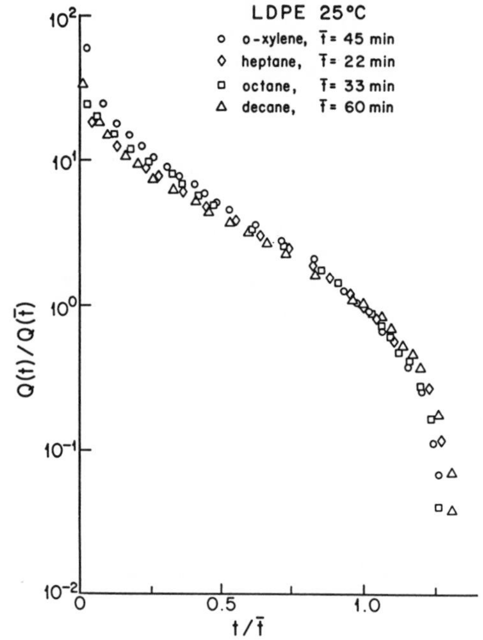

Figure 4: Normalized desorption curves for hydrocarbons in LDPE

Near the ordinate, $Q(t) \sim t^{-\alpha}$ and at the final stage of desorption $Q(t) \sim t^{-\beta}$. For the systems illustrated in Figure 5, $\alpha \approx 0.6$ and $\beta \approx 5.4$. We assume that the CTRW model can be applied to our experiments. The penetrants are continuously evacuated from the sample surface (the case of absorbing boundary). We therefore use the following equation for the propagator P(l,t) - the probability of a walker (diffusant molecule) being at plane l, at time t, if it started at plane l_0 at time t=0 (we assume one-dimensional diffusion on a simple cubic lattice with L cells in the x-direction), [23]

$$P(l,t) = G(l - l_0, t) - \int_0^t G(l, t - s)\, F(L - l_0, s)\, ds \tag{32}$$

Here G(l,t) is the free-space propagator (no absorbing boundaries) and $F(L-l_o,t)$ is the first passage-time distribution function for the transition l_o -> L (L being the position of the absorbing boundary). Since

$\sum_{l=1}^{L} G(l - l_0, t) = 1$, one can calculate from equ. (32), the fraction $\varphi(\tau)$ of walkers which have survived until time τ,

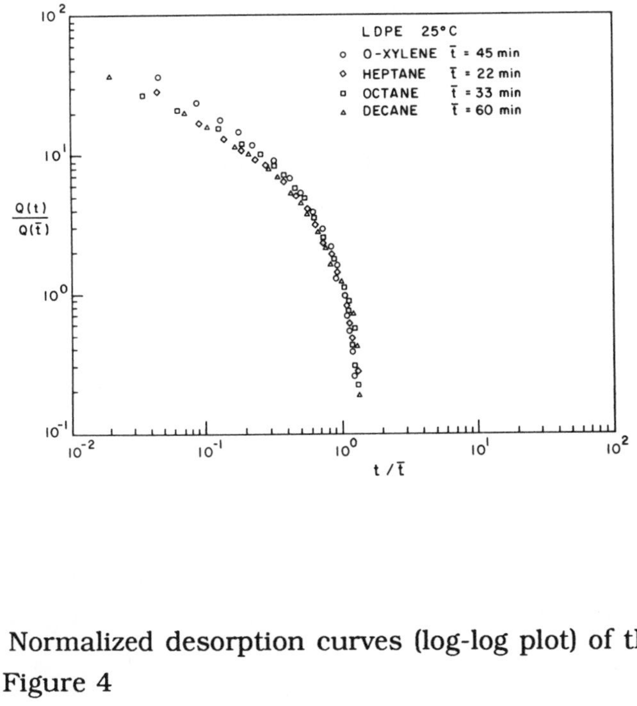

Figure 5: Normalized desorption curves (log-log plot) of the systems in Figure 4

$$\varphi(\tau) = 1 - \int_0^\tau F(L - l_0, s)\, ds \tag{33}$$

Since from the desorption experiments one can calculate $\varphi(\tau) = Q(\tau)/Q(o)$, it follows that

$$F(L - l_o, \tau) = -\frac{1}{Q(o)}\frac{dQ(\tau)}{d\tau} \tag{34}$$

Equation (32) also yields the mean position of the propagating packet of a diffusant,

$$\langle l(t)\rangle = l_o + \langle l(t)\rangle_o + \frac{1}{Q(o)}\int_o^t \langle l(t-s)\rangle_o \frac{dQ(s)}{ds}\,ds \tag{35}$$

where $\langle l(t)\rangle = \sum_{l=1}^{L} lP(l,t)$ and $\langle l(t)\rangle_o = \sum_{l=1}^{L} lG(l,t)$.

Since the flux of penetrant J(t) is proportional to $\frac{d\langle l(t)\rangle}{dt}$, it is possible to estimate the form of J(t), from equ. (35). For Gaussian packets [23],

$\langle l(t)\rangle_o = \frac{\bar{l}}{\bar{\bar{t}}}\,t$, where $\bar{l}$ is the mean displacement for a single jump and $\bar{\bar{t}}$ is the mean time between jumps. Let us note that for Gaussian propagation the jumping-time distribution function $\psi(t)$, has the simple exponential form, $\psi_1(t) = \lambda e^{-\lambda t}$.

For example, the flux of a Gaussian packet in a free space (no absorbing boundaries) is constant $J_o \sim \bar{l}/\bar{\bar{t}}$, while in the presence of absorbing boundaries $J \sim \frac{\bar{l}}{\bar{\bar{t}}}\frac{Q(t)}{Q(o)}$, where Q(t) is the amount of penetrant in a sample at time t. The Gaussian transport is characterized by the existence of the first two moments of G(l,t). This condition is satisfied, for large t, by such $\psi(t)$ which yields a finite mean time between jumps. A prototype of such a jumping-time distribution function, $\psi(t) = \lambda(\lambda t)^{\lambda\bar{t}-1} e^{-\lambda t}/\Gamma(\lambda\bar{t})$, has been studied [23]. It is shown there, that different time parameters, λ, correspond to different dispersions of Gaussian packets (i.e. different constant diffusion coefficients). Studying a different class of jumping-

time distribution functions, particularly $\Psi_1(t) = 4\lambda e^{\lambda t}\, i^2 \mathrm{erfc}(\sqrt{\lambda t})$, Shlesinger [37] obtained the following interesting results.
Note that this class of jumping-time distribution functions have an extended tail decaying asymptotically

as $$\Psi(t) \sim \left[A t^{1+q}\,\Gamma(1-q)\right]^{-1}, \quad 0 < q < 1,\ t \to \infty \tag{36}$$

Using Tauberian theorems, Shlesinger [37] obtained asymptotic expressions for $\langle l(t)\rangle_0$ and $\langle l(t)\rangle$, i.e. for the mean position of walkers in the packet:

$$\langle l(t)\rangle_0 \sim \bar{l}\, A t^{q}/\Gamma(q+1) \tag{37}$$

and $$\langle l(t)\rangle \sim C - \bar{l}\, B/t^{q}\,\Gamma(q+1) \tag{38}$$

where C and B are parameters depending on the used lattice model. This means that the flux associated with a moving packet is given by:

$$J(t) \sim \begin{cases} t^{-(1-q)}, & \langle l\rangle << L \\ t^{-(1+q)}, & \langle l\rangle \sim L \end{cases} \tag{39}$$

We can see that the flux J(t), in these cases, undergoes a transition from one fractional power law into another one, after some characteristic time (which is given by the properties of the particular system, i.e. by $\psi(t)$). A similar type of behavior is illustrated by Figure 5. Detailed calculations for $\psi_2(t) = 4\lambda e^{\tau} i^2 \mathrm{erfc}(\tau^{1/2}),\ \tau = \lambda t$ can be found in [36]. This distribution function has no finite positive integral moments. This fact results in an unusual character of the propagating packet. The mean position of the propagating packet for $t >> 1$ is:

$$\langle l\rangle_0 \sim \bar{l}\,(\tau/\pi)^{1/2} \tag{40}$$

and the flux J(t) is proportional to $\tau^{-1/2}$,

$$J_0(t) \sim \frac{1}{2}\,\bar{l}\,(\pi\tau)^{-1/2} \tag{41}$$

The peak in the packet remains near its starting position. However, the mean advances with $\tau^{1/2}$, hence the ratio of dispersion (σ) to the mean displacement is independent of time t. For Gaussian packets this ratio is proportional to $\tau^{-1/2}$.

The first passage-time distribution function ($F(L-l_o,t)$ has been determined [23] for a simple cubic lattice from the relation:

$$F(L-l_o,t) = L^{-1}\left[\tilde{G}(L-l_o,\psi^*(u)) / \tilde{G}(0,\psi^*(u))\right] \qquad (42)$$

where L^{-1} is the inverse Laplace transform, $\tilde{G}(l,t)$ is a random-walk generating function and $\psi^*(u)$ is the Laplace transform of the jumping-time distribution $\psi(t)$ $\left(\psi^*(u) = \int_0^\infty e^{-ut}\psi(t)dt\right)$. Approximately,

$$\tilde{G}(L-l_o,\psi^*) / \tilde{G}(l,\psi^*) \simeq \alpha_1^{L-l_o} + (1-\alpha_1^L)\,\alpha_2^{-l_o} \qquad (43)$$

where

$$\alpha_1 = 1-\epsilon(\psi^*),\ \alpha_2 = K(1+\epsilon(\psi^*)),\ K>1 \qquad (44)$$

and

$$\epsilon(z) = A_1(1-z) + A_2(1-z)^2 + \ldots \qquad (45)$$

Parameters K and A_i depend on the lattice model of the transport process [23] (i.e. transition probabilities and a bias parameter which models the driving force of the transport). Combining equs. (34) and (42) one can write:

$$Q(t) = Q(o) - Q(o)\int_0^t L^{-1}\left[\frac{\tilde{G}(L-l_o,\psi^*)}{\tilde{G}(0,\psi^*)}\right]dt \qquad (46)$$

Here, an experimentally accessible quantity Q(t) is related to the jumping-time distribution function $\psi(t)$, which models the properties of the system. Theoretically by solving the inverse problem, one can obtain $\psi(t)$. Practically, one can study asymptotic regions of the transport process and obtain a nonclassical flux exponent [42,23,36].

During the process of diffusion in a polymeric sample, the diffusing molecules may introduce various local conformational abnormalities (a field of local strains) into the system of polymer chains. As a result of vacating the position of such a local disturbance, the penetrant molecule will cause a disturbance at some other position in the network. After some time, the neighborhood of the disturbed segment will relax. According to this picture, the migration of penetrant molecules may cause a mechanical relaxation in a polymer. The fraction of segments surviving the "attack" of the diffusing molecules can be estimated from the CTRW model. If a typical unperturbed segment is at the origin of the periodic space lattice, we have to determine the flux of walkers at the origin. Let f(r,t) be the probability density, at time t, that a walker, originally at r reaches the origin for the first time. The probability that the walker has not reached the origin in the time interval (0,t) is then

$1-\int_0^t f(r,t')\,dt'$. The probability that none of the n walkers has reached the origin is the survival probability $\bar{P}(t)$ of the undisturbed segments. This quantity is given as follows:

$$\bar{P}(t) = \sum_{r_1=1}^{N} \cdots \sum_{r_n=1}^{N} \prod_{i=1}^{n} \left[1-\int_0^t f(r_i,t')dt'\right] \bar{p}(r_1, r_2, \ldots, r_n) \tag{47}$$

Here, n is the number of walkers, N is the number of lattice points and $\bar{p}(r_1, \ldots, r_n)$ is the initial probability density function of walker positions. With walkers originally randomly distributed, $\bar{p} = N^{-n}$, n and N large and c = n/N, the last equation yields

$$\bar{P}(t) = \exp\left[-c\int_0^t J(t')\,dt'\right] \tag{48}$$

where the flux of walkers is given by:

$$J(t) = \sum_{r\neq 0} f(r,t) \tag{49}$$

The relaxation modulus, G(t), should be proportional to the rate of change of the survival probability $\bar{P}(t)$ of undisturbed segments. Then writing

$$G(t) = -k\,\frac{d\bar{P}(t)}{dt},\ k \equiv \text{const} \tag{50}$$

one observes that:

$$G(t) = ckJ(t)\exp\left(-c\int_0^t J(t')\,dt'\right) \tag{51}$$

and

$$J(t) = \frac{G(t)}{c\left[k - \int_0^t G(t')\,dt'\right]} \qquad (52)$$

Thus for some systems it would be possible to relate the diffusional and rheological characteristics. If for example $J(t) \sim t^{\alpha-1}$, then equ. (51) yields $G(t) \sim t^{\alpha-1} \exp(-c\,t^{\alpha})$, which is the form of the stretched exponential relaxation function rediscovered after more than a century [19,27,21].

Concluding Remarks

In this work, we have mentioned three possible approaches to the problem of diffusion in polymeric materials. We have chosen the methods of continuum mechanics (classical and modern or "rational"), experimental and kinetic. This short review is not meant to be complete; many ideas have been omitted. One has to be aware that diffusion plays an important role in many subfields of physics and is being extensively studied in mathematics, computer science and in a variety of disciplines of the natural sciences. Since polymers are very complex systems, we hope that some features of diffusion in polymers are of a general enough nature to better help us understand this important physical process.

Acknowledgement

D. De Kee acknowledges with thanks, the financial support through DSS contract #W7714-8-5558/01-SS.

Nomenclature

A_r	area
b	barodiffusion ratio
$\underset{\sim}{b}$	body force
c	concentration
c_p	specific heat
$D(D_0, \bar{D}, D_T)$	diffusion coefficient (at zero concentration, average, thermodynamic)
$\bar{f}$	free energy (specific)
f_o	volume fraction of "empty space"
$F(L-l_o,t)$	first passage-time distribution function
g	Gibbs potential (per unit mass)

$G(l,t)$	free space propagator
$\tilde{G}(l,t)$	random walk generating function
$G(t)$	relaxation modulus
$\underset{\sim}{j}, J_i (J(t))$	flux of mass (walkers)
$\underset{\sim}{k}$	interaction force
l	thickness in section 2, otherwise $l=l_1$
$l_i=1,2,\ldots L_i$	components of dimensionless lattice vector
$\bar{l}$	mean displacement for a simple jump
$\underset{\sim}{L}(L_{ij})$	Onsager's (kinetic) coefficients
$L[f] = f^*(u)$	Laplace transform
m	mass
$M_t(M_\infty)$	mass of penetrant at time t (equilibrium)
N	number of lattice points
p	pressure
$P(l,t)$	propagator in the presence of absorbing boundary
$\bar{P}(t)$	survival probability of undisturbed segments
$\bar{p}$	initial probability density function of walker positions
$\underset{\sim}{q}$	heat flux
$Q(t)$	amount of diffusant
R	gas constant
$S(s)$	entropy (specific)
$t(\bar{\bar{t}})$	time (mean time between jumps)
$T(T_g)$	temperature (glass transition)
$\underset{\sim}{T}$	stress tensor
u	Laplace transform parameter
$\bar{u}$	internal energy
$V(v)$	volume (specific)
$\underset{\sim}{X}(X_i)$	thermodynamic force
$\underset{\sim}{w}$	diffusion velocity
$\bar{\beta}_i$	internal parameters
$\Gamma()$	Gamma function
λ	rate constant
μ	chemical potential

χ	thermal conductivity coefficient
$\varphi(t)$	fraction of walkers surviving until time t
$\psi(t)$	jumping-time distribution function
ρ	density
$\tau=\lambda t$	dimensionless time
υ	thermodiffusion ratio
ΔE_a	activation energy

$$i^n \mathrm{erfc}\, x = (2/\pi^{1/2} n!) \int_x^\infty (t - x)\, e^{-t^2}\, dt$$

<>	average
$\underset{\sim}{\nabla}$	gradient operator
$\underset{\sim}{a}$	vector or tensor a

References

1. Alfrey, T., Gurnee, F.E. and Lloyd, W.G., J. Polym. Sci., C., 12, 249 (1966).
2. Asfour, A., D. De Kee, D. and Stastna, J., Report TP002.D88, University of Windsor (1988).
3. Bagley, E. and Long, F.A., J. Am. Chem. Soc., 77, 2172 (1955).
4. Barrer, R.M. and Chio, H.T., J. Polym. Sci. C., 10, 111 (1965).
5. Bowen, R.M., in Continuum Physics, Vol. III, A.C. Eringen, ed., Academic Press, New York (1976).
6. Coleman, B.D. and Gurtin, M.E., J. Chem. Phys., 47, 597 (1967).
7. Coleman, B.D. and Noll, W., Arch. Rat. Mech. Anal., 13, 167 (1963).
8. Crank, J., The Mathematics of Diffusion, Clarendon Press, Oxford (1975).
9. Crank, J. and Park, G.S., Trans. Faraday Soc., 47, 1072 (1951).
10. Daynes, H.A., Proc. Roy. Soc. A., 94, 286 (1920).
11. De Groot, S.R. and Mazur, P., Nonequilibrium Thermodynamics, North-Holland, Amsterdam (1962).
12. Frisch, H.L., J. Chem. Phys., 37, 2408 (1962).
13. Frisch, H.L., Polym. Eng. Sci., 20, 2 (1980).
14. Fujita, H., Fortschr. Hochpolym. Forsch., 3, 1 (1961).
15. Fujita, H., Kishimoto, H. and Odani, H., Progr. Theor. Phys., 10, 210 (1959).
16. Hopfenberg, H.B., J. Membrane Sci., 3, 215 (1978).
17. Jacques, C.H.M., Hopfenberg, H.B. and Stannet, V.T., in Permeability of Plastic Films and Coatings, ed., Hopfenberg, H.B., Plenum Press, New York (1965).
18. Kenkre, V.M., Montroll, E.W. and Shlesinger, M.F., J. Stat. Phys., 9, 45 (1973).
19. Kohlrausch, F., Pogg. Ann. Phys., 119, 352 (1832).
20. Lewis, T.J., Polymer, 19, 285 (1953).

21. Li, K.L., Inglefield, P.T., Jones, A.A., Bendler, J.T. and English, A.D., Macromolecules, 21, 3110 (1988).
22. Mandelbrot, B.B., The Fractal Geometry of Nature, Freeman, New York (1982).
23. Montroll, E.W. and Scher, H., J. Stat. Phys., 9, 101 (1973).
24. Müller, I., Thermodynamik Grundlagen der Materialtheorie. Bertelsman Universitätsverlag, Düsseldorf (1973).
25. Onsager, L., Phys. Rev., 37, 405 and 38, 2265 (1931).
26. Park, G.S. and Saleem, M., J. Membr. Sci., 18, 177 (1984).
27. Park, G.S., J. Polym. Sci., 11, 97 (1953).
28. Petropoulos, J.H., J. Polym. Sci., Polym. Phys. Ed., 22, 1855 (1984).
29. Prager, S., Bagley, E. and Long, F.A., J. Am. Chem. Soc., 75, 1255 (1953).
30. Rogers, W.A., Buritz, R.S. and Alpert, J., J. Appl. Phys., 25, 868 (1954).
31. Saleem, M., Ph.D. Thesis, University of Wales, U.K. (1977).
32. Saleem, M., Asfour, A.A., De Kee, D. and Harrison, B., J. Appl. Polym. Sci., 37, 617 (1989).
33. Asfour, A.A., Saleem, M., De Kee, D. and Harrison, B., J. Appl. Polym. Sci., 38, 1503 (1989).
34. Samohyl, I., Nguyen, X.Q. and Sipek, M., Coll. Czech. Chem. Comm., 50, 2346 (1985).
35. Samohyl, I., Nguyen, X.Q. and Sipek, M., Coll. Czech. Chem. Comm., 50, 2364 (1985).
36. Scher, H. and Montroll, E.W., Phys. Rev. B, 12, 2455 (1975).
37. Shlesinger, M., J. Stat. Phys., 10, 421 (1974).
38. Smith, D.M. and Keller, J.F., Ind. Eng. Chem. Fund., 24, 497 (1985).
39. Thomas, N.L. and Windle, A.H., Polymer, 23, 529 (1982).
40. Truesdell, C.A., A First Course in Rational Continuum Mechanics, Johns Hopkins University, Baltimore (1972).
41. Vrentas, J.S. and Duda, J.L., J. Polym. Sci., Phys. ed., 15, 403 and 417 (1977).
42. Williams, G.O., Frisch, H.L. and Ogawa, H., J. Colloid Interface Sci., 123, 448 (1987).

Acknowledgements: We are indebted to John Wiley and Sons Inc. for permission to reproduce figures one and two, (Saleem et al. 1989) [32]

D.A. HILL AND D.S. SOANE

Rheology of dispersions of rodlike particles in viscoelastic fluids

1. INTRODUCTION

Fiber filled thermoplastics and thermosets have found a wide range of applications as light weight high performance structural elements. Their formation involves complex rheology, and many factors influence the processability of these materials. Fiber geometry, concentration, orientation distribution as well as matrix rheological characteristics are all important parameters. Thus, fundamental understanding of the interrelations among microscopic parameters and bulk rheological properties is critical not only for improved characterization but also for enhanced processability.

To date, much of the theoretical work on suspensions of rodlike particles has focused on Newtonian matrices. Following the early work of Jeffery [1] on the motion of spheroidal particles in creeping flow, a large number of theoretical investigations has appeared. Most of this development, for Brownian and non-Brownian particle suspensions, has been summarized by Brenner [2]. For non-Brownian particles additional major contributions are those of Batchelor [3] who developed a general mathematical formalism by which constitutive equations for suspensions of particles of arbitrary shape and concentration can be derived. Dinh and Armstrong [4] developed a constitutive model predicting transient rheological properties of semi-dilute suspensions, and Hinch and Leal [5] (among others) the steady shear flow orientation distribution function of dilute, weakly Brownian fibers. Furthermore, Evans [6] demonstrated the equivalence between the single particle theory of Jeffery and Ericksen's continuum theory of the Transversely Isotropic Liquid [7]. The ability of the TIL constitutive equation to describe successfully the rheology of dilute fiber suspensions in complex geometries with no adjustable parameters was recently demonstrated by Lipscomb and coworkers [8].

Despite the abundance of literature on suspensions in Newtonian fluids, relatively little attention has been devoted to systems in which the suspending fluid is Non-Newtonian viscoelastic. In particular, to date all of the analyses available for the motion of particles in Non-Newtonian matrices are concerned exclusively with second order fluids. A second order fluid has a shear rate independent viscosity but non-zero first and second normal stress differences. These are all desirable features if one wishes to model (even approximately) "real" non-Newtonian suspensions. On the other hand, a second order fluid is also inelastic, in the sense that it possesses no memory. Thus, the use of this model to describe transient (creeping) flows of viscoelastic liquids may lead to highly inaccurate predictions. In fact, in most transient flows of polymeric liquids it is the memory of the fluid which dictates the richness of features exhibited by the system. It is precisely this point that we wish to address (approximately) in the present paper, with the hope of capturing, at least qualitatively, the main physics of the problem. The following brief overview is designed to outline the state of affaires in the modeling of non-Newtonian suspensions. Particle motion in a second order fluid has been examined recently by Kim [9] and earlier by Brunn [10]. For dilute spheres Mifflin [11] calculated the energy dissipation in a second order fluid, and found that it could be either smaller or larger than the Newtonian case depending on the type of flow. Suspensions of spherical particles in second order fluids have also been considered by Kaloni and Stastna [12], who showed that second order terms have no effect on the effective shear viscosity in steady shear flow. Perhaps, one of the most important pioneering contributions to the field of non-Newtonian suspensions of slender particles is that of Leal [13,14], who studied the motion of slender rodlike particles in a Rivlin-Ericksen second order fluid in steady shear flow. Jeffery's theory predicts that isolated spheroidal particles immersed in a Newtonian fluid in steady shear flow will rotate periodically and indefinitely along fixed elliptical orbits centered around the local vorticity axis of the flow. In their periodic revolution, particles of large aspect ratios ($r_a >> 1$) spend a time of order $t_p \sim r_a/\dot{\gamma}$ (t_p=period of rotation and $\dot{\gamma}$=shear rate) along the plane of shear, and then quickly "flip" over in a time of order $t_f \sim 1/\dot{\gamma}$. During the flipping motion the particles rotate "affinely", i.e., at a rate dictated by the local vorticity of the flow. By calculating the rate of change of particle orientation in simple shear, Leal demonstrated that in a second order fluid the prevailing second normal stress differences induce a drift in the particle orbits towards an orientation parallel to the vorticity axis. The characteristic time for the drift, t_d, is of order $t_d \sim 8\eta/(\dot{\gamma}^2\psi_2)$, where η is the viscosity, and ψ_2 the second normal stress coefficient of the fluid (in the limit of zero shear). The results of Leal's analysis were later used by Cohen et. al. [15] to predict the steady state orientation distribution function of dilute, slender spheroids in simple shear flow, under the influence of weak elasticity and weak Brownian motion. These authors showed that weak elastic torques can lead (at sufficiently long times) to considerable deviations of the orientation distribution of the fibers from the Newtonian profile. Neglecting the effects of fluid elasticity on the stress profile along the particles, i.e., considering only contributions to the suspension stress due to the biased orientation distribution, the authors demonstrated the existence of a second shear thinning mechanism, distinct from that of the pure matrix. This shear thinning stems from the competition between fluid elasticity and Brownian rotational diffusion. The calculation, however, is limited to steady state, a condition which (depending on the magnitude of the second normal stresses and rotational diffusivity) may require a very long time to achieve in some cases.

In certain processing applications, such as molding, the characteristic time scale for flow is of order $t_f \sim \dot{\gamma}^{-1}$, which also corresponds to the "flip over" time of the fibers. Here, a measure of the "elastic" orbital drift occurring during an affine rotation period can be taken as the ratio t_d/t_f, i.e.:

$$\frac{t_d}{t_f} \sim \frac{8\,\eta\,\dot{\gamma}}{\dot{\gamma}^2\,\Psi_2} \sim \frac{8\,T_{12}}{(T_{22} - T_{33})} \qquad (1)$$

where T_{12} and $(T_{22}$-$T_{33})$ are the shear stress and second normal stress difference of the fluid respectively. For ordinary polymer melts at low Deborah numbers, a conservative estimate of t_d/t_f can be easily obtained from literature data. Surprisingly, we find [11] that, even for Deborah numbers up to $Dh \sim O(1)$, values for t_d/t_f of order ~ 100 are not uncommon. Since t_d/t_f scales as ~ $1/Dh$, for $Dh \leq 1$ the fibers will undergo a negligible orbital drift during an "affine rotation" period ($t_f \sim \dot{\gamma}^{-1}$) (obviously, for fluids having $(T_{22}$-$T_{33})=0$ no drift at all would occur). Thus, in transient flows for which $t_d/t_f >> 1$, the orientation distribution of ensembles of dilute rods will evolve as if the suspending fluid were essentially Newtonian. The rheology of such (dilute) suspensions will reflect primarily the viscoelastic nature of the stress profile along the rods. Contributions to the total stress due to distortions in the orientation distribution function, are expected to be only of minor importance.

The present paper studies the effects of matrix viscoelasticity on the rheology of a dilute suspension of large aspect ratio rodlike particles. Different from the case considered by Cohen et. al. [15], we are primarily concerned with the rheological consequences of a viscoelastic stress profile along the rods. The suspending medium is assumed here to behave as a "Maxwell" fluid in the limit of linear viscoelasticity. We recognize this linearity condition to be a serious limitation of our approach. This simplification, however, allows us to obtain manageable analytical expressions for a system which, otherwise, would require the solution of a full non-linear problem. Note that in this case the absence of secondary normal stress differences is only a necessary condition to ensure zero particle orbital drift. There is no proof that for a Maxwell fluid the above condition is also a sufficient one. Nevertheless, we shall assume here that orbital drifts in the particle trajectories can be neglected. (This assumption can be, in part, justified by considering the fact that a Maxwell fluid is characterized by a constant (steady) shear viscosity and first normal stress coefficient, and zero second normal stress differences (i.e., $t_f/t_d=0$).) From an experimental point of view the assumption of negligible orbital drift should be reasonable if $t_f/t_d << 1$. Hinging on these approximations we derive analytical expressions for the hydrodynamic stresses on the particles based on physically plausible simplifying assumptions, and we compute bulk rheological properties using Newtonian orientation distributions (errors arising from this approximation are expected to be at most of order t_f/t_d). It must be pointed out that the results of our analysis are strictly valid only for Dh<<1 and r_a>>1, and that analytical complexities preclude us from obtaining an explicit form for the constitutive equation for the suspension. Indeed, our calculations of the particle stress are limited to shear flows in which the initial orientation distribution of the fibers at the inception of flow is of complete alignment. Because of our many approximations and simplifying assumptions, the results of the present analysis are to be evaluated primarily from a qualitative point of view. For this reason, even though the analysis is valid in principle only for Dh<<1, in order to clearly expose the qualitative features our calculations, most of the results are presented for values of the Deborah numbers up to order unity.

This paper is structured as follows. First, a derivation of the macroscopic excess-stress tensor due to the presence of the particles in the fluid is presented, based on the microscopic dynamics of fiber-flow interaction. We claim no originality for the results of this section, which have been previously developed in great detail by Evans [6] at first and subsequently by Dinh and Armstrong [4] for the case of a Newtonian matrix. However, the derivations are repeated for clarity and to avoid possible ambiguities which could arise because of the viscoelastic nature of the suspending fluid. Second, an analytical expression for the hydrodynamic stress profile along the fiber will be derived for a Maxwell fluid. Some results for stress profiles in simple shear flow as functions of rod orientation and Deborah number will be presented. Third, the stress profile will be repeatedly integrated to obtain a tension profile at first and then the average tension in the fibers. The latter will finally be used to calculate the stress in flows of assigned kinematics.

2. THEORY

Model development is subject to the following assumptions:

(1) We consider a homogeneous suspension of monodisperse cylindrical rods of length L and diameter d, with an aspect ratio r_a=L/d>>1. As the rods do not interact with each other we are concerned with the case ϕr_a^2<<1, ϕ being the volume fraction of the rods.

(2) No external torque-inducing fields act on the system, and Brownian motion is negligible.

(3) Inertia is unimportant for both the suspending medium and the particles, which are also neutrally buoyant.

(4) The flow field is homogeneous.

(5) Particle motion is affine.

(6) In the constitutive equation for the matrix only linear terms are important.

2.1 The Orientation Distribution Function

Let **p** be a unit vector specifying the orientation of a rod (Fig.1). In a statistical sense, the orientation of the particles is uniquely described by a distribution function, χ, such that $\chi(\mathbf{p},t)\,d\mathbf{p}$ is the fraction of particles with orientation between **p**, and **p**+d**p** (d**p** is a differential solid angle around **p**). The following restrictions apply to χ:

$$\chi(\mathbf{p}) = \chi(-\mathbf{p}) \qquad \text{(symmetry)} \qquad (2)$$

$$\oint \chi(\mathbf{p},t)\,d\mathbf{p} = 1 \qquad \text{(normalization)} \qquad (3)$$

The ensemble average (< >) of a quantity **Q** can be calculated as follows:

$$\langle Q\rangle(t) = \oint Q(\mathbf{p})\,\chi(\mathbf{p},t)\,d\mathbf{p} \qquad (4)$$

In Eqs.3 and 4 the integration extends over the entire phase (angular) space.

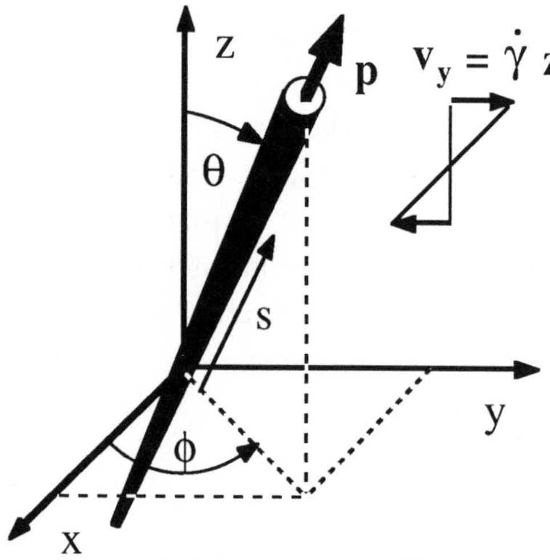

Fig.1 The coordinate system; **p** is a unit vector describing the orientation of the rod and s is the curvilinear coordinate along the major axis with origin at the center of mass of the particle. The velocity field is of simple shear along the y direction.

A conservation statement in phase space can be written for the ensemble of particles:

$$\frac{\partial \chi}{\partial t} = -\frac{\partial}{\partial \mathbf{p}}\,\dot{\mathbf{p}}\,\chi \qquad (5)$$

where $\dot{\mathbf{p}}$ is the angular velocity of a rod of orientation **p**, $\partial/\partial t$ the time derivative, and $\partial/\partial\mathbf{p}$ the phase space gradient operator. For affine particle rotation $\dot{\mathbf{p}}$ coincides with the "Newtonian" angular velocity. In a homogeneous flow field of velocity gradient $\mathbf{K}=\nabla\mathbf{V}$, Jeffery's equations of motion for spheroidal particles of large aspect ratio (r_a>>1) reduce to [4]:

$$\dot{\mathbf{p}} = \mathbf{K}\cdot\mathbf{p} - \mathbf{K}:\mathbf{ppp} \tag{6}$$

In simple shear flow, Eq.6 predicts that a particle will rotate affinely and ultimately align with the flow without any further rotation; in fact, for r_a>>1 the actual angular velocity of the particles in the aligned configuration is of order $|\dot{\mathbf{p}}| \sim O(\dot{\gamma}/r_a) << 1$. Note that since $|\dot{\mathbf{p}}| \propto |\mathbf{K}|$, $\mathbf{p}$ in Eq.6 is a function only of the total deformation.

Equations 5 and 6 posses analytical solutions. The general solution of Eq.6 is [4, 6]:

$$\mathbf{p} = \frac{\mathbf{E}\bullet\mathbf{p_o}}{(\mathbf{E}^{\dagger}\bullet\mathbf{E}:\mathbf{p_o p_o})^{1/2}} \tag{7}$$

where $\mathbf{E}$ is the deformation gradient tensor ($\mathbf{E} = \partial\mathbf{x}/\partial\mathbf{x}_o$, $\mathbf{x}_o$ and $\mathbf{x}$ are the initial and actual position vectors of a fluid element respectively), and $\mathbf{p}_o$ is the initial orientation of the particle. Dinh and Armstrong [4] solved Eq.5, subject to Eqs.2 and 3, for an initial random orientation distribution. The solution is:

$$\chi(\mathbf{p},t) = \frac{1}{4\pi}(\Delta^{\dagger}\cdot\Delta:\mathbf{pp})^{-3/2} \tag{8}$$

where $\Delta=\mathbf{E}^{-1}$. Note that the symbols $\mathbf{p}$ appearing in Eq.7 and 8 have different meanings. In Eq.7, $\mathbf{p}$ indicates the orientation of a specific rod as a function of deformation (or time), whereas in Eq.8, $\mathbf{p}$ is the coordinate of an arbitrary point in phase-space.

Another simple solution of Eq.5 can be obtained for the case in which the rods are all initially oriented along the same direction, $\mathbf{p}_o$. Such an orientation distribution could be induced, for example, by a prolonged exposure of the suspension to an external uniform electric or magnetic field. This strategy, however, would be practical only for suspending fluids of low viscosity. For very viscous matrices, repeated extensional flow of the suspension would be a much more efficient way of aligning the fibers. In either of these cases, the prevailing "initial" orientation distributions would closely resemble a Dirac δ function, i.e.:

$$\chi(\mathbf{p},0) = \delta(\mathbf{p}-\mathbf{p}_o) \tag{9}$$

Obviously, the solution of Eq.5 for this initial configuration is:

$$\chi(\mathbf{p},t) = \delta(\mathbf{p}-\mathbf{p}(t)) \tag{10}$$

where $\mathbf{p}(t)$ is given by Eq.6.

Once a kinematic is assigned, ensemble averages of orientation-dependent variables can be easily computed by means of Eqs.4, 8 and 10. The derivation of a relationship between the macroscopic stress and orientation-dependent microscopic hydrodynamic forces is the subject of the next section.

2.2 The Stress Tensor

The overall (deviatoric) stress, τ, exhibited by the suspension is the sum of matrix and particle contributions:

$$\tau = \tau_m + \tau_p \tag{11}$$

In the absence of inter-particle forces, and for negligible Brownian (entropic) forces, the particle contribution, τ_p, is purely hydrodynamic. The calculation of the hydrodynamic particle stress has been considered in detail by Evans [6] at first and later by Dinh and Armstrong [4], for rodlike particles in Newtonian matrices. By reducing Batchelor's multiparticle problem to a single particle one, these authors obtained a relation giving the hydrodynamic particle stress as a function of the suspension microstructure. Application of the above results to our case of a viscoelastic matrix is not straightforward, because in Batchelor's theory the linearity of the Newtonian constitutive equation was exploited. We note, however, that the basic idea behind the above derivations is that the hydrodynamic forces exerted by the fluid on a particle (and *vice versa*) are strongly anisotropic in nature. Because particle rotation is affine, at any point along the rod the component of the velocity of the fluid relative to the rod (orthogonal to **p**) is identically zero. Since the drag force is proportional to the relative velocity, its orthogonal component is zero as well. The only non-zero component of the hydrodynamic force is that in the **p** direction. This force arises primarily because of the shearing motion of the fluid relative to the rod, and thus is a functional of the shear stress profile along the particle. (Stated differently, the rod sustains either tension or compression due to the shear deformation at the fluid-rod interface. The exact tensile or compressive force profile along the rod is what we are after now.) This force is important because in creeping flow (i.e., for negligible inertia), the active hydrodynamic force exerted by the fluid on the rod is equal and opposite to the reaction of the rod. Thus, there is a direct correspondence between the hydrodynamic force on the particle and the particle stress τ_p. Although originally formulated for a Newtonian matrix, the above concepts are equally valid for a viscoelastic fluid, provided that the assumption of linear elasticity is made. The complication, here, is given by the normal stresses, which, in principle, could affect both the magnitude and the direction of the hydrodynamic force on the rod. This complexity, however, can be resolved as follows. The constitutive model adopted here is a special case of the eight constants Oldroyd model given by Bird et. al.[16]. In particular, with reference to their notation, the values of the phenomenological constants are: $\lambda_1=\mu_1=\lambda$, and $\lambda_2=\mu_2=\nu_2=\mu_0=\nu_1=0$. This model is also equivalent to Denn's "modified Maxwell model" [16] with $\mu_2=0$. For simple unsteady shear flow along the "1" direction, with the vorticity in the "3" direction, the constitutive equation gives:

$$
\begin{aligned}
&T_{13} = T_{23} = 0 \\
&(1 + \lambda D) T_{12} = -\eta_o \dot{\gamma} \\
&(1 + \lambda D) T_{11} = 2 T_{12} \dot{\gamma} \lambda \\
&(1 + \lambda D) T_{22} = (1 + \lambda D) T_{33} = 0
\end{aligned}
\qquad (12)
$$

where T_{ij} (= **T**) is the deviatoric stress, D the material derivative, and $\gamma = \partial v_1/\partial x_2$ (**v** = $(v_1,0,0)$ being the fluid velocity). Note that for an initially stress-free material $T_{22}=T_{33}=0$. Consider now a coordinate system along the surface of a rod, such that the 1 and 2 directions correspond to the direction of **p**, and to the (local) outward normal to the surface respectively. Not too far from the surface the flow is well approximated by a shear flow (with, $v_2=v_3=0$) and, if non-linear terms in the constitutive equation are neglected, Eqs.12 can be used to calculate the stress. The incremental force, **df**, on a surface element of area dA and outward normal $\mathbf{e}_2$ is

$$\mathbf{df} = \mathbf{T} \cdot \mathbf{dA} = \mathbf{T} \cdot \mathbf{e}_2 \, dA \qquad (13)$$

which, since $T_{22}=T_{32}=0$, gives

$$\mathbf{df} = T_{12} \mathbf{e}_1 \, dA = T_{12} \mathbf{p} \, dA \qquad (14)$$

Thus, normal stresses (as well as the isotropic pressure) do not contribute to **df**. Within our assumptions, the total hydrodynamic drag force on the rod depends uniquely on the shear stress and, as it would be for a Newtonian fluid, is directed along **p**.

Calculation of the particle hydrodynamic stress at this point is straightforward, and follows, albeit somewhat differently, the analysis in Ref. [4]. Consider a test rod of length L, diameter d and orientation **p**, in a homogeneous flow field of assigned kinematic, $\mathbf{K}=\nabla\mathbf{v}$. Let s be the curvilinear coordinate of a point along the major axis of the rod relative to its center of mass ($R \geq s \geq -R$, with $R=L/2$) (Fig.1), and $\boldsymbol{t}$ be a vector specifying the tension in the rod. The tension would be the force required to keep an imaginary detached section of the rod in dynamic equilibrium under the action of external drag forces. Obviously, from the arguments above, $\boldsymbol{t}$ and **p** are parallel, i.e.,

$$\boldsymbol{t} = t \, \mathbf{p} \qquad (15)$$

Because the resultant of the hydrodynamic forces on a rod segment depends on the length of the segment considered, so does the magnitude of the tension, t, which is zero at $s=\pm R$. It is convenient then to define a linear stress (or a force per unit length), $\boldsymbol{\sigma}$, such that:

$$d\boldsymbol{t} = \boldsymbol{\sigma} \, ds = \sigma \, ds \, \mathbf{p} \qquad (16)$$

where $d\boldsymbol{t}$ is the incremental tension between s and s+ds, and $\sigma=|\boldsymbol{\sigma}|$. The linear stress σ, is obtained by integration of the hydrodynamic stress along the perimeter of the cross section of the rod:

$$\sigma = \frac{d}{2} \int_0^{2\pi} T_{12} \, d\theta \qquad (17)$$

where d is the diameter of the rod. The tension in a rod segment between s and R is then:

$$t(s) = \int_{s}^{R} \sigma(s')\, ds' \tag{18}$$

This is also the force exerted by the segment on the fluid.

In order to evaluate the particle stress we consider a surface element **dA** within a suspension in which n particles per unit volume are present. The incremental contribution, $\mathbf{df}_A$, to the total force on **dA** exerted by the rods with orientation between **p** and **p+dp**, contained within a differential volume dV=**dA** · ds' (positioned at a distance **s'** (=s' **p**) from **dA**) is:

$$\mathbf{df}_A = \boldsymbol{t}(\mathbf{s}')\, n\, \chi(\mathbf{p})\, d\mathbf{p}\, d\mathbf{s}' \cdot d\mathbf{A} \tag{19}$$

(note that the tension t depends on s'). The incremental particle stress $d\tau_p$ is then (see also Eq.12):

$$d\tau_p = n\, \chi(\mathbf{p})\, d\mathbf{p}\, \boldsymbol{t}(\mathbf{s}')\, d\mathbf{s}' \tag{20}$$

or, from Eq.16 and 18:

$$d\tau_p = n\, \mathbf{pp}\, \chi(\mathbf{p})\, d\mathbf{p} \int_{s'}^{R} \sigma(s'')\, ds''\, ds' \tag{21}$$

which can be formally integrated over the entire phase space and total particle length to give:

$$\tau_p = n \oint \mathbf{pp}\, \chi(\mathbf{p}) \left\{ \int_{-R}^{R} \left[\int_{s'}^{R} \sigma(s'')\, ds'' \right] ds' \right\} d\mathbf{p} \tag{22}$$

This is traditionally referred to as the Kramers expression for the particle stress [16]. Note that in Eq.22, σ is not only a function of s'', but of **p**, time, and shear rate as well. Equation 22 links directly microstructure and particle stress, and it is valid as long as $\boldsymbol{t}$ and **p** are parallel. Once an equation for σ (i.e., for T_{12} in Eq.21) is prescribed, for a given orientation distribution function, Eq.22 can be used to calculate the particle stress either analytically or numerically.

As an example of application of Eq.22 we consider the case of a Newtonian matrix, for which the stress on the surface of the rod is proportional to the **p** component of the local fluid-rod relative velocity, $(\mathbf{v}_r)_\mathbf{p}$, giving

$$\sigma = -\, \xi_{//}\, (\mathbf{v}_r)_\mathbf{p} \tag{23}$$

where $\xi_{//}=2\pi\eta/\ln(r_a)$ is the hydrodynamic friction coefficient in a direction parallel to **p** (η is the viscosity of the matrix), and

$$(\mathbf{v}_r)_\mathbf{p} = \mathbf{v}_r \cdot \mathbf{p} = \mathbf{K} : \mathbf{pp}\, s \tag{24}$$

Substitution of Eq.23 and 24 into Eq.22 and subsequent integration gives:

$$\tau_p = -\frac{n\, \xi_{//}\, L^3}{12}\, \mathbf{K} : \oint \mathbf{pppp}\, \chi(\mathbf{p},t)\, d\mathbf{p} \tag{25}$$

which is the result obtained by Evans [6] and Dinh and Armstrong [4].

The above discussion lays the foundation for the extension of Dinh and Armstrong's model to accommodate matrix viscoelasticity. For purely viscous matrices the linear stress σ is history invariant, i.e., at any given orientation **p**, σ depends uniquely on the local kinematics of the flow. On the other hand, viscoelastic fluids are characterized by a fading memory, i.e., by their (partial) ability to remember previous kinematics. Thus, in viscoelastic matrices the magnitude and sign of the hydrodynamic stress on the rod will not only depend upon the present but also on past configurations as dictated by the history of deformation. The calculation the viscoelastic stress profile on an isolated rod is considered in the next section.

2.3 The Viscoelastic Stress Profile on the Rod

We consider a slender rod of orientation **p** immersed in a Maxwell fluid in creeping flow. We assume that in a homogeneous flow field of velocity gradient **K** the motion of the rod can be described by Eq.6. In order to analyze this fluid dynamic problem, we choose $\{\rho,\Theta,s\}$ to be a cylindrical coordinate system with origin at the center of mass of the rod and with the s axis parallel to **p**. For negligible inertia this "co-rotating" frame of reference is completely equivalent to any stationary laboratory frame. Because the rod is slender, the surrounding velocity field is within a good approximation "Θ"-independent with $v_\Theta=0$, i.e.:

$$\mathbf{v} = (v_\rho(\rho,s),\ 0\ ,\ v_s(\rho,s)) \tag{26}$$

For an incompressible matrix continuity requires:

$$\frac{1}{\rho}\frac{\partial}{\partial\rho}\rho\, v_\rho + \frac{\partial}{\partial s} v_s = 0 \tag{27}$$

Since the hydrodynamic problem is characterized by only two length scales, L and d, we must have O(ρ)~d and O(s)~L (O(x) stands for "order-of-magnitude of x"). This allows us to obtain an order-of-magnitude estimate of v_ρ from Eq.27:

$$O(v_\rho) \sim r_a^{-1}\, O(v_s) \tag{28}$$

which, for $r_a \gg 1$ implies $O(v_\rho) \ll O(v_s)$, i.e., the flow is predominantly of shear in character. Thus, the Cauchy momentum equations and the constitutive equation simplify to

$$\frac{\partial P}{\partial\rho} = -\frac{\partial T_{\rho s}}{\partial s} \tag{29}$$

$$\frac{\partial P}{\partial s} = -\left(\frac{1}{\rho}\frac{\partial}{\partial\rho}\rho\, T_{s\rho} + \frac{\partial}{\partial s} T_{ss}\right) \tag{30}$$

$$(1 + \lambda D)\, T_{\rho s} = -\,\eta\,\dot{\gamma} \tag{31}$$

$$(1 + \lambda D)\, T_{ss} = 2\, T_{\rho s}\, \lambda\, \dot{\gamma} \tag{32}$$

where D is the substantial derivative and $\dot{\gamma}=\partial v_s/\partial\rho$. For $r_a>>1$ Eqs.29 and 30 can be further simplified as follows. From Eq.29 we estimate:

$$O(P) \approx r_a^{-1}\, O(T_{\rho s}) \tag{33}$$

Thus, in Eq.30:

$$\frac{O(\partial P/\partial s)}{O(1/\rho\, \partial/\partial\rho(\rho T_{s\rho}))} \approx r_a^{-2} << 1 \tag{34}$$

To estimate the magnitude of the second term on the right-hand-side of Eq.30 relative to the first one, we use Eq.32 which gives:

$$O(T_{ss}) \leq 2\, O(T_{s\rho}\, Dh) \tag{35}$$

where $Dh = \lambda\,\dot{\gamma}$ is the Deborah number. Therefore:

$$\frac{O(\partial T_{ss}/\partial s)}{O(1/\rho\partial/\partial\rho(\rho T_{s\rho}))} \leq 2\, r_a^{-1}\, Dh \tag{36}$$

Thus, for $Dh/r_a<<1$ the momentum equations reduce to:

$$\frac{1}{\rho}\frac{\partial}{\partial\rho}\, \rho\, T_{s\rho} = 0 \tag{37}$$

where $T_{s\rho}$ is given by Eq.31. Equation 37 can be integrated to give:

$$T_{s\rho}(\rho,s,t) = \frac{\rho_o}{\rho}\, T_{s\rho}(\rho_o,s,t) \tag{38}$$

where $T_{s\rho}(\rho_o,s,t)$ is the (s,t)-dependent shear stress at the surface, and $\rho_o=d/2$ is the radius of the cross section of the rod. The combination of Eq.38 and Eq.31 gives:

$$-\eta\,\frac{\partial v_s}{\partial\rho} = \frac{\rho_o}{\rho}(1+\lambda\, D)\, T_{s\rho}(\rho_o,s,t) \tag{39}$$

Equation 39 can be integrated subject to the following conditions:

$$v_S = 0 \qquad \text{for} \qquad \rho = \rho_o \tag{40}$$
$$v_S \rightarrow v_S^*(\mathbf{p},s,t) \qquad \text{for} \qquad \rho \rightarrow \rho^*$$

where v_s^* is the fluid velocity at a distance ρ^* sufficiently far away from the surface. Information on these quantities will be derived below shortly. Integration of Eq.39 subject to Eq.40 gives:

$$(1+\lambda\, D)\, T_{s\rho}(\rho_o,s,t) = -\,\frac{v_s^*(\mathbf{p},s,t)\,\eta}{\rho_o \ln(\rho^*/\rho_o)} \tag{41}$$

Integration of Eq.41 according to Eq.17 yields an expression σ:

$$(1+\lambda\, D)\, \sigma(s,t) = -\,\frac{2\,\pi\,\eta}{\ln(\rho^*/\rho_o)}\, v_s^*(\mathbf{p},s,t) \tag{42}$$

In order to evaluate ρ^* and v_s^* we note that in the limit of $\lambda\rightarrow 0$, i.e., for a Newtonian fluid (see Eqs. 23 and 24):

$$\sigma = -\frac{2\pi\eta}{\ln(r_a)}\ \mathbf{K{:}pp}\ s \tag{43}$$

Thus :

$$(1 + \lambda\, D)\ \sigma(s,t) = -\frac{2\pi\eta}{\ln(r_a)}\ \mathbf{K{:}pp}\ s \tag{44}$$

(this result could be obtained, equivalently, by setting in Eq.42 $v_s^* = (v_r)_p$ and recognizing that $\rho^*/\rho_o \sim r_a$). In contrast to the Newtonian case, σ is now a function of the history of kinematics. Evaluation of σ at time t at any point s along the rod requires knowledge of the history of deformation of the fluid element which at the specific time t contacts the rod at s.

Equation 49 can be integrated subject to the initial condition that σ=0 for t=0, giving

$$\sigma(t,s) = -\frac{2\pi\eta}{\ln(r_a)}\ e^{-t/\lambda} \int_0^{t/\lambda} e^{t'/\lambda}\ (\mathbf{K{:}pp}\ s)\ \ d(t'/\lambda) \tag{45}$$

where, again, time t must be measured from the viewpoint of the fluid element, i.e., relative to a <u>material</u> frame of reference (within this framework (**K:pp** s) is a function of t').

Equation 45 allows to calculate σ at any point along the rod once the function (**K:pp** s) is specified. For the particular case of shear flow this is done in the next section.

2.4 The Linear Stress in Shear Flow

We consider, here, the special case of homogeneous, steady shear flow. We choose as reference frame the spherical coordinate system illustrated in Fig.1, and write:

$$\mathbf{p} = (\sin\theta\ \cos\phi,\ \sin\theta\ \sin\phi,\ \cos\theta) \tag{46}$$

thus

$$\mathbf{K} : \mathbf{pp} = \dot{\gamma}\ \ \sin\theta \cos\theta \sin\phi \tag{47}$$

where $\dot{\gamma} = \partial v_y/\partial z$. The equations of motion of the rod (Eq.6) become:

$$d\theta/d\gamma = \cos^2\theta\ \sin\phi \tag{48}$$

$$d\phi/d\gamma = \cot\theta\ \cos\phi$$

where $d\gamma = \dot{\gamma}\ dt$ is the incremental shear deformation of the matrix between t and t+dt. Subject to the initial condition that γ=0 at ϕ=0, Eqs 48 can be solved to give:

$$\tan\theta\ \cos\phi = C \tag{49}$$

$$\tan\phi = \gamma/C$$

where C is the orbit constant of the particle, and γ ($\infty > \gamma > -\infty$) is the total deformation measured with respect to the zx plane (ϕ=0). Equations 49 give the evolution of rod orientation (along a specific orbit, C) as a function of the macroscopic deformation. Such knowledge is essential for relating the position of a given fluid element relative to the particle to the particle orientation. As mentioned previously, integration of Eq.45 requires knowledge of the history of deformation of the fluid elements in the proximity of the rod, i.e., in principle of the entire velocity field surrounding the rod. Because of the complexity of the viscoelastic fluid-dynamic problem, such information is very difficult to obtain either analytically, or even by numerical means. To resolve this difficulty, we recall that, according to the affine rotation assumption, on average, the motion of fluid elements in the vicinity of the rod with respect to the coordinate system ($L/2 \geq s \geq -L/2$) on the rod is described by:

$$\frac{ds}{dt} = \mathbf{K}:\mathbf{pp}\, s \tag{50}$$

which, again, stems from the fact that the component of the fluid-rod relative velocity in a direction orthogonal to the rod axis is zero. As illustrated in Fig.2, the above statement implies that (with respect to the stationary frame of reference in Fig.2) on average the fluid elements move on trajectories characterized by z=constant. Equivalently, if $\mathbf{s}_o = s_o \mathbf{p}_o$ and $\mathbf{s}_1 = s_1 \mathbf{p}_1$ are the (axial) positions of a fluid element relative to the rod at time t_o and t_1 respectively, the following relation holds ($\mathbf{e}_z$ is the unit vector of the z axis):

$$\mathbf{s}_o \cdot \mathbf{e}_z = \mathbf{s}_1 \cdot \mathbf{e}_z \tag{51}$$

which, in our coordinate system (Fig.2, $\mathbf{p} \cdot \mathbf{e}_z = \cos\theta$) becomes:

$$s_o \cos\theta_o = s_1 \cos\theta_1 \tag{52}$$

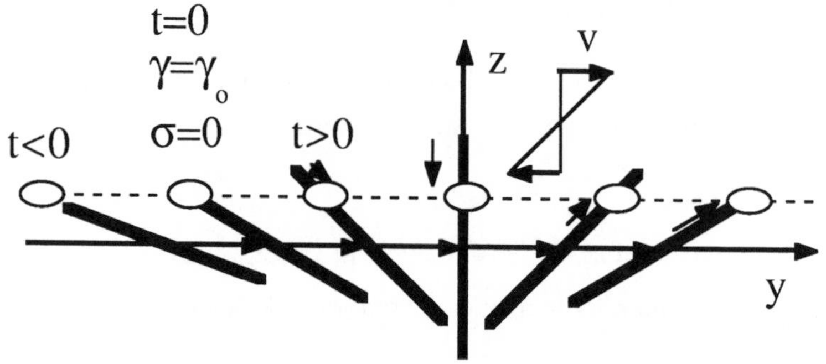

Fig.2 The motion of a fluid element (ellipse) relative to a test rod. The particle is followed from a fixed coordinate frame. Each of the segments represents a snap-shot of the testrod at a particular time t. The arrows indicate the direction of the shearing motion of the fluid relative to the rod, which gives rise to the stress (σ) in the matrix. The fluid element is stress-free for $t \leq 0$.

Equations 52 and 49 give the position of a fluid element relative to the rod at any time, provided an initial coordinate is known; they allow to determine the time (or deformation) at which a given fluid element first contacted the rod (i.e., $s = \pm L/2$), i.e., the time (t=0) at which σ=0 in Eq.45. The problem can be formulated as follows: given at time t a fluid element of position s relative to a rod of orientation **p**, find the time t_o and orientation $\mathbf{p}_o$ such that $s_o = \pm L/2$. From Eqs.52 and 49 the solution is straightforward. Given the orientation **p** (i.e., θ and

ϕ) from Eqs.49 the orbit constant, C, and the deformation coordinate, γ, can be determined. At a designated distance from the rod's center of mass, s, Eq.52 gives the angle θ_o (where $s_o = \pm L/2$); ϕ_o is then determined from the first of Eqs.49. The geometrical interpretation of the above procedure is schematically illustrated in Fig.3.

Following these considerations, combination of Eqs.45, 47, and 52 gives

$$\sigma(t,s) = -\xi_{//} s \cos\theta\, e^{-t/\lambda} \int_0^{t/\lambda} e^{t'/\lambda}\, \dot{\gamma} \sin\theta' \sin\phi'\, d(t'/\lambda) \qquad (53)$$

which, for $\dot{\gamma}$ = constant, can be further rearranged in terms the deformation, γ, as follows:

$$\sigma = -\dot{\gamma}\, \xi_{//} s \cos\theta\, e^{-t/\lambda} \int_0^{t/\lambda} e^{t'/\lambda} \frac{\gamma'}{\sqrt{(1 + C^2 + \gamma'^2)}}\, d(t'/\lambda) \qquad (54)$$

where we have used the identities (see Eq.49):

$$\tan\theta \sin\phi = \gamma \qquad (55)$$

$$\cos\theta = (1 + C^2 + \gamma^2)^{-1/2} \qquad (\text{for } \theta \le \pi/2)$$

For a constant rate of shear a relationship between the (material-element's) time t and the deformation γ is easily obtained as:

$$\gamma = \gamma_0 + \dot{\gamma}\, t \qquad (56)$$

where γ_0 is the value of the deformation at time t=0, i.e., at the time when the fluid element first came in contact with the rod end. Using Eq.56, changing variables, and rearranging Eq.54 can be rewritten as:

$$\sigma = -\dot{\gamma}\, \xi_{//} s \cos\theta \int_0^{\frac{(\gamma-\gamma_0)}{Dh}} e^{-\alpha} \frac{(\gamma - Dh\,\alpha)}{\sqrt{(1 + C^2 + (\gamma - Dh\,\alpha)^2)}}\, d\alpha \qquad (57)$$

where $Dh = \lambda \dot{\gamma}$ is the Deborah number. In this form it is easy to recover the Newtonian case (Eq.43) as $Dh \rightarrow 0$. Together with Eqs.49 and 52, Eq.57 can be used to calculate the stress at any coordinate, s, along a rod of given orientation $\mathbf{p} \equiv (\theta, \phi)$. Note that, because the orientation of the rod corresponding to zero time-of-contact changes with s, γ_0 in Eq.57 is a function of s (see also Fig.3). Also note that, the integrand is not a function of s; the only function of s in the integral appears as a limit of integration. This property will be exploited to evaluate the integrals of σ.

It is convenient to non-dimensionalize the curvilinear coordinate s with respect to the characteristic longitudinal dimension R=L/2. We define:

$$x = s/R = s/(L/2) \qquad (58)$$

which allows us to rewrite Eq.57 as:

$$\sigma = -\dot{\gamma}\, \xi_{//} R \left(x \cos\theta \int_0^{h(x,Dh,\gamma)} e^{-\alpha}\, g(\alpha, Dh, \gamma)\, d\alpha \right) \qquad (59)$$

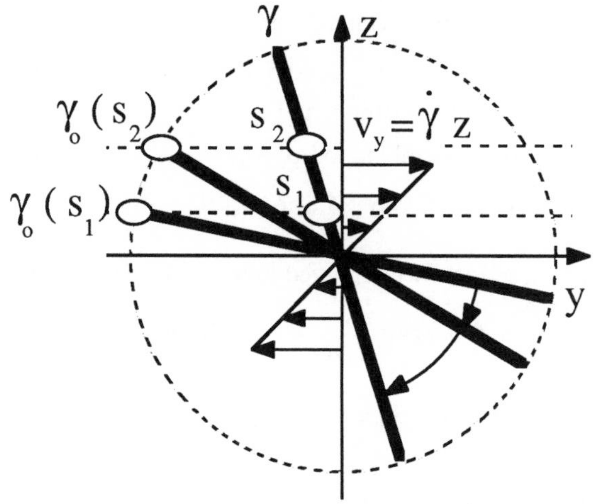

Fig.3 The relative motion of a fluid element and a rod with respect to a frame which translates with the rod. Each segment represents a snap-shot of the rod at a particular time t (or deformation γ). The fluid elements follow (on average) trajectories characterized by z=constant. Note that, the orientation of the rod ($\gamma_0(s)$) corresponding to the initial contact of a given fluid element depends on the axial position s of that element. This concept is schematically illustrated for two different coordinates: s_1 and s_2.

where:

$$h(x, Dh, \gamma) = \frac{(\gamma - \gamma_0(x))}{Dh} \qquad (60)$$

$$g(\alpha, Dh, \gamma) = \frac{(\gamma - Dh\,\alpha)}{\sqrt{(1 + C^2 + (\gamma - Dh\,\alpha)^2)}}$$

Once an orientation $\mathbf{p} \equiv (\theta,\phi)$ is assigned, calculation of σ at a given coordinate x ($-1<x<1$) proceeds as follows. First, C and γ are determined from Eqs.54. The angles θ_o and ϕ_o are calculated from:

$$\cos\theta_o = x \cos\theta \qquad (61)$$
$$\cos\phi_o = C/\tan\theta_o$$

Next, γ_0 is given by:

$$\gamma_0 = -\left((x \cos\theta)^{-2} - 1 - C^2\right)^{1/2} \qquad (62)$$

Finally, σ is calculated by numerical integration from Eq.59. Note that with respect to our coordinate system (with $v_y = \dot\gamma z$) $\gamma_0 \le 0$ always. Because the motion of the rods is assumed to start from a completely aligned orientation, the fluid elements can meet the rod for the first time only in phase space regions where $\gamma \le 0$. Consistent with this concept Eq.62 gives: $\gamma_0 \rightarrow -|\gamma|$ for $x \rightarrow 1$, and $\gamma_0 \rightarrow -\infty$ for $x \rightarrow 0$. Typical normalized linear-stress profiles for different values of the Deborah number are given in Fig.5 (for $\theta=45^o$ $\phi=-45^o$) and Fig.6 (for $\theta=45^o$ $\phi=45^o$) (see Fig.4 for rod orientation). The curves are for a rod originally aligned along the y axis (i.e., $\gamma\rightarrow -\infty$), where there is no stress on the rod (i.e., $\sigma=0$ for every x). The following features may be noted: (1) as Dh tends to zero the profiles approach consistently the Newtonian behavior. In this limit, for $\gamma<0$ ($\phi<0$), the rod is being compressed by the fluid ($\sigma<0$), while for $\gamma>0$ ($\phi<0$) the rod is in traction ($\sigma>0$). (2) As Dh increases the area under the normalized linear-stress curves decreases in absolute value, i.e., the force and rate of energy dissipation (compared to the Newtonian case) decrease. (3) A very interesting feature emerges in Fig.6 where for Dh~1 the linear-stress profile undergoes a change of sign along the rod, being still negative near the rod center. This indicates that in some regions along the particle the fluid has not yet completely forgotten the previous "compression phase". This prediction is reasonable in that it agrees with what one would expect as a result of first pushing and then pulling (at a fixed rate) a rod into and from a highly viscoelastic fluid such as a molten thermoplastic. During the penetration half of the experiment some of the energy given to the rod is stored in the fluid as elastic energy. When the direction of motion of the rod is suddenly reversed (pull), part of the stored energy is released and given back to the object. Thus, even though the relative motion has been reversed the force required to keep the object in motion at a certain speed can still be temporarily oriented in the direction of penetration (i.e., the rod is pushed out by the fluid).

In order to calculate the particle stress in Eq.22, the linear stress must be integrated twice over the particle length and then again twice over the angular distribution of rod orientation. Because the linear stress itself is given as an integral, calculation of τ_p requires in general evaluation of five nested integrals. A "brute-force" approach to this problem by numerical methods is clearly an extremely inefficient (if not impossible) strategy to pursue. Fortunately, in our case the problem can be simplified considerably by exploiting the fact that in Eq.59 the kernel is independent of x.

The tension in a rod segment between x (=s/(L/2)) and 1 is given by:

$$t(\mathrm{x}) = \frac{L}{2}\int_{\mathrm{x}}^{1} \sigma \; d\mathrm{x}' \qquad (63)$$

Retaining explicitly only the x functionality in Eq.59, Eq.63 can be written as:

$$t(\mathrm{x}) = -\frac{L^2}{4}\dot{\gamma}\; \xi_{//}\cos\theta \int_{\mathrm{x}}^{1} \mathrm{x}' \int_{0}^{h(\mathrm{x}')} g(\alpha)\, d\alpha \; d\mathrm{x}' \qquad (64)$$

This equation can be integrated by parts with the aid of Leibnitz's rule.

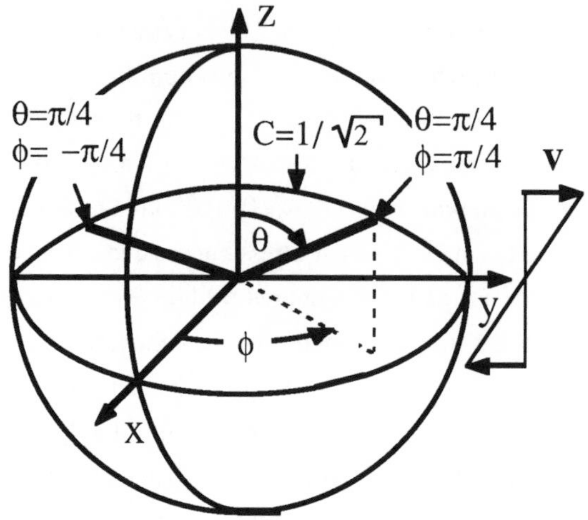

Fig.4 The rod orientations corresponding to the stress profiles given in Figs.5 and 6. The curve marked C=1/√2 is a particular orbit.

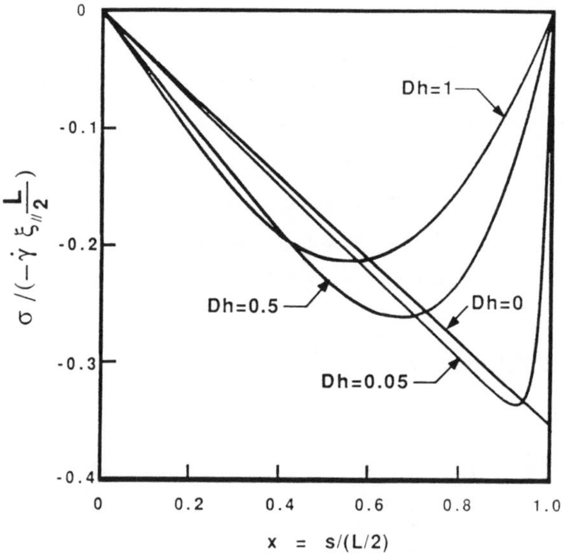

Fig.5 The normalized linear-stress (σ) vs. dimensionless curvilinear coordinate $x=s/(L/2)$ (Eq.59) ($x=0$ and $x=1$ indicate the rod center and tip respectively) along a rod of orientation of ($\theta=\pi/4$, $\phi=-\pi/4$), at different values of the Deborah number, Dh.

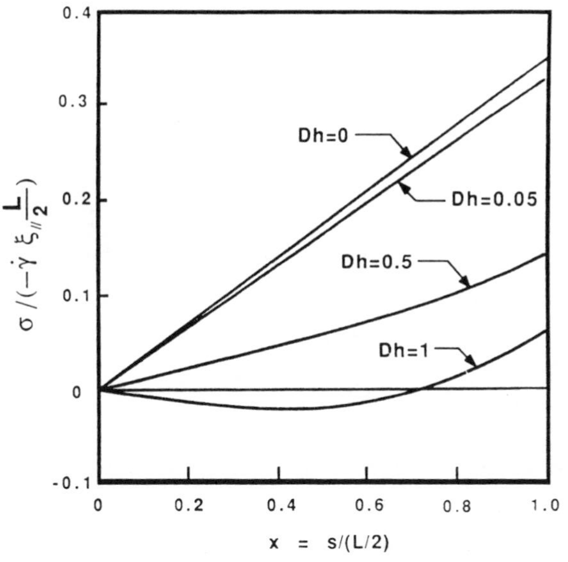

Fig.6 The normalized linear-stress (σ) vs. dimensionless curvilinear coordinate x=s/(L/2) (Eq.59) (x=0 and x=1 indicate the rod center and tip respectively) along a rod of orientation of ($\theta=\pi/4$, $\phi=\pi/4$), at different values of the Deborah number, Dh.

We obtain:

$$\frac{-8\,t\,(x)}{(L^2\xi_{//}\dot{\gamma}\cos\theta)} = \int_0^{h(1)} g(\alpha)\,d\alpha \;-\; x^2\int_0^{h(x)} g(\alpha)\,d\alpha \;+\; -\int_{h(x)}^{h(1)} x^2(\alpha)\,g(\alpha)\,d\alpha \qquad (65)$$

where $g(\alpha)$ and $h(x)$ are given by Eq.60, and:

$$x(\alpha) = \left(\cos\theta\,\left(1 + C^2 + (\gamma - Dh\,\alpha)^2\right)^{1/2}\right)^{-1} \qquad (66)$$

Evaluation of the the particle stress (Eq.22) requires further integration of Eq.65 over the total particle length, i.e.:

$$\int_{-L/2}^{L/2} t\,(s)\,ds = L\int_0^1 t\,(x)\,dx \qquad (67)$$

where the integral on the right hand side could be considered as a (spatial) average of the tension in the fiber. Here, again integration by parts and Leibniz rule can be used to obtain:

$$\frac{\int_{-L/2}^{L/2} t\,(s)\,ds}{\left(-\frac{L^3}{12}\dot{\gamma}\,\xi_{//}\cos\theta\right)} = \int_0^{h(1)} g(\alpha)\,d\alpha - \int_{h(0)}^{h(1)} x^3(\alpha)\,g(\alpha)\,d\alpha \qquad (68)$$

where:

$$h(0) = \frac{(\gamma - \gamma_o(0))}{Dh} = \frac{(\gamma - (-\infty))}{Dh} = \infty \tag{69}$$

$$h(1) = \frac{(\gamma - \gamma_o(1))}{Dh} = \frac{(\gamma - (-|\gamma|))}{Dh} = \frac{(\gamma + |\gamma|)}{Dh} \tag{70}$$

(Note that, from Eq.70, h(1)=0 for $\gamma<0$, and h(1)= 2γ/Dh for $\gamma>0$.) Equations 68 to 70 are consistent with our assumption of zero stress at an initial orientation corresponding to ($\theta=\pi/2$, $\phi=-\pi/2$) (i.e., $\gamma=-\infty$) (i.e., the rod motion must start from a completely aligned orientation). In the Newtonian case Eq.68 correctly gives:

$$\int_0^1 t\,(x)\,dx = (-\frac{L^2}{12}\dot{\gamma}\,\xi_{//})\,\sin\theta\,\cos\theta\,\sin\phi \tag{71}$$

Figure 7 shows the (normalized) average tension versus azimuthal angular position (ϕ), for a rod orbiting along the trajectory C=1/$\sqrt{2}$ (see also Fig.4), for three values of the Deborah number. As Dh increases: (1) the profiles become increasingly more asymmetric;

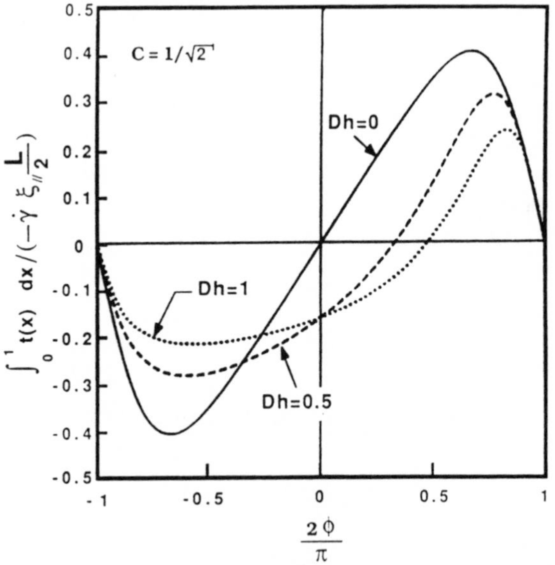

Fig.7 The normalized average tension (Eq.67) vs. normalized azimuthal angular position (ϕ) for a rod orbiting along the trajectory C=1/$\sqrt{2}$, for three values of Dh. The original and final orientation of the rod correspond to perfect alignment with the flow (i.e., orthogonal to the vorticity vector).

(2) the average tension in the fiber decreases (with respect to the Newtonian case); (3) the point of zero tension shifts to the right. This behavior is, again, consistent with the fact that

elastic energy is being stored for $\gamma<0$ ("compression phase"), and later released with partial memory in regions where $\gamma>0$. Note that sufficiently close to complete alignment (i.e., around $\phi=\pm\pi/2$) the fluid behaves as if it were Newtonian, due to the small rod angular velocity.

Once the average tension as a function of particle orientation and the orientation distribution function are known the particle stress can be calculated from Eqs.22 and 68. Let (in Eq.68):

$$G(Dh, \mathbf{p}) = \cos\theta \left(\int_0^{h(1)} g(\alpha)\, d\alpha - \int_{h(0)}^{h(1)} x^3(\alpha)\, g(\alpha)\, d\alpha \right) \tag{72}$$

then, Eqs.22 and 68 can be formally combined to give:

$$\tau_p = - \frac{n L^3 \xi_{//} \dot{\gamma}}{12} \oint \mathbf{pp}\ G(Dh,\mathbf{p})\ \chi(\mathbf{p})\ d\mathbf{p} \tag{73}$$

Note that, under our assumptions, the front factor in Eq.73 (viscoelastic) and Eq.25 (Newtonian) are the same, and that the particle stress normalized by the front factor is exclusively a function of the Deborah number and deformation. Also note that now the evaluation of τ_p requires only a triple numerical integration. The results of such calculations for the particular kinematics chosen here are discussed in the next section.

3. RESULTS AND DISCUSSION

Equation 68 has a very limited range of applicability, being restricted to the case of constant shear rate and for an initially complete alignment of the rod. Nevertheless, there are two simple cases which, while satisfying the above conditions, are amenable to both experimental verification, and (relatively) easy theoretical treatment. We consider two separate experiments, differing from each other only by the initial orientation distribution of the fibers. In the first experiment (Case A) the fibers are all initially aligned parallel to each other and orthogonal to the plane of shear (Fig.8a) (methods for obtaining this orientation have been discussed previously), while in the second one (Case B) the initial orientation distribution is random (Fig.8b). In both cases, we consider the following sequence of actions. First, the suspension is slowly sheared "backwards" as in Fig.8c, until the fibers are all approximately aligned parallel to the plane of shear. For negligible Brownian motion, very small rates of deformation (i.e., negligible inertia and elasticity), and neutrally buoyant fibers this motion is reversible. In other words by reversing the direction of flow it is always possible to bring the orientation distribution back to its initial configuration, at least conceptually. Next, when the fibers have reached the desired (high) degree of alignment, the motion is stopped. Finally the flow is suddenly reversed with a step change in shear rate, and the stress is measured in time (Fig.8d). For sufficiently large initial deformations the start-up transient for the matrix can be

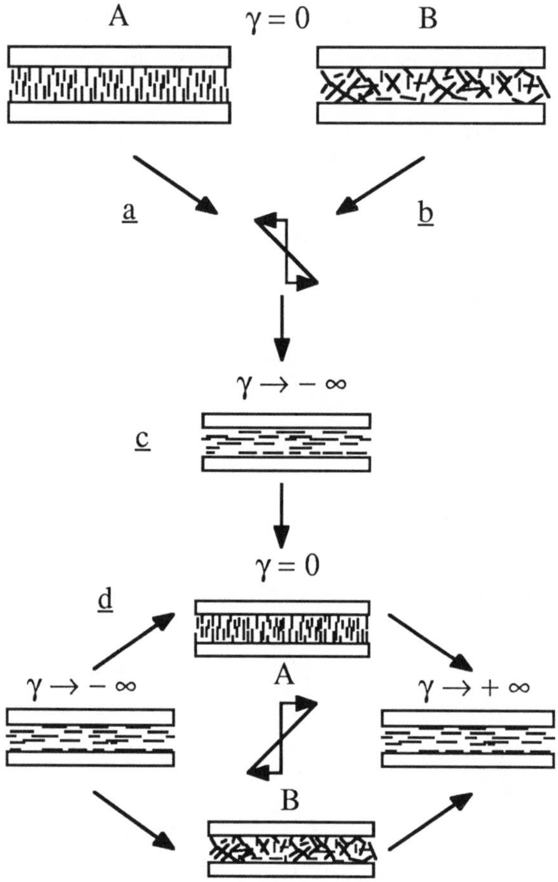

Fig.8 Schematics of experiments A and B described in the text. These experiments differ only by the initial orientation distribution of the fibers (see top). In Case A the fibers are originally all perpendicular to the plane of shear, while in Case B the orientation distribution is random. In both cases, by slowly back-shearing the suspension, an almost complete alignment ($\gamma \rightarrow -\infty$) of the fibers is induced at first. The flow is then suddenly reversed with a step change in shear rate and the stress is measured.

easily decoupled from the transient associated with the reorientation of the fibers. From Eq.11 the transient particle stress can thus be obtained. Case A and B described above have been chosen here for the following reasons. Case B has been already studied by Dinh and Armstrong [4] for Newtonian matrices and thus is useful for comparison of theoretical results for particles in purely viscous and viscoelastic fluids. From an experimental point of view, however, the absence of randomizing Brownian motion considerably complicates the task of obtaining a truly random initial orientation distribution of the fibers. On the other hand, the orientation distribution of Case A is both well characterized (see Eq.10) and easily reproducible experimentally. Thus in experiment A, the orientation distribution dependence and the hydrodynamic stress profile dependence of τ_p, can be conveniently decoupled, and the effect of orientation and hydrodynamic force on the particles can be examined separately.

Consider Case A first. Here, all the rods rotate simultaneously along the same orbit of constant C=0 (i.e., they rotate within the zy plane). The orientation distribution function for this case is given in Eq.10 with **p**(t) (or **p**(γ)) given by Eq.7. From Eq.73 we obtain the particle stress as:

$$\tau_p = -\frac{n L^3 \xi_{//} \dot{\gamma}}{12} \ \mathbf{p}(\gamma)\mathbf{p}(\gamma) \ G(Dh,\mathbf{p}(\gamma)) \tag{74}$$

where γ is the total deformation (related to the orientation of the particles (θ,ϕ) by Eqs.55). It is convenient to rewrite Eq.74 in terms of transient material functions, defined as:

$$\begin{aligned} \eta_p^+ &= -\tau_{12p}/\dot{\gamma} \\ \Psi_{1p}^+ &= -(\tau_{11p}-\tau_{22p})/\dot{\gamma}^2 \\ \Psi_{2p}^+ &= -(\tau_{22p}-\tau_{33p})/\dot{\gamma}^2 \end{aligned} \tag{75}$$

where η_p^+, Ψ_{1p}^+, and Ψ_{2p}^+ are transient (particle) viscosity, first, and second normal stress coefficient respectively. If we let for simplicity $\Gamma = nL^3\xi_{//}/12$, then from Eqs.75, 74 and 57 we obtain (in general, for any value of C):

$$\eta_p^+/\Gamma = G(Dh,\gamma)\ \gamma/(1+C^2+\gamma^2) \tag{76}$$

$$\Psi_{1p}^+\dot{\gamma}/\Gamma = G(Dh,\gamma)\ (\gamma^2-1)/(1+C^2+\gamma^2) \tag{77}$$

$$\Psi_{2p}^+\dot{\gamma}/\Gamma = G(Dh,\gamma)\ (1-C^2)/(1+C^2+\gamma^2) \tag{78}$$

In the present case (case A) C=0. For a Newtonian matrix, function G(Dh, γ) becomes:

$$G(Dh=0,\gamma) = \gamma/(1+C^2+\gamma^2) \tag{79}$$

giving

$$\eta_p^+/\Gamma = \gamma^2/(1+C^2+\gamma^2)^2 \tag{80}$$

$$\Psi_{1p}^+\dot{\gamma}/\Gamma = \gamma\ (\gamma^2-1)/(1+C^2+\gamma^2)^2 \tag{81}$$

$$\Psi_{2p}^+\dot{\gamma}/\Gamma = \gamma\ (1-C^2)/(1+C^2+\gamma^2)^2 \tag{82}$$

Note that the signs of η_p+ and Ψ^+_{1p} are independent of C, while Ψ^+_{2p} changes sign at C=±1. It is interesting to examine the behavior of the above functions for small and large values of γ. For γ→0 Eqs.80, 81 and 82 give:

$$\eta_p^+/\Gamma \to \gamma^2/(1+C^2)^2 \qquad (83)$$

$$\Psi_{1p}^+\dot{\gamma}/\Gamma \to -\gamma/(1+C^2)^2 \qquad (84)$$

$$\Psi_{2p}^+\dot{\gamma}/\Gamma \to \gamma\,(1-C^2)/(1+C^2)^2 \qquad (85)$$

while, for γ→±∞:

$$\eta_p^+/\Gamma \to \gamma^{-2} \qquad (86)$$

$$\Psi_{1p}^+\dot{\gamma}/\Gamma \to \gamma^{-1} \qquad (87)$$

$$\Psi_{2p}^+\dot{\gamma}/\Gamma \to (1-C^2)\,\gamma^{-3} \qquad (88)$$

Note that for C=0, while Ψ^+_{1p} undergoes a change of sign at γ=±1, Ψ^+_{2p} is always positive for γ>0. Also note that all the transient material functions exhibit power law behavior at large γ, and that the viscosity and second normal stress coefficient decay faster than the first normal stress coefficient.

For Dh≥0, Eqs.76 to 78 have been solved numerically. The results are plotted in Fig.9 (G), and Fig.10 (η_p^+), Fig.11 (Ψ^+_{1p}), Fig.12 (Ψ^+_{2p}), for a Newtonian matrix and for a

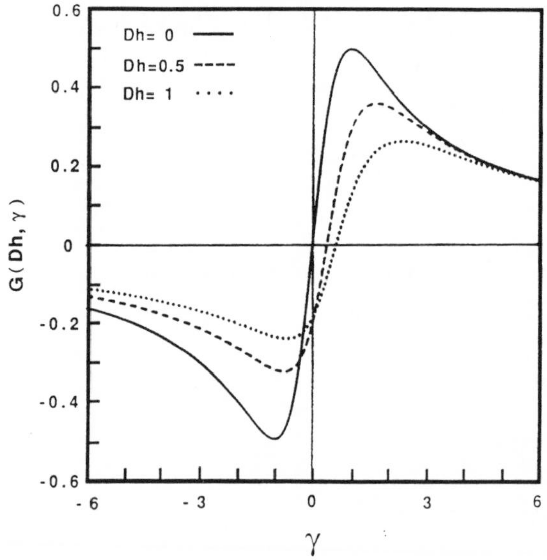

Fig.9 Case A. Function G(Dh, γ) (equivalent to a normalized average tension in the fiber, see Eqs.72, 68, and 67) against the deformation γ, for Dh=0, 0.5 and 1.0.

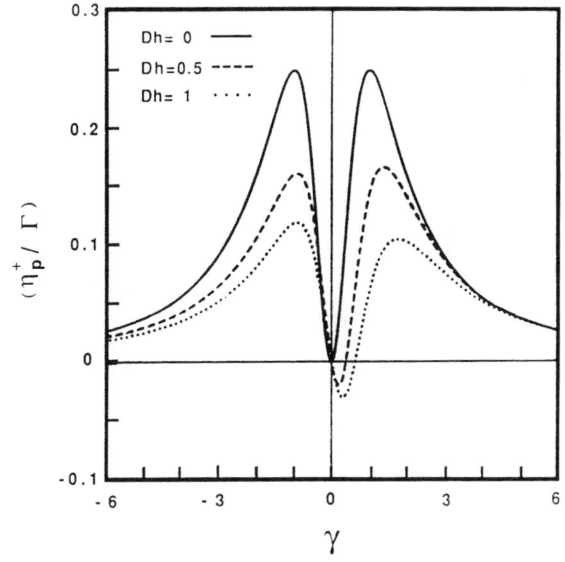

Fig.10 Case A. Normalized particle viscosity $\eta_p{}^+$ against deformation γ, for Dh=0, 0.5, 1.0.

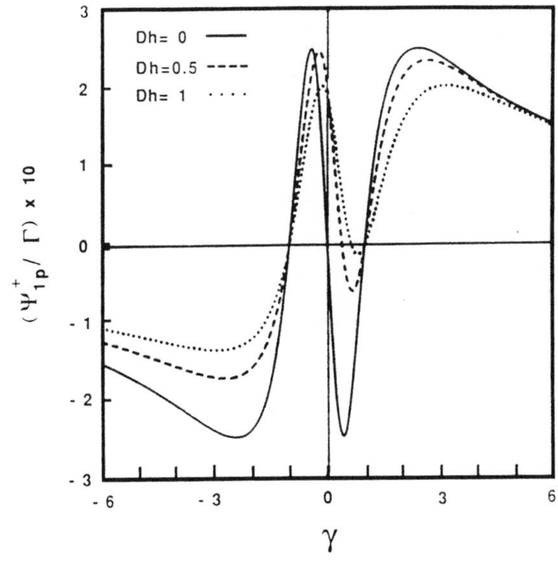

Fig.11 Case A. Normalized particle first normal stress coefficient $\Psi_{p1}{}^+$ against deformation γ, for Dh=0, 0.5, 1.0.

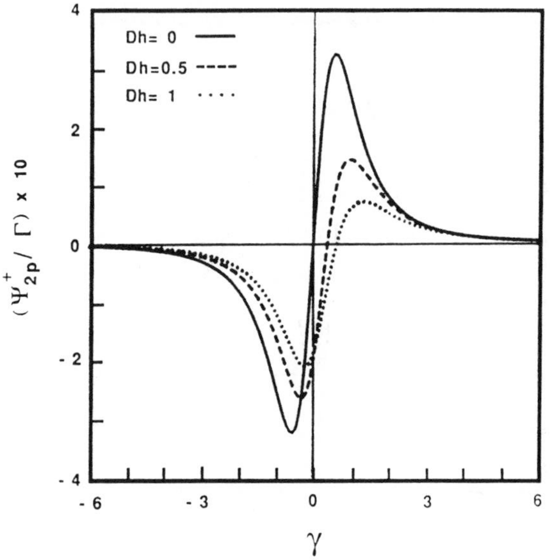

Fig.12 Case A. Normalized particle second normal stress coefficient $\Psi_{p2}{}^{+}$ against deformation γ, for Dh=0, 0.5, 1.0.

viscoelastic fluid at two different values of the Deborah number. In the Newtonian case (Dh=0) large overshoots arise as the fibers align along the principal axis of strain. Also, because at θ=0 the rods are orthogonal to the flow field, as $\gamma \to 0$ all material functions approach zero. The situation changes for Dh>0; as the Deborah number increases, memory effects become more pronounced. Indeed, even though the matrix is non-shear-thinning, the magnitudes of the material functions decrease, indicating a decreased dissipation. Also note that as the Deborah number is increased the curves become progressively more distorted. This distortion is particularly evident in the normal stress coefficients Figs.11 and 12, for which the value of the "zero-stress" deformation (γ=0 for Newtonian) progressively shifts to positive values of γ. Thus, at γ=0, even though the fibers are all oriented perpendicular to the plane of shear (i.e., orthogonal to the solid shearing surfaces), appreciable normal forces can still be measured for Dh>0. The most remarkable memory effect, however, is associated with the transient (particle) viscosity, $\eta_p{}^{+}$ (Fig.10), which becomes negative over a finite interval of γ. Thus, over a finite range of deformations (or times) the total (matrix+particle) transient shear stress is not only smaller than the corresponding Newtonian one, but is also smaller than the final steady state stress exhibited by the suspension when all the fibers are aligned parallel to the plane of shear (note that, as in [4] we neglect the particle thickness, so that the particles are invisible to the flow when they are lying parallel to the plane of shear). This phenomenon is again attributable to the fact that the elastic energy stored by the fluid for $\gamma<0$ is subsequently released for $\gamma>0$. Finally, it may be noted that (different from the Newtonian case) for Dh>0 integration of Ψ^{+}_{1p} and Ψ^{+}_{2p} from $\gamma=-\infty$ to $\gamma=+\infty$ yields non-zero values. This feature could, perhaps, be used experimentally to detect the anomalies in the material functions predicted by the present model.

In Case B the orientation distribution is given by Eq.11, which in our coordinate system becomes:

$$\chi(\gamma,\theta,\phi) = \frac{1}{4\pi}(1+\gamma^2\cos^2\theta - 2\,\gamma\,\sin\theta\,\cos\theta\,\sin\phi)^{-3/2} \tag{89}$$

Here again, as in Case A, the flow is started from a large negative value of γ (in principle from a completely aligned orientation of the fibers). In this case, however, the orientation distribution becomes completely random at $\gamma=0$. Note that, in principle, this type of experiment is different from the one described by Dinh and Armstrong [4] in which the flow is started at $\gamma=0$ (i.e., from a random distribution) and continues for $\gamma>0$. For a viscoelastic matrix, for a same orientation distribution and Deborah number, different stresses can be produced depending upon the initial orientation distribution of the fibers. For a Newtonian matrix, however, it makes no difference whether the flow is started from an almost completely aligned orientation or from a random distribution; the particle stress depends exclusively on γ. Thus, only for a Newtonian matrix (i.e.,Dh=0) there is no difference between Case B and Dinh and Armstrong's experiment. The particle stress in this case was evaluated numerically from Eq.72 with $\chi(\mathbf{p})$ given by Eq.89. The calculated material functions are given in Fig.13 (η_p^+), Fig.14 (Ψ_{1p}^+), and Fig.15 (Ψ_{2p}^+) for Dh=0 and Dh=1. For Dh=0 the results are identical to those reported by Dinh and Armstrong [4], and, thus, the corresponding analysis will not be repeated here (in the comparison it is useful to note that η_p^+ is an even function of γ while Ψ_{1p}^+ and Ψ_{2p}^+ are odd). For Dh>0, the curves exhibit the same qualitative trends as in Case A. As the Deborah increases from zero to unity: (1) the material functions decrease in magnitude, (2) the curves "shift" to the right to positive values of γ, and (3) at $\gamma=0$ (random orientation) both first and second normal stress differences are non-zero for Dh>0. In contrast to Case A, however, here the particle viscosity is never negative (at least up to Dh=1); also, for comparable values of Dh, the distortion of the curves is much less severe than for Case A . This is because the magnitude of the viscoelastic distortion for each rod greatly depends on its trajectory and orientation. For instance, particles which orbit very close to the plane of shear experience an essentially Newtonian environment, while fibers having the trajectory C=0 (as in case A) interact much more strongly with the flow. In a system where a variety of orientations coexist, the effects

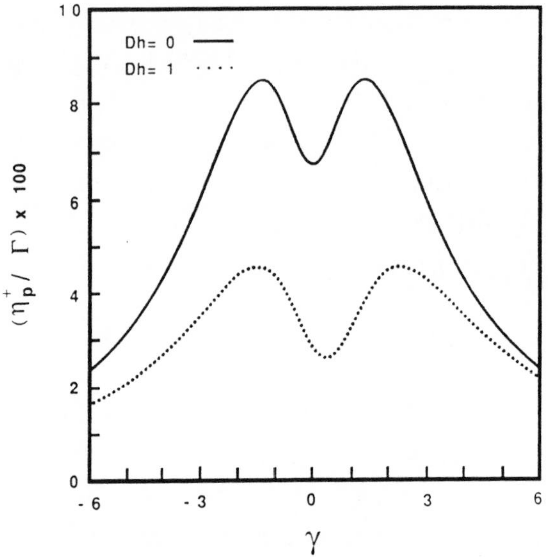

Fig.13 Case B. Normalized particle viscosity $\eta_p{}^+$ against deformation γ, for Dh=0 and 1.0.

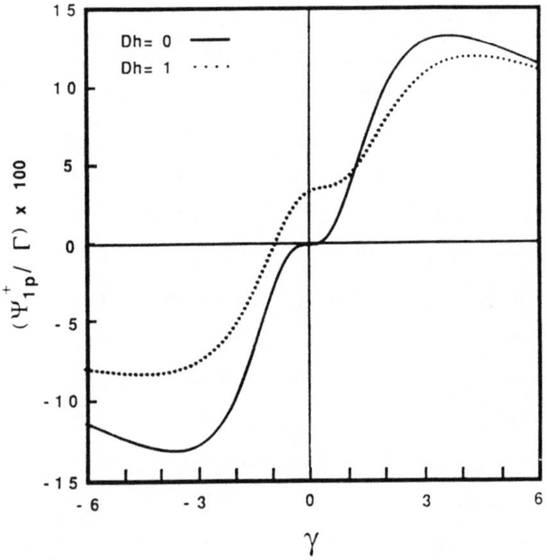

Fig.14 Case B. Normalized particle first normal stress coefficient $\Psi_{p1}{}^+$ against deformation γ, for Dh=0 and 1.0.

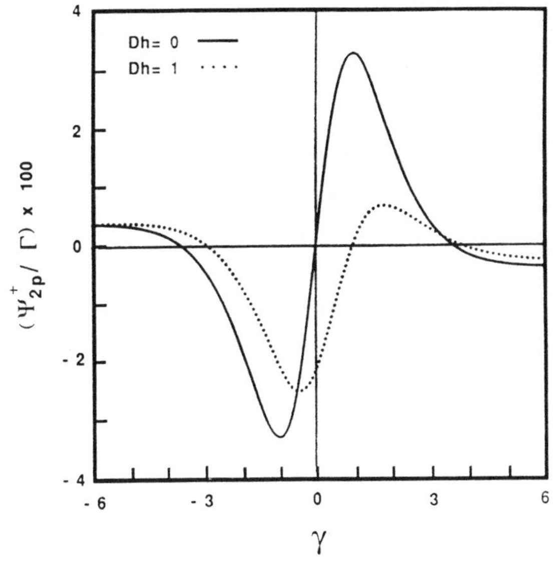

Fig.15 Case B. Normalized particle second normal stress coefficient Ψ_{p2}^{+} against deformation γ, for Dh=0 and 1.0.

of elasticity are then smeared out by the broad distribution. This broadness is also responsible for the disappearance of the negative region in the particle viscosity.

4. CONCLUDING REMARKS

We have attempted to extend Dinh and Armstrong's model for Newtonian, dilute fiber suspensions to account for matrix viscoelasticity. Our treatment is subject to two key assumptions: (1) on the time scale of the experiment no appreciable elastic orbital drift occurrs, and (2) the matrix behaves as a Maxwell fluid in the limit of linear viscoelasticity. The first of these assumptions can be partially justified if the matrix has negligible (or zero) second normal stress differences, while the second one (linearity), however poor it may be, is necessary to keep the algebra manageable. Such simplifications allowed us to calculate the hydrodynamic force on a fiber in a viscoelastic fluid as a function of particle orientation and Deborah number. The assumption of negligible orbital drift allowed us to use "Newtonian" transient orientation distributions to calculate the particle stress in two particular flows of assigned kinematics.

For Newtonian suspensions the instantaneous values of the transient (particle) material functions are controlled exclusively by the total matrix deformation γ, in agreement with the analysis of Dihn and Armstrong. On the other hand, for viscoelastic matrices calculation of the transient particle stress requires the additional knowledge of: (1) the Deborah number, and (2) the initial orientation distribution of the particles.

Despite its extremely simplified structure, the model does show a combination of interesting memory effects, including a negative (particle) viscosity during startup. The predictions are reasonable, in that they agree with our intuition on very simple experiments on viscoelastic systems.

NOMENCLATURE

A	area element
C	particle orbit constant
d	rod diameter
D	material derivative
Dh	Deborah number
$\mathbf{e}_i$	(i=1, 2) unit vectors
$\mathbf{E}$	deformation gradient tensor
$\mathbf{f}$	force on area element
$\mathbf{K}$	velocity gradient tensor
L	rod length
n	rod number concentration
$\mathbf{p}$	rod directional unit vector
$\dot{\mathbf{p}}$	rod angular velocity
P	pressure
r_a	rod aspect ratio
R	rod semi-length
s	curvilinear coordinate along rod major axis
t	time
t_d	elastic-orbital-drift time
t_f	rod "flip over" time
t_p	rod period of rotation around Jeffery orbit
$\boldsymbol{t}$, t	the tension in the rod, and its magnitude
T_{ij}	stress in the fluid
$\mathbf{v}$	velocity field
$\mathbf{v}_r$	fluid-rod relative velocity
x	dimensionless axial coordinate along the rod
Δ	inverse of velocity gradient tensor ($=\mathbf{E}^{-1}$)
χ	rod orientation distribution function
ϕ	rod volume fraction
γ	shear deformation
$\dot{\gamma}$	shear rate
η	matrix viscosity
$\eta_p{}^+$	transient particle viscosity

Ψ^{+}_{1p} transient particle first normal stress coefficient
Ψ^{+}_{2p} transient particle second normal stress coefficient
ψ_2 matrix second normal stress coefficient
λ matrix relaxation time
(θ,ϕ) angular phase-space coordinates
Θ cylindrical angular coordinate around a rod
ρ cylindrical radial coordinate from rod axis
σ linear stress (force per unit length) on the rod
τ suspension deviatoric stress
τ_m matrix contribution to the stress
τ_p particle stress
$\xi_{//}$ hydrodynamic friction coefficient parallel to the rod

REFERENCES

[1] G.B Jeffery, Proc. Roy. Soc., **A102**, 161, (1922)
[2] H. Brenner, Int. J. Multiphase Flow, **1**, 195, (1974)
[3] G.K. Batchelor, J. Fluid Mech. ,**41**, 545, (1970)
[4] S.M. Dinh, R.C. Armstrong, J. Rheol., **28**(3), 207, (1984)
[5] E.J. Hinch, L.G. Leal, J. Fluid Mech., **52**(4), 683, (1972)
[6] J.G. Evans, Ph.D. Dissertation, Cambridge University, (1975)
[7] J.L. Ericksen, Kolloid Z., **173**(2), 117, (1960)
[8] G.G. Lipscomb II, M.M. Denn, D.U. Hur, D.V. Boger, J. Non-Newtonian Fluid Mech., **26**, 297, (1988)
[9] S. Kim, J. Non-Newtonian Fluid Mech., **21**, 255, (1986)
[10] P. Brunn, J. Non-Newtonian Fluid Mech., **7**, 271, (1980)
[11] R.T. Mifflin, J. Non-Newtonian Fluid Mech., **17**, 267, (1985)
[12] P.N. Kaloni, V. Stastna, Polym. Eng. Sci., **23**, 465, (1983)
[13] L.G. Leal, J. Fluid Mech., **69**, 305, (1975)
[14] L.G. Leal, J. Non-Newtonian Fluid Mech., **5**, 33, (1979)
[15] C. Cohen, B. Chung, W. Stasiak, Rheol. Acta, **26**, 217, (1987)
[16] R.B. Bird, R.C. Armstrong, O. Hassager, "Dynamics of Polymeric Liquids", vol.1, Fluid Mechanics, p.367 (Eqs. 8.1-3 to 8.1-6), John Wiley and Sons, New York, (1977)

This work was supported by The Director, Office of Energy Research, Office of Basic Energy Sciences, Material Science Division of the U.S. Department of Energy under Contract No. DE-AC03-76SF00098.

Davide A. Hill and David S. Soane
Center for Advanced Materials
Lawrence Berkeley Laboratory
and
Department of Chemical Engineering
University of California at Berkeley
Berkeley, CA 94720

P.J. CARREAU, M. GRMELA AND A. ROLLIN

Rheological models for rigid and worklike macromolecules

1. Introduction

Most polymeric systems such as high polymer weight melts, blends, composites and solutions are known to exhibit complex rheological behaviour. Constitutive equations or rheological models are needed for various purposes. First, they can be very useful for understanding and clarifying the strange flow phenomena encountered in polymer processing: extrudate swell, vortices, instabilities, etc.. Secondly, parameters contained in rheological models can be advantageously used to correlate data and obtain master curves incorporating such effects as molecular weight, molecular weight distribution, polymer concentration, chain flexibility, solvent properties, etc. Thirdly, a ultimate objective in developing constitutive equations is to obtain an exact expression for the stress tensor in terms of the thermo-mechanical history of any given polymeric system.

Three basic approaches can be followed in writing down a constitutive relation. One can use the principles of continuum mechanics to propose admissible forms and assess these forms with appropriate sets of experimental data. A more promising and rewarding route is, however, through the so-called molecular theories. These theories should lead to a number of meaningful parameters that can be used not only for correlating data but for extrapolation and formulation of new products. Molecular theories leading to constitutive equations applicable to large deformation flows can be subdivided into four main categories: i) the phase space and configuration space kinetic theories such as those proposed by Kirkwood [18] and refined and extended by Bird and co-workers [6], ii) the network theories derived for polymeric liquids from solid rubber theories by Lodge [22] and by Yamamoto [27] and extended by Carreau [7], Macdonald [24] as well as many others, iii) reptation theories, based on the original concept of Edwards [12] or de Gennes [10], proposed by Doi and Edwards [11], extended to large deformation flows by Marrucci [23] and put into the phase space kinetic framework by Curtiss and Bird [9] and iv) the rheological models based on the conformation tensor, as those proposed by Giesekus [13] and Leonov [21]. Useful reviews of the first three categories of molecular theories have been presented by Bird [5] and a recent review by Carreau and Grmela [8] includes the conformation tensor models. The third route, combining the previous ones, has been proposed by Grmela [14] and by Grmela and Carreau [15]. It improves the first route by allowing to use molecular state variables as for example distribution functions and the second route by guaranteeing the compatibility of the governing equations of the model with thermodynamics and by providing a general formula for the extra stress tensor.

This contribution is part of a series of papers on conformation tensor models based on the third route. Ait Kadi et al. [1] introduced a FENE-Charged potential to account for the "coil-stretch" transformation of macromolecules and explain rheological data obtained for polyacrylamide solutions in aqueous solvents containing various quantities of salt. Ajji et al. [2] used two conformation tensors to describe polymer melts as a blend of free macromolecular chains and entangled chains in a network. With that model, they could account for changes in the entanglement structure due to a controlled solution precipitation treatment. In this work we review the case of rigid polymeric chains (Grmela and Carreau [15], Grmela [17]) and focus on semi-flexible chains with applications to liquid crystalline polymers and polymers reinforced with fibers. The modification needed to account for the semi-flexibility of chains has been introduced by Grmela and Chhon Ly [16].

2. The Model

We follow the method proposed by Grmela [14] and Grmela and Carreau [15]. The state variable chosen to characterize the polymeric chains or the fibers is a second-order conformation tensor $\underline{\underline{c}}$. The conformation tensor as used by Giesekus [13] and Leonov [21] can be viewed in reference to a distribution function $\psi(\underline{r},t)$ as the second moment of the end-to-end vector $\underline{R}$ (see Figure 1), i.e.

$$\underline{\underline{c}}(\underline{r},t) = < \underline{R}\ \underline{R} > = \int \psi(\underline{r},\underline{R},t)\ \underline{R}\ \underline{R}\ d\ r \qquad (1)$$

$\underline{\underline{c}}(\underline{r},t)$ is taken as a positive definite and symmetric tensor. It represents the average conformation of a structure under any flow conditions.

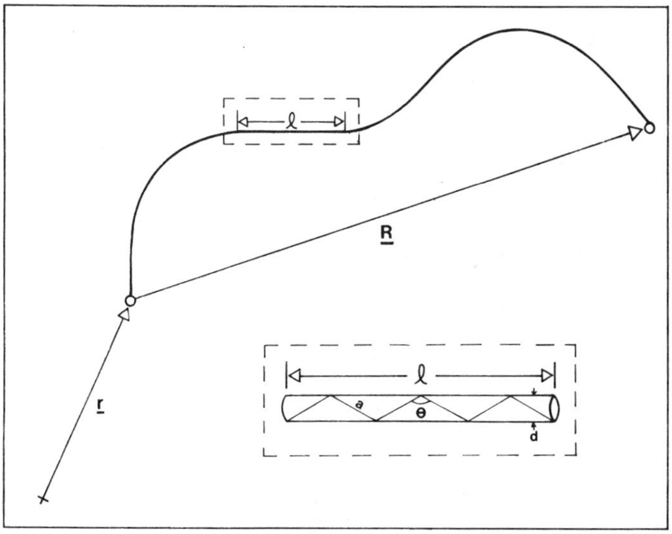

Figure 1 **Sketch of a wormlike molecule**

We will assume that the flow is locally homogeneous and take $\underline{\underline{c}}$ independent of the position vector $\underline{r}$. We will also restrict ourselves to isothermal and affine flows. The most simple time evolution of the structure that will respect the Clausius-Duhem inequality (or positive entropy production) is the following:

$$\frac{D\underline{\underline{c}}(t)}{Dt} = (\nabla\underline{v}^{+} \cdot \underline{\underline{c}}) + (\underline{\underline{c}} \cdot \nabla\underline{v}) - \Lambda\ (\underline{\underline{c}})\ \underline{\underline{c}} \cdot \frac{dA(\underline{\underline{c}})}{d\underline{\underline{c}}} \qquad (2)$$

where D/Dt is the substantial derivative, $\Lambda\ (\underline{\underline{c}})$ is the mobility, a function of the conformation tensor and possibly a tensorial quantity to account for non-isotropic effect; $A(\underline{\underline{c}})$ is the Helmholtz free energy, which needs to be specified. The first three terms of Equation (2) are the upper convective derivative of the conformation tensor. This is the convective part or reversible part of the evolution equation. The last term is the diffusive or dissipative term. More complex evolution equations for $\underline{\underline{c}}$ include effects for slip or non-affine deformations (Grmela and Carreau [15], Ait Kadi et al. [1]), internal viscosity, etc..

The extra stress tensor is related to the free energy by (see Grmela [14]):

$$\underline{\underline{\sigma}} = -2n\, \underline{\underline{c}} \cdot \frac{dA(\underline{\underline{c}})}{d\underline{\underline{c}}} \tag{3}$$

where n is the chain density.

The Helmholtz free energy of a chain is taken as the sum of an elastic (intramolecular energy) contribution and an entropic term, i.e.

$$A(\underline{\underline{c}}) = E(\underline{\underline{c}}) - TS(\underline{\underline{c}}) \tag{4}$$

In this paper, we shall consider one choice of the intramolecular energy $E(\underline{\underline{c}})$ (corresponding physically to chains that are in average <u>rigid</u> [inextensible] and two choices of the entropy (corresponding physically to unflexible [rod-like] and semi-flexible chains). The mobility $\Lambda(\underline{\underline{c}})$ will be, for the sake of simplicity, considered to be a positive constant independent of $\underline{\underline{c}}$.

2.1 Rigid-in-Average and Rod-like chains

First we focus on rigid-in-average and inflexible chains. We shall take into account the rigidity by introducing the constraint

$$\mathrm{tr}\ \underline{\underline{c}} = R_0{}^2 \tag{5}$$

where R_0 is the total length of a rod-like chain or the end-to-end distance of a chain made of rigid segments and tr is the trace of the tensor. By using the Lagrange multiplier method we replace the free energy (4) by

$$A\ (\underline{\underline{c}}) = \frac{1}{2}\ k_B T\beta(\mathrm{tr}\ \underline{\underline{c}} - R_0{}^2) - T\ S\ (\underline{\underline{c}}) \tag{6}$$

where the Lagrange multiplier β is obtained from

$$\frac{dF}{dt} = \frac{d}{dt}\ (\mathrm{tr}\ \underline{\underline{c}} - R_0{}^2) = 0 \tag{7}$$

expressing the requirement that the constraint (5) holds for all times. For the rod-like chains, the entropy $S(\underline{\underline{c}})$ is the Boltzmann entropy (Sarti and Marrucci [26], Grmela and Carreau, [15])

$$S(\underline{\underline{c}}) = -\frac{1}{2}\ k_B\ T\ \ln\det \underline{\underline{c}} \tag{8}$$

where k_B is the Boltzmann constant and det $\underline{\underline{c}}$ means the determinant of $\underline{\underline{c}}$.

Equations (2), (3), (6), (7) and (8) specify the complete rheological model. In general, these equations cannot be written as a single constitutive equation, for example a single equation determining the stress tensor.

Solving the set of equations in terms of β, we obtain:

$$\underline{\underline{\sigma}}(\underline{\underline{c}}) = nk_B T\ (\underline{\underline{\delta}} - \beta \underline{\underline{c}}) \tag{9}$$

$$\frac{D\underline{\underline{c}}}{Dt} = \nabla \underline{v}^{+} \cdot \underline{\underline{c}} + \underline{\underline{c}} \cdot \nabla \underline{v} + \frac{1}{2} k_B T \Lambda\ (\underline{\underline{\delta}} - \beta \underline{\underline{c}}) \tag{10}$$

where the Lagrange multiplier is obtained from (7) and (10):

$$\beta = \frac{4 \mathrm{tr}\ (\underline{\underline{c}} \cdot \nabla \underline{v}^{+})}{k_B T \Lambda \mathrm{tr}\ \underline{\underline{c}}} + \frac{3}{\mathrm{tr}\ \underline{\underline{c}}} \tag{11}$$

We will consider here only simple shear flows,

$$\nabla \underline{v}^{+} = \begin{bmatrix} 0 & \dot{\gamma} & 0 \\ 0 & 0 & 0 \\ 0 & 0 & 0 \end{bmatrix} \tag{12}$$

and $\underline{\underline{c}}$ can be written as

$$\underline{\underline{c}}^{*} = \underline{\underline{c}} / R_0^2 = \begin{bmatrix} \frac{1}{2}(C+N) & C_{12} & 0 \\ C_{12} & \frac{1}{2}(C-N) & 0 \\ 0 & 0 & 1-C \end{bmatrix} \tag{13}$$

where C, N and C_{12} are the three dimensionless characteristic values of the tensor $\underline{\underline{c}}$. Equation (10) yields the following three equations:

$$\frac{\partial C_{12}}{\partial t} = \frac{1}{2}(C - N)\dot{\gamma} - \frac{\beta^{*}}{\lambda} C_{12} \tag{14}$$

$$\frac{\partial N}{\partial t} = 2\ C_{12}\ \dot{\gamma} - \frac{\beta^{*}}{\lambda} N \tag{15}$$

$$\frac{\partial C}{\partial t} = 2\ C_{12}\ \dot{\gamma} + \frac{1}{\lambda}(2 - \beta^{*} C) \tag{16}$$

where β^{*} is given by

$$\beta^{*} = \beta R_0^2 = \frac{4\dot{\gamma}\ C_{12}\ R_0^2}{k_B T \Lambda} + 3 = 2\dot{\gamma}\ \lambda\ C_{12} + 3 \tag{17}$$

and the time constant is defined by

$$\lambda = 2R_0^2 / k_B T \Lambda \tag{18}$$

The four unknowns in Equations (14) to (17), C_{12}, N, C and β, can be solved numerically taking for example isotropic conditions for the initial values of the conformation tensor. The components for the stress tensor are then obtained from Equation (9).

The governing equation of the model can be solved analytically for steady shear and elongational flows and for start-up and relaxation in elongational flows (Grmela and Carreau [15], Grmela [17]). For steady simple <u>shear flow</u>, simple algebra leads to the following results:

$$\lambda\dot{\gamma} = \left[\frac{1 - \eta/\eta_0}{\frac{2}{27}(\eta/\eta_0)^3}\right]^{1/2} \tag{19}$$

where η is the shear viscosity and η_0 is the zero-shear viscosity obtained by taking $\beta^* = 3$ as $\dot{\gamma} \to 0$, i.e.

$$\eta_0 = \lim_{\dot{\gamma}\to 0} - \sigma_{12}/\dot{\gamma} = \frac{1}{3} nk_B T\lambda \tag{20}$$

The primary and the secondary normal stress coefficients are respectively given by

$$\begin{aligned} \psi_1 &= -(\sigma_{11} - \sigma_{22})\dot{\gamma}^2 \\ &= \frac{2\eta}{nk_B T} = \frac{2}{3}\left(\frac{\eta}{\eta_0}\right)^2 \eta_o\lambda \end{aligned} \tag{21}$$

and

$$\psi_2 = 0 \tag{22}$$

These last results have been obtained by Grmela and Carreau [15] in a more general framework.

For high shear rates, the model predicts unique power-law curves, i.e.

$$\eta \; \alpha \; (\lambda\dot{\gamma})^{-2/3}$$

and (23)

$$\psi_1 \; \alpha \; (\lambda\dot{\gamma})^{-4/3}$$

We notice that this result is significantly different from that obtained by Bird et al. [6] for rigid dumbbells in the context of the configuration phase space kinetic theory, which does not admit a closed form solution. The power-law exponents for the viscosity and the primary normal stress coefficient were found to be respectively - 1/3 and -2/3, i.e. half of the values obtained for the rigid-in-average model. It is not clear to us why such differences in the high-shear rate behaviour are observed. Nevertheless, both models yield single master curves for the steady-shear viscosity and primary normal stress coefficient. With one single adjustable parameter, λ, these models are obviously not flexible enough. In the next paragraph, we introduce the results of Khokhlov and Semenov [20, 21] for semi-flexible polymer chains.

2.2 Semi-Flexible Chains

Again we consider the polymer chains to be rigid in average, but we allow for some flexibility in the chains. We choose the wormlike model with Kuhn segments of length ℓ, as shown in Figure 1.

The results of Khokhlov and Semenov [20, 21] obtained for entropy of chains made of sub-segments vibrating about an angle θ close to 2π can be written as:

$$S = \frac{1}{2} k_B T \left[\ln \det \left(\frac{\underline{\underline{c}}}{R_o^2}\right) - b \operatorname{tr} \left(\left(\frac{\underline{\underline{c}}}{R_o^2}\right)^{-1}\right) \right] \tag{24}$$

where b is a parameter related to the chain flexibility that we take as

$$b = L/\ell \tag{25}$$

where L is the contour length of the polymer chains. We note that if $b \ll 1$, the chains are rigid. On the contrary, if $b \gg 1$, the chains are quite flexible.

Equation (6) for the free energy becomes:

$$A(t) = \frac{1}{2} k_B T \left[\beta F(\underline{\underline{c}}) - a \ln \det \left(\frac{\underline{\underline{c}}}{R_o^2}\right) + b \operatorname{tr} \left(\left(\frac{\underline{\underline{c}}}{R_o^2}\right)^{-1}\right) \right] \tag{26}$$

In Equation (26) we have introduced also the parameter a, which weights the influence of the standard contribution of the Brownian motion with respect to the effect of the chain flexibility. The physical justification is the following. Brownian motion is of importance for small or molecular size particles. For larger particles such as crystallites, fillers, fibers, the contribution of the Brownian motion to the free energy will become more and more negligible as the particles size increases.

To obtain the components of the stress tensor and material functions under various flow situations, we follow the procedure outlined in the previous paragraph. However, the solutions are no longer analytical and a numerical scheme has to be used. The zero-shear viscosity is not affected by the chain flexibility and it is expressed by Equation (20). On the other hand, the zero-shear primary normal coefficient is now expressed by:

$$\psi_{10} = \frac{2nk_B T\lambda^2}{9(a+12b)} = \frac{2\eta_0 \lambda}{3(a+12b)} \tag{27}$$

As expected, both contributions to the Brownian motion affects the material's elasticity as well as the shear-thinning properties. Figure 2 shows steady-shear viscosity master curves for different values of the parameters a and b.

It is interesting to note that with increasing chain flexibility (increasing value of b), the onset of shear thinning appears at higher reduced shear rate, $\lambda\dot{\gamma}$. Also, with increasing flexibility, the viscosity becomes less shear-thinning: the slope in the power-law region increases from -2/3 for rigid chains to approximately -1/2 for semi-flexible chains.

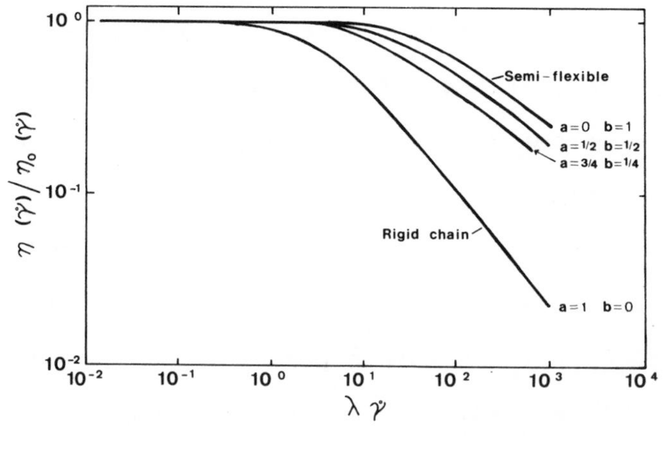

Figure 2 <u>**Effect of chain flexibility on the reduced shear viscosity**</u>

The primary normal stress coefficient describes similar master curves as shown in Figure 3.

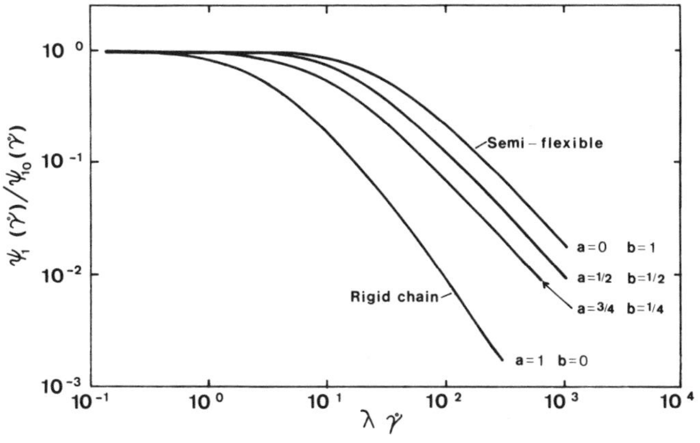

Figure 3 **Effect of chain flexibility on the reduced primary normal stress coefficient**

The results obtained numerically when flexibility is considered, appear to verify the relationship (21) between the primary normal stress coefficient and the shear viscosity. Hence, the power-law slopes for ψ_1 varies from -4/3 for rigid chains to -1 for semi-flexible chains.

The transient behaviour in shear flow and the elongational properties predicted by the model are of considerable interest. Figure 4 compares the model predictions of the shear stress growth function for the rigid and semi-flexible chains. The growth function is defined by:

$$\eta^{+} = -\sigma_{12}(t,\dot{\gamma})/\dot{\gamma} \tag{28}$$

where σ_{12} is the shear stress measured after imposing a sudden and constant shear rate.

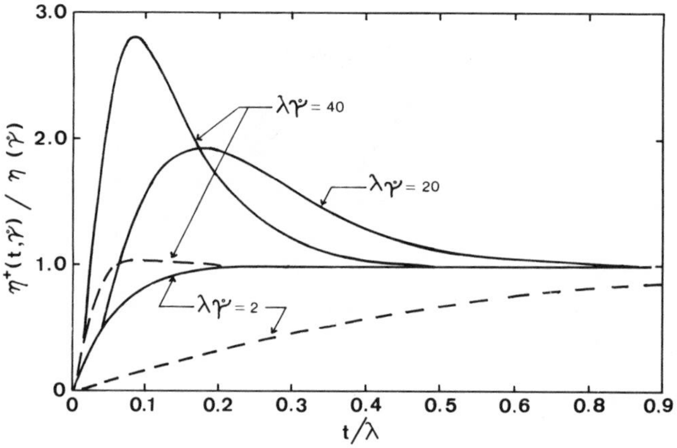

Figure 4 __**Effect of chain flexibility on the shear stress growth predictions**__

——————— rigid chains (a=1, b=0)

--------- semi-flexible chains (a=0, b=1)

We notice that the growth function increases much faster for the rigid chain than for the semi-flexible ones. Moreover, the model predicts large stress overshoots at large reduced shear rates, $\lambda\dot{\gamma}$, whereas the overshoots are eliminated (or considerably reduced) when chain flexibility is incorporated. The predictions for the rigid chains are, at least qualitatively, in agreement with the measurements done on thermotropic polymers [4].

In line with the stress growth observations, the stresses after steady-shear flow are found to relax much more rapidly for semi-flexible chains than for rigid ones [25]. Also the model predicts that the uniaxial elongational viscosity increases for rigid chains at a considerably lower value of the reduced elongational rate. Rigid (rod-type) chains are oriented sooner in the flow compared to the semi-flexible chains.

3. Assessment of the model

For a preliminary evaluation of the model we use the data of Baird and Ballman [3], obtained on a Rheometrics Spectrometer. Two types of polymers were investigated. The first one was a polyester (PPT) of molecular weight equal to 27600 and 32000 kg/kmol in sulphuric acid. The chains of this polymer system are believed to be rigid. The second system consisted of Nylon 6,6 (M_w = 25100, 35200 and 42300 kg/kmol) also in sulphuric acid. The polymeric chains of these Nylon solutions are somewhat flexible. In all cases, the polymer concentration was kept at 0.117g/mL and the rheological measurements were made at 22°C.

Figure 5 compares the viscosity data for both sets of solutions with the model predictions. The corresponding comparison for the primary normal stress coefficient is shown in Figure 6. The parameters are reported in Table 1.

Table 1 **Parameters for two sets of polymer solutions in H_2SO_4 at 0.117 g/mL and 22°C (data from Baird and Ballman [3])**

Polymer	M_w kg/kmol	η_o Pa.s	λ s	a -	b -
PPT	27600	520	0.149	0.2	0
"	32000	1725	0.573	0.2	0
Nylon 6,6	25100	2.00	$5.20 \cdot 10^{-4}$	0.2	-
"	35200	3.40	$1.24 \cdot 10^{-3}$	0.2	-
"	42300	6.00	$2.64 \cdot 10^{-3}$	0.2	-

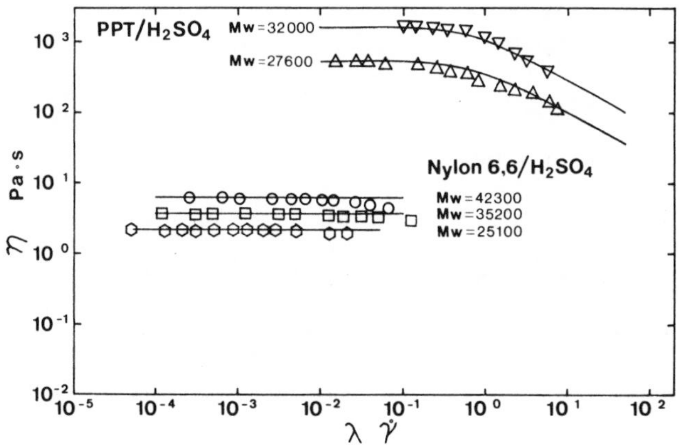

Figure 5 **Viscosity of PPT and Nylon 6,6 solutions in H_2SO_4.**
Data from Baird and Ballman [3]
— Model predictions

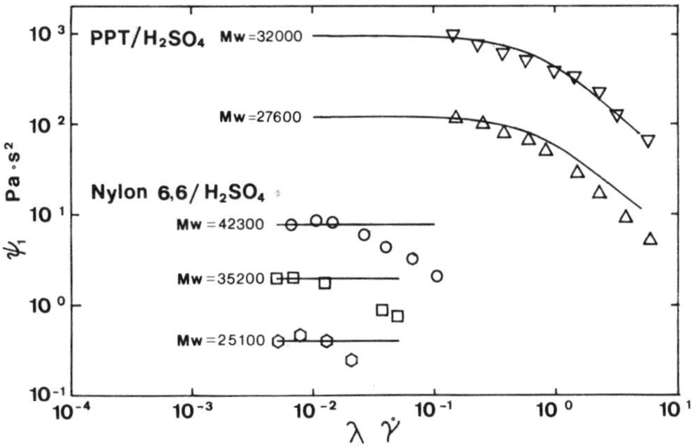

Figure 6 <u>Primary normal stress coefficient of PPT and Nylon 6,6 solutions in H_2SO_4</u>
- Data from Baird and Ballman[3]
—— Model predictions

First, it is interesting to note that the viscosity, the primary normal stress coefficient and hence the time constant, λ, for the rigid PPT polymeric chains are much higher than for the semi-flexible Nylon chains. Since the same concentration and the same range of molecular weight were used, it is obvious that chain rigidity is responsible for the high values. In the model, this is accounted for in the definition of the time constant (Eq. 18). It is reasonable to expect large end-to-end distance, R_0, and lower mobility, Λ, for the more rigid chains. It is also probable that particle-particle interactions and chain entanglements play an important role and reduce the mobility. Notice from Table 1 that λ is strongly dependent on the molecular weight whereas it should be a linear function of the molecular weight if the solutions were dilute ones.

Overall, the model predictions for the PPT solutions are excellent. The parameter b was set equal to zero and the parameter a was found to be equal to 0.2. It is reasonable to expect a reduction of the influence of the Brownian motion since, as discussed above, particle-particle interactions are highly dominant. The onset of shear thinning in the case of the Nylon 6,6 solutions is observed at values of the reduced shear rate, $\lambda\dot{\gamma}$, smaller than 0.1. This is much lower than predicted by the model in any cases. Hence no value for the parameter b could be obtained and the low shear-rate data could be fitted by using a=0.2 and b=0. Accounting for the viscosity of the solvent could possibly improve the fit of the model with respect to the shear-thinning effects.

4. Concluding Remarks

In conclusion we recapitulate the advantages of the models presented above.

(i) The models are intrinsically consistent and compatible with thermodynamics. This is because they possess the structure identified on the molecular level of description.

(ii) The models are simple. They could also be incorporated into numerical schemes for the calculations of flows arising in polymer processing.

(iii) To the best of our knowledge they are the only models that take into account the flexibility of macromolecular chains in large deformation flows according to this preliminary assessment.

(iv) Predictions of the models agree qualitatively with results of observations.

These models obtained here for rigid and semi-flexible chains are excellent candidates for describing the rheological behaviour of thermoplastics reinforced with fibers. The model for rigid-in-average chains predicts qualitatively well the steady and transient shear data of fiber composites. In a forthcoming publication, an extension of these models for two-phase systems is proposed. The model is based on two conformation tensors, one describing the fibers and the other one the polymeric matrix. The matrix is considered as made of flexible free chains or consists of a network of entangled chains (see reference [2]). We also discuss the effect of including a non isotropic mobility.

Acknowledgement

We wish to acknowledge the financial support received from the National Science and Engineering Research Council of Canada.

Nomenclature

a	parameter weighing the Boltzmann contribution to the Brownian motion in Eq. (26)
A	Helmholtz free energy
b	parameter weighing the Lifshitz contribution to the Brownian motion in Eq. (24)
$\underline{\underline{c}}$	conformation tensor defined by Eq.(1)
C	dimensionless normal component of $\underline{\underline{c}}$
C_{12}	dimensionless shear component of $\underline{\underline{c}}$
det	determinant of a tensor
E	intramolecular energy
F	internal energy defined by Eq. (7)
k_B	Boltzmann's constant
ln	logarithm
ℓ	Kuhn's length (see Figure 1)
L	contour length of a chain
M_w	molecular weight
n	number density of particles or macromolecules
N	dimensionless normal component of $\underline{\underline{c}}$
$\underline{r}$	position vector
$\underline{R}$	end-to-end vector of a chain
R_0	maximum extensibility of a chain
S	entropy
tr	trace of a tensor
T	absolute temperature
$\underline{v}$	velocity vector

Greek letters

$\dot{\gamma}$	shear rate
$\underline{\underline{\delta}}$	unit tensor
$\nabla \underline{v}^{+}$	transpose of velocity gradient tensor
η	shear viscosity
η_{0}	zero-shear viscosity
η^{+}	stress growth shear function defined by Eq. (28)
λ	time constant defined by Eq. (18)
Λ	mobility of a chain
$\underline{\underline{\sigma}}$	stress tensor
ψ	distribution function
ψ_{1}	primary normal stress coefficient
ψ_{2}	secondary normal stress coefficient

Operator

D/Dt	substantial or material derivative

References

1. A. Ait Kadi, M. Grmela and P.J. Carreau, Rheol. Acta, 27, 241 (1988).

2. A. Ajji, P.J. Carreau, M. Grmela and H.P. Schreiber, J. Rheol., 33, 401 (1989).

3. D.G. Baird and R.L. Ballman, J. Rheol., 23, 505 (1979)

4. D.G. Baird, A. Gotis and G. Viola, in Polymeric Liquid Crystal, edited by A. Blumstein, Plenum Press, N.Y. (1983)

5. R.B. Bird, J. Rheol, 26, 277 (1982); Chem. Eng. Commun., 16, 175 (1982).

6. R.B. Bird, C.F. Curtiss, R.C. Armstrong and J. Hassager, Dynamics of Polymeric Liquids, Vol. 2, Kinetic Theory -second edition, Wiley, N.Y. (1987).

7. P.J. Carreau, Trans. Soc. Rheol., 16, 99 (1972).

8. P.J. Carreau and M. Grmela, in "Recent Advances in Structured Continua", edited by D. De Kee and P.N. Kaloni, Pitman Publishing Company, Chap. 4, p. 105 (1986).

9. C.F. Curtiss and R.B. Bird, J. Chem. Phys., 74, 2016, 2026 (1981).

10. P.G. de Gennes, J. Chem. Phys., 60, 5030 (1974).

11. M. Doi and S.F. Edwards, J. Chem. Soc. Faraday, Trans. II, 74, 1789, 1802, 1818 (1978).

12. S.F. Edwards, Proc. Phys. Soc., London, 92, 9 (1967).

13. M. Giesekus, J. Non-Newt. Fluid Mech., 11, 69 (1982).

14. M. Grmela, Physics Letters, 111A, 36 and 41 (1985); Physica D, 21, 179 (1986).

15. M. Grmela and P.J. Carreau, J. Non-Newt. Fluid Mech., 23, 27 (1987).

16. M. Grmela and Chhon Ly, Phys. Lett. A, 120, 281 (1987).

17. M. Grmela, J. Rheol., 33, 207 (1989).

18. J.G. Kirkwood in P. Auer (Ed.), Documents in Modern Physics, Gordon and Breach, N.Y. (1967).

19. A.R. Khokhlov and A.N. Semenov, Macromolecules, 17, 2678 (1984).

20. A.R. Khokhlov and A.N. Semenov, J. Stat. Phys., 38, 161 (1985).

21. A.I. Leonov, Rheol. Acta, 15, 85 (1976).

22. A.S. Lodge, Rheol. Acta, 7, 379 (1968).

23. G. Marrucci, in Advances in Transport Processes, Vol. V, A.S. Majumdar and R.A. Mashelkar Eds., Wiley, N.Y. (1984).

24. I.F. Macdonald, Rheol. Acta, 14, 801, 899, 906 (1975).

25. A. Rollin, "Modélisation rhéologique de solutions diluées; macromolécules semiflexibles versus inflexibles", A.M. Sc. Thesis, Ecole Polytechnique of Montreal (1988)

26. G.S. Sarti and G. Marrucci, Chem. Eng. Sci., 28, 1053 (1973).

27. M. Yamamoto, J. Phys. Soc. Jpn, 11, 413 (1956).

Authors and Address

Pierre J. Carreau, Miroslav Grmela and André Rollin
Centre de recherche appliquée sur les polymères (CRASP)
Ecole Polytechnique de Montréal
Case postale 6079, Succursale A
Montréal (Québec)
H3C 3A7
Canada

J.M. DEALY

Nonlinear viscoelasticity of polymeric liquids

Introduction

The storage of elastic energy in an incompressible material implies anisotropy and thus a type of structure. The notion of linear viscoelastic behavior arises only as a mathematical limiting case as the strain amplitude or strain rate approaches zero. The observation of a finite zone of linear behavior depends on the specific criterion adopted for linearity and on the resolution of the experimental technique. In other words, the establishment of a linear regime of behavior is a matter of definition and signal to noise ratio, not of polymer physics.

Nevertheless, the concept of linear viscoelasticity has proven useful as a basis for material characterization and as a point of departure for the description of nonlinear behavior. This type of rheological response can be described entirely within the framework of the Boltzmann superposition principle, which can be expressed as follows:

$$\tau_{ij}(t) = \int_{-\infty}^{t} G(t-t')\dot{\gamma}_{ij}dt' \qquad (1)$$

where: τ_{ij} = component of the extra stress tensor

$G(t-t')$ = shear stress relaxation modulus

γ_{ij} = component of the rate of deformation tensor

Within this regime of behavior, a complete description of the rheological behavior of the material is provided by giving the relaxation modulus function. Alternative representations, which are, in principle equivalent, include the spectrum function, the loss modulus function, $G''(\omega)$ and the storage modulus function, $G'(\omega)$ [1]. Once one of these characterizing functions has been determined, the response of the material to any deformation can be calculated.

Outside the linear regime, the description of viscoelastic behavior becomes much more complicated. There is no unifying theory to replace the Boltzmann superposition principle, and it is not possible to generalize the results of various types of experiment. Thus, the response to a deformation depends on its size, rate and kinematics, and each experiment yields more or less unique information.

This lack of a general theory of nonlinear behavior limits our ability to relate nonlinear properties to molecular structure. In addition, it makes it difficult to model the behavior of polymeric liquids in manufacturing processes.

Theories of Nonlinear Viscoelasticity

For the reasons cited above, there has been a substantial effort to develop at least approximate, empirical constitutive equations to describe the nonlinear viscoelasticity of polymeric liquids. Much of this work has been reviewed by Larson [2]. It would be particularly useful to have a model that related rheological behavior to molecular structure, and for this reason, the theory of Doi and Edwards [3] has attracted considerable attention in recent years. However, in order to put this theory in the form of an explicit constitutive equation, it is necessary to make certain simplifying assumptions that reduce its regime of validity to rather small deviations from linear behavior.

An empirical equation that has proven useful, at least as a basis for the presentation of experimental results, is the following generalization of the Boltzmann superposition principle.

$$\tau_{ij}(t) = \int_{-\infty}^{t} m(t-t')h(I)[\beta B_{ij}(t,t')+(1-\beta)C_{ij}(t,t')]dt' \qquad (2)$$

where: $m(t-t')$ = memory function

$h(I)$ = damping function, a scalar function of the invariants of the Finger and Cauchy strains.

B_{ij} = component of the Finger tensor

C_{ij} = component of the Cauchy tensor

The memory function is directly related to the relaxation modulus of the Boltzmann superposition principle, which can be expressed as follows:

$$m(t-t') = \frac{dG(t-t')}{dt'} \tag{3}$$

Thus, the memory function can be obtained from any linear viscoelastic characterizing function for the material of interest. For the generalized Maxwell model representation of linear behavior, it is a sum of exponentials.

Many of the empirical constitutive equations that have been proposed to describe the nonlinear behavior of polymeric liquids are of the general type of Equation (2), including the Doi-Edwards constitutive equation and the BKZ equation. This particular form has been used by Tanner and Luo [4] and by Wagner [5].

One of the important implications of Equation (2) is that the relaxation modulus is a separable function of time and strain. For example, for a step shear strain experiment, (3) predicts that:

$$G(t,\gamma) = G(t)h(\gamma) \tag{4}$$

Thus, one convenient test of the model is to see if nonlinear step strain data determined at various levels of strain are superposable by a vertical shift on a plot of G versus t. If superposibility is observed, the shift factor gives the damping function $h(\gamma)$.

Unfortunately, it is not possible to infer the complete dependence of the damping function on the Finger and Cauchy tensors on the basis of any single type of experiment. All that can be done is to try to formulate an empirical function h(I) that is consistent with results from several types of experiment. For example, Wagner [5] has proposed a form that he found to be adequate to describe the multiaxial extensional flow results of Desmarmels and Meissner [6].

It is obviously essential in any thorough study of the nonlinear viscoelastic behavior of a material to obtain data for several types of deformation. The state of the art in regard to experimental methods for doing this is described in detail by Dealy [7] and is summarized in the following section.

Shearing Flows

Three rheologically significant quantities can, in principle, be measured in a simple shear flow: the shear stress and two normal stress differences. Within the linear regime, the normal stress differences are zero. For steady simple shear, or more generally, for any viscometric flow, these quantities are functions of shear rate but not of time. For transient flows, however, they are also functions of time. For example, in a step strain experiment, where the strain amplitude, γ, is outside the range of linear behavior, three material functions of t and γ can, in principle, be determined. Many other strain histories have been used to study the nonlinear behavior of polymeric liquids, including large amplitude oscillatory shear, start up and cessation of steady shear, and exponential shear.

However, commercially available rheometers have been found to be of limited use in nonlinear studies. Pressure flow rheometers (capillary or slit) can only be used for steady shear experiments. In a capillary instrument, only the shear stress can be measured. While methods have been proposed for inferring the first normal stress difference from slit flow measurements [8,9], the validity of these methods has not yet been established.

Rotational rheometers, usually of the cone-plate type, have been used to obtain most of the nonlinear shear data now in the literature, which have been recently summarized [10]. However, edge effects and flow instabilities limit the use of rotational rheometers to rather small departures from the linear regime. Moreover, it is particularly difficult to obtain reliable normal stress data, especially at high temperatures [11].

In order to overcome the limitations of rotational rheometers, there has been a growing use in recent years of sliding cylinder and sliding plate rheometers [12]. However, at large shear strains, these instruments also are subject to significant edge and/or end effects. In order to eliminate errors due to end and edge effects in sliding plate rheometers, a shear stress transducer has been developed [13]. This device senses the shear stress exerted by the fluid on only a small area of a plate, and by positioning it such that it is never near to an edge, only the region of the sample in which the strain is uniform contributes to the stress measurement. Other advantages of the use of a shear stress transducer are that the quantity measured does not include instrument friction and that degradation that occurs at the edges of the sample has no effect on the measurement.

Sliding plate rheometers equipped with shear stress transducers have been used to study large amplitude oscillatory shear [14], exponential shear [15] and bidirectional shear [16]. In the last mentioned study, the shear stress transducer was designed to respond to stresses in two orthogonal directions. A sliding plate rheometer equipped with a shear stress transducer has also proven useful in the study of the wall slip of molten thermoplastics [17].

It is not possible to determine normal stress differences

by means of mechanical measurements in a sliding plate rheometer, because the deformation at the ends and edges of the sample is nonuniform and uncontrolled. Birefringence can be used, however, to infer normal stress differences by means of the stress optical law. This is particularly convenient for the measurement of the stress difference, $\sigma_{11} - \sigma_{33}$, which is related to the attenuation of light in the x_2 direction, i.e., the direction normal to the shear planes.

Extensional Flows

It is considerably more difficult to generate uniform, controlled extensional deformations than simple shear deformations, because the sample cannot be supported by instrument fixtures. While a number of techniques have been designed to meet this challenge [7], great care and skill are required in their use, and strain rates are generally limited to quite low values. Furthermore, unless the material under study has a tendency to become more resistant to deformation as the strain rate is increased, it is very difficult to generate a uniform deformation well into the region of nonlinear behavior. It has been proposed that converging pressure flow and extrudate drawing be used to obtain information about the response of polymeric liquids to extensional flows, but the deformation is not uniform in these flows, and well-defined material functions cannot be inferred from the results [10]. For these reasons, much less is known about the response to extensional flows than is known about shear properties.

Summary

There is a strong interest in nonlinear behavior on the part of both polymer scientists and plastics engineers. However, advances in our understanding of this subject are limited on the one hand by the lack of a unifying theory and on the other by the major difficulties that are involved in carrying out definitive experiments. Empirical constitutive equations are useful for categorizing response types and as a basis for the presentation of experimental results. However, it is not possible to establish with any confidence the range of validity of such a model.

The use of sliding plate rheometers is leading to a significant advance in our empirical knowledge of nonlinear behavior in shear, although measurements of the normal stress differences are still problematic. Experimental difficulties still severely limit our ability to study the response of polymeric liquids to extensional modes of deformation.

Thus, there remain many major challenges in the development of theories and experimental techniques in the area of nonlinear viscoelasticity.

References

1. N. W. Tschoegl, The Phenomenological Theory of Linear Viscoelastic Behavior, Springer-Verlag, Berlin, 1989.

2. R. G. Larson, Constitutive Equations for Polymer Melts and Solutions, Butterworths, Boston, London, 1988.

3. M. Doi and S. F. Edwards, The Theory of Polymer Dynamics, Oxford Science Publishers, Oxford, 1986.

4. Y. L. Luo and R. I. Tanner, Int. J. Num. Meth. Eng., 25, 9 (1988).

5. M. H. Wagner, "A Constitutional Analysis of Multiaxial Elongations of Polyisobutylene", Paper II.1.2, 61st Ann. Mtg., Soc. of Rheol., Montreal, 1989.

6. A. Desmarmels and J. Meissner, Coll. Polym. Sci., 264, 829 (1986).

7. J. M. Dealy, Rheometers for Molten Plastics, Van Nostrand Reinhold, New York, 1982.

8. C. D. Han, "Slit Rheometry", Chapter 2 of Rheological Measurement, ed. by A. A. Collyer and D. W. Clegg, Elsevier Applied Science Publishers, London, 1988.

9. A. S. Lodge, "Normal Stress Differences from Pressure Hole Measurements", Chapter 11 of Rheological Measurement, ed. by A. A. Collyer and D. W. Clegg, Elsevier Applied Science Publishers, London, 1988.
10. J. M. Dealy and K. F. Wissbrun, Melt Rheology and its Role in Plastics Processing, Van Nostrand Reinhold, New York, 1990.
11. J. Meissner, R. W. Garbella and J. Hostettler, J. Rheol, 33, 843 (1989).
12. J. M. Dealy and A. J. Giacomin, "Sliding Plate and Sliding Cylinder Rheometers", Chapter 12 of Rheological Measurement, ed. by A. A. Collyer and D. W. Clegg, Elsevier Applied Science Publishers, London, 1988.
13. J. M. Dealy, "Method of Measuring Shear Stress", U.S. Patent 4,463,928; August 1984.
14. A. J. Giacomin, T. Samurkas and J. M. Dealy, Polym. Eng. Sci., 29, 499 (1989).
15. S. R. Doshi and J. M. Dealy, J. Rheol., 31, 563 (1987).
16. J. Meissner, "Improved Conventional Methods and New Approaches in Polymer Melt Rheology", Paper II.1.2, 61st Ann. Mtg., Soc. of Rheol., Montreal, 1989.
17. S. Hatzikiriakos and J. M. Dealy, 61st Ann. Mtg., Soc. of Rheol., 31, 563 (1987).

NONLINEAR VISCOELASTICITY OF POLYMERIC LIQUIDS

John M. Dealy
Department of Chemical Engineering
McGill University
Montreal, Canada
H3A 2A7

P.N. KALONI AND A.M. SIDDIQUI

Generalized helical flow of a simple fluid

1. Introduction

The problem of a helical flow in viscoelastic fluids in which the annular mass of the fluid is contained between two rotating coaxial cylinders and a constant pressure gradient acts parallel to the axis of the cylinder, has been considered by several authors. Rivlin [7], who first introduced the term helical, gave expressions for the stress components, etc., for Rivlin-Ericksen fluids [8] but never solved the resulting equations. A few years later, Fredrickson [4] devised a trial and error method to solve the non-linear equations of [7]. Coleman and Noll [1] gave solution of this problem in a simple fluid but in terms of viscosity and normal stress functions. Dierckes and Schowalter [2] did measurements of pressure drop and volumetric flow rate in a helical flow in a 3% measurements of pressure drop and volumetric flow rate in a helical flow in a 3% polysobutylene dissolved in decalin and found excellent agreement with a power-law fluid theory. Savins and Wallick [9] employed a numerical method to calculate viscosity profiles, discharge rates and torque, etc., for an Oldroyd type fluid in a helical flow. These authors specifically considered the coupling effects between rotational motion and axial motion and, amongst other things, found that the axial discharge rate in a helical flow is increased in comparison to a purely annular flow field.

The purpose of the present paper is to give a complete analytical solution of the spiral flow problem in a simple fluid. Following the perturbation scheme of Joseph and Fosdick [5], Fosdick and Kao [3] and Rajagopal, Kaloni and Tao [6] we consider the successive approximation of the simple fluid up to fourth order. One advantage in this manner, as opposed to considering the order or grade fluids as exact models, is that one avoids the problem of prescribing additional boundary conditions in this approach. We determine explicit expressions for the velocity, pressure, etc., up to fourth order and compute torque, volume discharge rate and normal-stress differences. In comparison to the linearly viscous fluid case we find that, if we take $(\beta_2+\beta_3)>0$, the volume discharge rate is decreased and the torque lowered, while for $(\beta_2+\beta_3)<0$, we find the effects to be of completely reverse order. We also find that even though there is coupling between the rotational and translational velocity components the velocity do not themselves depend upon the expressions involved in normal stress functions.

2. Basic Equations

It can be shown that Cauchy stress $\mathbf{T}$ in an incompressible simple fluid can be expressed as [11]

$$\mathbf{T}(\mathbf{x},t) = -p\mathbf{I} + \mathop{\mathbf{H}}_{s=0}^{\infty} (\mathbf{C}_t(t-s)) \tag{1}$$

where $\mathop{\mathbf{H}}_{s=0}^{\infty}$ is the functional operator, $\mathbf{C}_t(t-s)$ is the relative deformation tensor, i.e., the history of the right Cauchy-Green tensor and p denotes the indeterminate pressure. Coleman and Noll (see Truesdell and Noll [11]), by introducing the concepts of retarded motions and fading memory, have developed consistent approximations to the constitutive equation (1). In such a case it follows that one can rewrite (1) as

$$\mathbf{T} = -p\mathbf{I} + \sum_{i=1}^{4} \mathbf{S}_i \tag{2}$$

where

$$\begin{aligned}
\mathbf{S}_1 &= \mu\mathbf{A}_1, \quad \mathbf{S}_2 = \alpha_1\mathbf{A}_2 + \alpha_2\mathbf{A}_1^2, \\
\mathbf{S}_3 &= \beta_1\mathbf{A}_3 + \beta_2(\mathbf{A}_1\mathbf{A}_2 + \mathbf{A}_2\mathbf{A}_1) + \beta_3(\mathrm{tr}\mathbf{A}_2)\mathbf{A}_1 , \\
\mathbf{S}_4 &= \gamma_1\mathbf{A}_4 + \gamma_2(\mathbf{A}_3\mathbf{A}_1\mathbf{A}_1\mathbf{A}_3) + \gamma_3\mathbf{A}_2^2 + \gamma_4(\mathbf{A}_2\mathbf{A}_1^2+\mathbf{A}_1\mathbf{A}_2^2) \\
&\quad + \gamma_5(\mathrm{tr}\mathbf{A}_2)\mathbf{A}_2 + \gamma_6(\mathrm{tr}\mathbf{A}_2)\mathbf{A}_1^2 + [\gamma_7\mathrm{tr}\mathbf{A}_3+\gamma_8\mathrm{tr}(\mathbf{A}_2\mathbf{A}_1 0]\mathbf{A}_1,
\end{aligned} \tag{3}$$

and where μ is the viscosity coefficient and α_1, α_2, $\beta_1,\ldots,\gamma_8$ are material constants. For steady motion the Rivlin-Ericksen tensors satisfy the recursion relation

$$\mathbf{A}_{\Gamma+1} = (\text{grad } \mathbf{A}_\Gamma)\mathbf{v} + \mathbf{A}_\Gamma \text{grad } \mathbf{v} + (\mathbf{A}_\Gamma \text{grad } \mathbf{v})^T,$$
$$\mathbf{A}_0 = \mathbf{I} \qquad \Gamma \geqq 1 \ , \tag{4}$$

where $\mathbf{v}$ is the velocity vector.

Equations (2)-(4) can also be obtained from (1) by considering motions which are slow and slowly varying, as is discussed by Saut and Joseph [10]. Thus, it appears that one can use above equations for motions which are not necessarily retarded. However, care has to be exercised not to make the problem a singular perturbation problem, as is pointed out in [6]. We also suggest the reader to look into this paper for other interesting discussions about the order fluid.

We shall assume that body forces are absent and motion is steady. The field equations then take the form

$$\text{div } \mathbf{v} = 0 \ ,$$
$$\rho(\text{grad } \mathbf{v})\mathbf{v} = -\text{grad } p + \text{div } \mathbf{S} \ . \tag{5}$$

We consider the motion of a simple fluid contained between two concentric circular pipes of radii a and b with a<b. The pipes rotate about their common axis with constant angular velocities Ω and $\lambda\Omega$, respectively. In addition the pipes may translate steadily, parallel to their common axis. We assume that the outer pipe translates with a velocity $\bar{U}$ relative to the inner one.

We use the cylindrical polar coordinate system (r,θ,z), in which the z-axis lies along the common axis of the pipes. The no-slip conditions require that

$$\mathbf{v} = \begin{cases} \Omega a \mathbf{e} & \text{at } r = a \ , \\ \lambda\Omega b \mathbf{e}_\theta + \bar{U}\mathbf{e}_z & \text{at } r = b \ , \end{cases} \tag{6}$$

where a is the radius of inner pipe and b is the radius of outer pipe and $\mathbf{e}_\theta$ and $\mathbf{e}_z$ are unit vectors in θ and z directions, respectively. For mathematical convenience we shall also assume that constant velocity $\bar{U} = U\Omega$. We can then consider the velocity field and pressure of the form:

$$\mathbf{v} = \mathbf{v}(r,z;\Omega),$$

$$p = p(r,z;\Omega), \tag{7}$$

Furthermore, we assume that (7) is sufficiently differentiable with respect to Ω to yield the series expansions of the type

$$\mathbf{v} = \sum_{i=1}^{n} \frac{1}{k!} \mathbf{v}^{(k)}\Omega^k + 0(\Omega^n),$$

$$p = \sum_{k=0}^{n} \frac{1}{k!} p^{(k)}\Omega^k + 0(\Omega^n), \tag{8}$$

where

$$(\,)^{(k)} = \frac{\partial^k}{\partial\Omega^k}(\,)\Bigg|_{\Omega=0},$$

and where $\mathbf{v}^{(k)}$ and $p^{(k)}$ are functions of position. Because of the axial symmetry it follows that

$$\mathbf{v}^{(k)} = u^{(k)}(r,z)\mathbf{e}_r + v^{(k)}(r,z)\mathbf{e}_\theta + w^{(k)}(r,z)\mathbf{e}_z \tag{9}$$

and that

$$p^{(k)} = p^{(k)}(r,z)\ . \tag{10}$$

Since each of the equations (5) is an identity in Ω, these may be differentiated any number of times with respect to Ω and evaluated at $\Omega=0$. Thus, we obtain

$$\operatorname{div} \mathbf{v}^{(k)} = 0 \tag{11}$$

$$\rho[(\operatorname{grad} \mathbf{v})\mathbf{v}]^{(k)} = -\operatorname{grad} p^{(k)} + \operatorname{div} \mathbf{S}^{(k)} \tag{12}$$

where $\mathbf{S}^{(k)}$ are given by (3).

From (8), (9) and (6), the boundary conditions for $\mathbf{v}^{(k)}$ have the form

$$u^{(1)} = 0,\quad w^{(1)} = 0,\quad v^{(1)} = a \quad \text{at } r = a,$$

$$u^{(1)} = 0,\quad w^{(1)} = U,\quad v^{(1)} = \lambda b \quad \text{at } r = b, \tag{13}$$

and for k>1,

$$u^{(k)} = 0, \quad w^{(k)} = 0, \quad v^{(k)} = 0, \qquad \text{at } r = a,$$
$$u^{(k)} = 0, \quad w^{(k)} = 0, \quad v(k) = 0, \qquad \text{at } r = b. \tag{14}$$

3. Solution

Following [5], [3] and [6] we now develop the perturbation solution in series in powers of Ω. This will be carried out by solving (11) and (12) for $\mathbf{v}^{(k)}$ and $p^{(k)}$, at each order $k = 1,2,3,4$, with the boundary conditions (13) and (14).

For $k = 1$, (11) and (12) have the form

$$\operatorname{div} \mathbf{v}^{(1)} = 0,$$
$$-\operatorname{grad} p^{(1)} + \operatorname{div} \mathbf{S}^{(1)} = 0, \tag{15}$$

where

$$\mathbf{S}^{(1)} = \mathbf{S}_1^{(1)} + \mathbf{S}_2^{(1)} + \mathbf{S}_3^{(1)} + \mathbf{S}_4^{(1)}.$$
$$\mathbf{S}_1^{(1)} = \mu \mathbf{A}_1^{(1)}, \quad \mathbf{S}_i^{(1)} = 0, \quad i \geqq 2. \tag{16}$$

On introducing the stream functions $\psi^{(1)}(r,z)$ such that

$$u^{(1)} = -\frac{1}{r}\frac{\partial \psi^{(1)}}{\partial z}, \qquad w^{(1)} = \frac{1}{r}\frac{\partial \psi^{(1)}}{\partial r}, \tag{17}$$

we find that (15) is identically satisfied. On substituting (16) in $(15)_2$ and eliminating the pressure we get

$$-\frac{\mu}{r} E^4 \psi^{(1)} = 0$$
$$\frac{\partial^2 v^{(1)}}{\partial r^2} + \frac{\partial}{\partial r}\left(\frac{v^{(1)}}{r}\right) + \frac{\partial^2 v^{(1)}}{\partial z^2} = 0. \tag{18}$$

The solution of (18) satisfying the boundary conditions (13) is

$$u^{(1)} = 0$$
$$v^{(1)} = \left(\frac{\lambda b^2 - a^2}{b^2 - a^2}\right) r + \frac{(1-\lambda) a^2 b^2}{(b^2 - a^2)} \frac{1}{r}$$

$$w^{(1)} = \frac{G}{4\mu}[a^2r^2 + \frac{(b^2-a^2)\ln(\frac{r}{a})}{\ln(b/a)}] + U\,\frac{\ln(r/a)}{\ln(b/a)}\,, \tag{19}$$

where

$$\frac{dp^{(1)}}{dz} = -G.$$

Also, the solution for $p^{(1)}$ is

$$p^{(1)} = C - Gz \tag{20}$$

where C is an arbitrary constant.

For k=2 the equations to be solved are

$$\text{div } \mathbf{v}^{(2)} = 0, \tag{21}$$

$$-\text{grad } p^{(2)} + \text{div } \mathbf{S}^{(2)} = 2\rho(\text{grad } \mathbf{v}^{(1)})\,\mathbf{v}^{(1)}, \tag{22}$$

where

$$\mathbf{S}^{(2)} = \mathbf{S}_1^{(2)} + \mathbf{S}_2^{(2)} + \mathbf{S}_3^{(2)} + \mathbf{S}_4^{(2)},$$

$$\mathbf{S}_1^{(2)} = \mu\mathbf{A}_1^{(2)},$$

$$\mathbf{S}_2^{(2)} = \alpha_1\mathbf{A}_2^{(2)} + 2\alpha_2(\mathbf{A}_1^{(1)})^2,$$

$$\mathbf{S}_i^{(2)} = 0,\ i\geqq 3, \tag{23}$$

and

$$\mathbf{A}_1^{(2)} = [\text{grad } \mathbf{v}^{(2)} + (\text{grad } \mathbf{v}^{(2)})^T]$$

$$\mathbf{A}_2^{(2)} = 2[(\text{grad } \mathbf{A}_1^{(1)})\mathbf{v}^{(1)} + \mathbf{A}_1^{(1)}\text{ grad } \mathbf{v}^{(1)} + (\mathbf{A}_1^{(1)}\text{ grad } \mathbf{v}^{(1)})^T]$$

$$\mathbf{A}_i^{(2)} = 0 \quad i\geqq 3$$

$$\mathbf{A}_1^{(1)} = \text{grad } \mathbf{v}^{(1)} + (\text{grad } \mathbf{v}^{(1)})^T. \tag{24}$$

With $\mathbf{v}^{(1)}$ and $p^{(1)}$ known, substitution of (23) and (24) in (21) and (22) leads to

$$\frac{\partial u^{(2)}}{\partial r} + \frac{u^{(2)}}{r} + \frac{\partial w^{(2)}}{\partial z} = 0,$$

$$-\frac{\partial p^{(2)}}{\partial r} + \mu[\frac{\partial^2 u^{(2)}}{\partial r^2} + \frac{\partial}{\partial r}(\frac{u^{(2)}}{r}) + \frac{\partial^2 u^{(2)}}{\partial z^2}] = \frac{-2(2\alpha_1+\alpha_2)}{r} \times$$

$$\times \frac{d}{dr}[r\{\frac{dv^{(1)}}{dr} - \frac{v^{(1)}}{r}\}^2 + (\frac{dw^{(1)}}{dr})^2\}] + \frac{2\alpha_2}{r}(\frac{dv^{(1)}}{dr}$$

$$- \frac{v^{(1)}}{r})^2 - \frac{2\rho}{r}(v^{(1)})^2,$$

$$-\frac{\partial p^{(2)}}{\partial z} + \mu[\frac{\partial^2 w^{(2)}}{\partial r^2} + \frac{1}{r}\frac{\partial w^{(2)}}{\partial r} + \frac{\partial^2 w^{(2)}}{\partial z^2}) = 0,$$

$$\mu[\frac{\partial^2 v^{(2)}}{\partial r^2} + \frac{\partial}{\partial r}(\frac{v^{(2)}}{r}) + \frac{\partial^2 v^{(2)}}{\partial z^2}] = 0. \quad (25)$$

The solution of (25) satisfying the boundary condition (14) is

$$u^{(2)} = v^{(2)} = w^{(2)} = 0. \quad (26)$$

$$p^{(2)} = 2(2\alpha_1+\alpha_2)[(\frac{dv^{(1)}}{dr} - \frac{v^{(1)}}{r})^2 + (\frac{dw^{(1)}}{dr})^2]$$

$$+ \int^r \frac{1}{r}[4\alpha_1\{(\frac{dv^{(1)}}{dr} - \frac{v^{(1)}}{r})^2 + (\frac{dw^{(1)}}{dr})^2\}$$

$$+ 2\alpha_2(\frac{dw^{(1)}}{dr})^2 + 2\rho(v^{(1)})^2]dr \quad (27)$$

where $v^{(1)}$ and $w^{(1)}$ are given by (19).

For k=3, the field equations become

$$\text{div } \mathbf{v}^{(3)} = 0, \quad (28)$$

$$- \text{grad } p^{(3)} + \text{div } \mathbf{S}^{(3)} = 3\ [(\text{grad } \mathbf{v}^{(2)})\mathbf{v}^{(1)}$$

$$+ (\text{grad } \mathbf{v}^{(1)})\mathbf{v}^{(2)}] \quad (29)$$

where

$$\mathbf{S}^{(3)} = \mathbf{S}_1^{(3)} + \mathbf{S}_2^{(3)} + \mathbf{S}_3^{(3)} + \mathbf{S}_4^{(3)},$$

$$\mathbf{S}_1^{(3)} = \mu\mathbf{A}_1^{(3)},$$

$$\mathbf{S}_2^{(3)} = \alpha_1\mathbf{A}_2^{(3)} + 3\alpha_2[\mathbf{A}_1^{(2)}\mathbf{A}_1^{(1)} + \mathbf{A}_1^{(1)}\mathbf{A}_1^{(2)}],$$

$$\mathbf{S}_3^{(3)} = \beta_1\mathbf{A}_3^{(3)} + 3[\beta_2(\mathbf{A}_2^{(2)}\mathbf{A}_1^{(1)} + \mathbf{A}_1^{(1)}\mathbf{A}_2^{(2)}) + \beta_3(\mathrm{tr}\mathbf{A}_2^{(2)})\mathbf{A}_1^{(1)}],$$

$$\mathbf{S}_i^{(3)} = 0 \qquad i \geqq 4, \tag{30}$$

with

$$\mathbf{A}_1^{(3)} = \mathrm{grad}\ \mathbf{v}^{(3)} + (\mathrm{grad}\ \mathbf{v}^{(3)})^T,$$

$$\mathbf{A}_2^{(3)} = 3[(\mathrm{grad}\ \mathbf{A}_1^{(2)})\mathbf{v}^{(1)} + \mathbf{A}_1^{(2)}\mathrm{grad}\ \mathbf{v}^{(1)} + (\mathbf{A}_1^{(2)}\mathrm{grad}\ \mathbf{v}^{(1)})^T$$
$$+ (\mathrm{grad}\ \mathbf{A}_1^{(1)})\mathbf{v}_1^{(2)} + \mathbf{A}_1^{(1)}\mathbf{v}^{(2)} + (\mathbf{A}_1^{(1)}\mathrm{grad}\ \mathbf{v}^{(2)})^T],$$

$$\mathbf{A}_3^{(3)} = 3[(\mathrm{grad}\ \mathbf{A}_2^{(2)}\mathbf{v}^{(1)} + \mathbf{A}_2^{(2)}\mathrm{grad}\ \mathbf{v}^{(1)} + (\mathbf{A}_2^{(2)}\mathrm{grad}\ \mathbf{v}^{(1)})^T],$$

$$\mathbf{A}_i^{(3)} = 0 \qquad i \geqq 4 \tag{31}$$

As before, we introduce a stream function $\psi^{(3)}$ such that

$$u^{(3)} = -\frac{1}{r}\frac{\partial\psi^{(3)}}{\partial z}, \qquad w^{(3)} = \frac{1}{r}\frac{\partial\psi^{(3)}}{\partial r}, \tag{32}$$

and find that (28) is identically satisfied. On substituting the values of $\mathbf{v}^{(1)}$, $\mathbf{v}^{(2)}$, $p^{(1)}$ and $p^{(2)}$ in (31) and then using (30) and (31) in (29) and eliminating the pressure $p^{(3)}$ between the resulting two equations we get

$$-\frac{\mu}{r}E^4\psi^{(3)} = 12(\beta_2+\beta_3)\frac{d}{dr}[\frac{1}{r}\frac{d}{dr}\{r\frac{dw^{(1)}}{dr}(\frac{dv^{(1)}}{dr} - \frac{v}{r})^2$$
$$+ (\frac{dw^{(1)}}{dr})\}], \tag{33}$$

$$\mu[\frac{\partial^2 v^{(3)}}{\partial r^2} + \frac{\partial}{\partial r}(\frac{v^{(3)}}{r}) + \frac{\partial^2 v^{(3)}}{\partial z^2}] = \frac{-12(\beta_2+\beta_3)}{r^2}\frac{d}{dr}[r^2.$$
$$.(\frac{dv^{(1)}}{dr} - \frac{v^{(1)}}{r})\{(\frac{dv^{(1)}}{dr} - \frac{v^{(1)}}{r})^2 + (\frac{dw^{(1)}}{dr})^2\}]. \tag{34}$$

On substituting the values of $v^{(1)}$ and $w^{(1)}$ on the right hand side of (33) and (34) from (19), the solutions for $v^{(3)}$ and $w^{(3)}$, satisfying the boundary conditions (14), turn out to be:

$$v^{(3)} = \frac{(\beta_2+\beta_3)}{\mu}\Big[16N^3\Big(\frac{1}{r^5} + \frac{b^2+a^2}{a^4b^4}r - \frac{b^4+a^4+a^2b^2}{a^4b^4}\frac{1}{r}\Big)$$

$$+ 6NB_1^2\Big\{\frac{1}{r^3} + \frac{1}{a^2b^2}r - \frac{a^2+b^2}{a^2b^2}\frac{1}{r}\Big\} + 24NB_o^2\Big\{-r\ln r$$

$$+ r\frac{b^2\ln b-a^2\ln a}{b^2-a^2} - \frac{a^2b^2\ln(\frac{b}{a})}{(b^2-a^2)r}\Big\}\Big], \tag{35}$$

$$w^{(3)} = \frac{(\beta_2+\beta_3)}{\mu}\Big[12N^2B_1\Big\{\frac{1}{r^4} - \frac{1}{a^4} + \frac{b^4-a^4}{a^4b^4} - \frac{\ln r/a}{\ln b/a}\Big\}$$

$$+ 6(B_1^3-4N^2B_o)\Big\{\frac{1}{r^2} - \frac{1}{a^2} + \frac{b^2-a^2}{a^2b^2} - \frac{\ln r/a}{\ln b/a}\Big\}$$

$$- 3B_o^3\Big\{a^4-r^4 + \frac{(b^4-a^4)\ln r/a}{\ln b/a}\Big\} + 18B_o^2B_1\Big\{a^2-r^2$$

$$+ \frac{(b^2-a^2)\ln r/a)}{\ln b/a}\Big\}\Big], \tag{36}$$

where

$$B_o = \frac{G}{2\mu}, \qquad N = \frac{(1-\lambda)a^2b^2}{(b^2-a^2)},$$

$$B_1 = \Big[\frac{B_o}{2}(b^2-a^2) + U\Big]\frac{1}{\ln(b/a)}. \tag{37}$$

The solution for $p^{(3)}$ is given by $p^{(3)} = C_3$ where C_3 is an arbitrary constant.

We note that first term in (35) is exactly the same as that reported in [5]. The remaining terms in (35) are, however, due to the generalized motion considered in this paper.

Finally, for k=4, the equations to be solved are:

$$\operatorname{div} \mathbf{v}^{(4)} = 0,$$

$$-\text{grad}\ p^{(4)} + \text{div}\ \mathbf{S}^{(4)} = \rho[4(\text{grad}\ \mathbf{v}^{(3)})\mathbf{v}^{(1)}$$

$$+ 4(\text{grad}\mathbf{v}^{(1)})\mathbf{v}^{(3)} + 6(\text{grad}\ \mathbf{v}^{(2)}\mathbf{v}^{(2)}]. \qquad (38)$$

where

$$\mathbf{S}^{(4)} = \mathbf{S}_1^{(4)} + \mathbf{S}_2^{(4)} + \mathbf{S}_3^{(4)} + \mathbf{S}_4^{(4)},$$

$$\mathbf{S}_1^{(4)} = \mu\mathbf{A}_1^{(4)},$$

$$\mathbf{S}_2^{(4)} = \alpha_1\mathbf{A}_2^{(4)} + \alpha_2[4(\mathbf{A}_1^{(1)}\mathbf{A}_1^{(3)} + \mathbf{A}_1^{(3)}\mathbf{A}_1^{(1)}) + 6\mathbf{A}_1^{(2)}\mathbf{A}_1^{(2)}],$$

$$\mathbf{S}_3^{(4)} = \beta_1\mathbf{A}_3^{(4)} + \beta_2[4(\mathbf{A}_1^{(1)}\mathbf{A}_2^{(3)} + \mathbf{A}_2^{(3)}\mathbf{A}_1^{(1)}) + 6(\mathbf{A}_1^{(2)}\mathbf{A}_2^{(2)}$$

$$+ \mathbf{A}_2^{(2)}\mathbf{A}_1^{(2)}) + \beta_3[4(\text{tr}\mathbf{A}_2^{(3)})\mathbf{A}_1^{(1)} + 6(\text{tr}\mathbf{A}_2^{(2)})\mathbf{A}_1^{(2)}],$$

$$\mathbf{S}_4^{(4)} = \gamma_1\mathbf{A}_4^{(4)} + 4\gamma_2[\mathbf{A}_3^{(3)}\mathbf{A}_1^{(1)} + \mathbf{A}_1^{(1)}\mathbf{A}_3^{(3)} + 6\gamma_3\mathbf{A}_2^{(2)}\mathbf{A}_2^{(2)}$$

$$+ 12\gamma_4[\mathbf{A}_2^{(2)}\mathbf{A}_1^{(1)}\mathbf{A}_1^{(1)} + \mathbf{A}_1^{(1)}\mathbf{A}_1^{(1)}\mathbf{A}_2{}^{(2)}]$$

$$+ 6\gamma_5(\text{tr}\mathbf{A}_2^{(2)})\mathbf{A}_2^{(2)} + 12\gamma_6(\text{tr}\mathbf{A}_2^{(2)})\mathbf{A}_1^{(1)}\mathbf{A}^{(1)}$$

$$+ 4\gamma_7(\text{tr}\mathbf{A}_3^{(3)})\mathbf{A}_1^{(1)} + 12\gamma_8(\text{tr}\mathbf{A}_1^{(2)}\mathbf{A}_1^{(1)}\mathbf{A}_1^{(1)}, \qquad (39)$$

with

$$\mathbf{A}_1^{(4)} = \text{grad}\ \mathbf{v}^{(4)} + (\text{grad}\ \mathbf{v}^{(4)})^T,$$

$$\mathbf{A}_2^{(4)} = [4\{(\text{grad}\ \mathbf{A}_1^{(3)})\ \mathbf{v}^{(1)} + \mathbf{A}_1^{(3)}\text{grad}\ \mathbf{v}^{(1)}$$

$$+ (\mathbf{A}_1^{(3)}\text{grad}\ \mathbf{v}^{(1)})^T\} + 4\{(\text{grad}\ \mathbf{A}_1^{(1)})\mathbf{v}^{(3)}$$

$$+ \mathbf{A}_1^{(1)}\text{grad}\ \mathbf{v}^{(3)} + (\mathbf{A}_1^{(1)}(\text{grad}\ \mathbf{v}^{(3)})^T\}$$

$$+ 6\{\mathbf{A}_1^{(2)}\text{grad}\ \mathbf{v}^{(2)} + (\mathbf{A}_1^{(2)}\text{grad}\ \mathbf{v}^{(2)})^T$$

$$+ (\text{grad}\ \mathbf{A}_1^{(2)})\ \mathbf{v}^{(2)}\}],$$

$$\mathbf{A}_3^{(4)} = [4\{(\text{grad } \mathbf{A}_2^{(3)})\, \mathbf{v}^{(1)} + \mathbf{A}_2^{(3)}\text{grad } \mathbf{v}^{(1)} + (\mathbf{A}_2^{(3)}\text{grad } \mathbf{v}^{(1)})^T\} + 6\{(\text{grad } \mathbf{A}_2^{(2)})\, \mathbf{v}^{(2)} + \mathbf{A}_2^{(2)}\text{grad } \mathbf{v}^{(2)} + (\mathbf{A}_2^{(2)}\text{grad } \mathbf{v}^{(2)})^T\}],$$

$$\mathbf{A}_4^{(4)} = [4\{(\text{grad } \mathbf{A}_3^{(3)})\, \mathbf{v}^{(1)} + (\mathbf{A}_3^{(3)}\text{grad } \mathbf{v}^{(1)}) + (\mathbf{A}_3^{(3)}\text{grad } \mathbf{v}^{(1)})^T\}] \,. \tag{40}$$

On employing the results obtained earlier, it is found that the last term on the right hand side of (38) and $\mathbf{S}_3^{(4)}$ become identically zero. A considerable simplification, in the manner followed earlier, leads to

$$\frac{\partial u^{(4)}}{\partial r} + \frac{u^{(4)}}{r} + \frac{\partial w^{(4)}}{\partial z} = 0 \,. \tag{41}$$

$$- \frac{\partial p^{(4)}}{\partial r} + \mu[\frac{\partial^2 u^{(4)}}{\partial r^2} + \frac{\partial}{\partial r}(\frac{u^{(4)}}{r}) + \frac{\partial^2 u^{(4)}}{\partial z^2}]$$

$$= \frac{-8(2\alpha_1+\alpha_2)}{r} \frac{d}{dr}[r\{(\frac{dv^{(1)}}{dr} - \frac{v^{(1)}}{r})(\frac{dv^{(3)}}{dr} - \frac{v^{(3)}}{r}) + \frac{dw^{(1)}}{dr} \frac{dw^{(3)}}{dr}]\} - 8\frac{\rho}{r}v^{(1)}v^{(3)}$$

$$+ \frac{8\alpha_2}{r} [(\frac{dv^{(1)}}{dr} - \frac{v^{(1)}}{r})(\frac{dv^{(3)}}{dr} - \frac{v^{(3)}}{r})$$

$$\frac{-96}{r}(\gamma_3+\gamma_4+\gamma_5+\gamma_{6/2})\frac{d}{dr}[r\{(\frac{dv^{(1)}}{dr} - \frac{v^{(1)}}{r})^2 + (\frac{dw^{(1)}}{dr})^2\}^2]$$

$$+ \frac{48\gamma_6}{r}[(\frac{dv^{(1)}}{dr} - \frac{v^{(1)}}{r})^2 \{(\frac{dv^{(1)}}{dr} - \frac{v^{(1)}}{r})^2 + \frac{dw^{(1)}}{dr})^2\}],$$

$$- \frac{\partial p^{(4)}}{\partial z} + \mu[\frac{\partial^2 w^{(4)}}{\partial r^2} + \frac{1}{r}\frac{\partial w^{(4)}}{\partial r} + \frac{\partial^2 w^{(4)}}{\partial z^2}] = 0,$$

$$\mu[\frac{\partial^2 v^{(4)}}{\partial r^2} + \frac{\partial}{\partial r}(\frac{v^{(4)}}{r}) + \frac{\partial^2 v^{(4)}}{\partial z^2}] = 0. \tag{42}$$

As in all earlier cases, we define the stream function $\psi^{(4)}(r,z)$ through

$$u^{(4)} = -\frac{1}{r}\frac{\partial\psi^{(4)}}{\partial z}, \quad w^{(4)} = \frac{1}{r}\frac{\partial\psi^{(4)}}{\partial r}, \tag{43}$$

and observe that (41) is satisfied identically. The rest of the equations in (42), after eliminating the pressure $p^{(4)}$, may be written as

$$-\frac{\mu}{r}E^4\psi^{(4)} = 0$$

$$\mu\left[\frac{\partial^2 v^{(4)}}{\partial r^2} + \frac{\partial}{\partial r}\left(\frac{v^{(4)}}{r}\right) + \frac{\partial^2 v^{(4)}}{\partial z^2}\right] = 0. \tag{44}$$

The solution of (44) satisfying (14) has the form

$$v^{(4)} = 0, \text{ i.e., } u^{(4)} = v^{(4)} = w^{(4)} = 0. \tag{45}$$

From (45) and (42), we get

$$\begin{aligned} p^{(4)} = {} & 8(2\alpha_1+\alpha_2)\left[\left(\frac{dv^{(1)}}{dr} - \frac{v^{(1)}}{4}\right)\left(\frac{dv^{(3)}}{dr} - \frac{v^{(3)}}{r}\right)\right. \\ & \left. + \frac{dw^{(1)}}{dr}\frac{dw^{(3)}}{dr}\right] + 96(\gamma_3+\gamma_4+\gamma_5\gamma_6/2)\left[\left(\frac{dv^{(1)}}{dr} - \frac{v^{(1)}}{r}\right)^2\right. \\ & \left. + \frac{dw^{(1)}}{dr}\right)^2\Big]^2 \\ & + \int^r \frac{1}{r}\left[16\alpha_1\left\{\left(\frac{dv^{(1)}}{dr} - \frac{v^{(1)}}{r}\right)\left(\frac{dv^{(3)}}{dr} - \frac{v^{(3)}}{r}\right) + \frac{dw^{(1)}}{dr}\frac{dw^{(3)}}{dr}\right\}\right. \\ & \left. + 8\alpha_2\left\{\frac{dw^{(1)}}{dr}\frac{dw^{(3)}}{dr}\right\}\right]dr + \int^r \frac{1}{r}\left[96\gamma_3+\gamma_4+\gamma_5)\left\{\left(\frac{dv^{(1)}}{dr} - \frac{v^{(1)}}{r}\right)^2\right.\right. \\ & \left. + \left(\frac{dw^{(1)}}{dr}\right)^2\right\}^2 + 48\gamma_6\left(\frac{dw^{(1)}}{dr}\right)^2\left\{\left(\frac{dv^{(1)}}{dr} - \frac{v^{(1)}}{r}\right)^2\right. \\ & \left.\left. + \left(\frac{dw^{(1)}}{dr}\right)^2\right\}\right]dr + \int^r \frac{1}{2}\left[8\rho v^{(1)}v^{(3)}\right]dr, \end{aligned} \tag{46}$$

where $v^{(1)}$, $w^{(1)}$, $v^{(3)}$ and $w^{(3)}$ are given by (20), (35) and (36).

On summarizing the results of the perturbation series through order four, we have:

$$u(r,z;\Omega) = 0$$

$$v(r,z;\Omega) = \Omega v^{(1)} + \frac{\Omega^3}{3!}v^{(3)} + 0(\Omega^5)$$

$$w(r,z;\Omega) = \Omega w^{(1)} + \frac{\Omega^3}{3!}w^{(3)} + 0(\Omega^5)$$

$$p(r,z;\Omega) = \Omega(-Gz+c) + \frac{\Omega^2}{2!}p^{(2)} + C_3 \frac{\Omega^3}{3!} + \frac{\Omega^4}{4!}p^{(4)} + 0(\Omega^5) \qquad (47)$$

where the fields $v^{(1)}$, $v^{(3)}$, $w^{(1)}$, $w^{(3)}$, $p^{(2)}$ and $p^{(4)}$ have been explicitly recorded earlier.

4. Torque, Normal Thrust, Volume Flux

In order to obtain the formulae for the torque and the thrust on the cylinders we substitute the power series expansion for the extra stress $\mathbf{S}$ and collect the information given in the earlier pages to obtain

$$\mathbf{T} = \Omega(-\mathbf{I}p^{(1)} + \mathbf{S}_1^{(1)}) + \frac{\Omega^2}{2!}(-\mathbf{I}p^{(2)} + \mathbf{S}_1^{(2)} + \mathbf{S}_2^{(2)}) + \frac{\Omega^3}{3!}(-\mathbf{I}p^{(3)} + \mathbf{S}_1^{(3)} + \mathbf{S}_2^{(3)}) + \mathbf{S}_3^{(3)} + \frac{\Omega^4}{4!}(-\mathbf{I}p^{(4)} + \mathbf{S}_1^{(4)} + \mathbf{S}_2^{(4)} + \mathbf{S}_3^{(4)} + \mathbf{S}_4^{(4)}) + 0(\Omega^5). \qquad (48)$$

(a) Torque

The Torque M per unit length to maintain the relative motion of bounding pipes is given by

$$M = 2\pi r^2 T_{r\theta}. \qquad (49)$$

On finding the value of $T_{r\theta}$ from (48), by computing the various S_i expressions from the known results of earlier sections, and then substituting in (49) we get

$$M = -4\pi\mu\Omega N + \left(\frac{\beta_2+\beta_3}{3}\right)\pi\Omega^3\left[48NB_1B_o - 48NB_o^2\,\frac{a^2b^2\ln(b/a)}{(b^2-a^2)} - 12NB_1^2\left(\frac{b^2+a^2}{a^2b^2}\right) - 32N^3\left(\frac{b^4+a^4a^2b^2}{a^4b^4}\right)\right] + 0(\Omega^5) \qquad (50)$$

(b) Normal Thrust Difference

We observe from the previous analysis that $\mathbf{S}_1^{(2)} = \mathbf{S}_3^{(3)} = \mathbf{S}_1^{(4)}$ $\mathbf{S}_3^{(4)} = 0$ and the normal components of $\mathbf{S}_1^{(1)}$, $\mathbf{S}_1^{(3)}$ and $\mathbf{S}_2^{(3)}$ are also zero. Using all this information and the results from the previous sections we find that the difference $\Delta\mathbf{T}_{rr} = T_{rr}\big|_{r=b} - T_{rr}\big|_{r=a}$ of the normal stresses at the outer and inner pipes is given by

$$\Delta\mathbf{T}_{rr} = -\frac{\Omega^2}{2!}\left[\int_a^b \frac{1}{r}\left\{4\alpha_1\left[\left(\frac{dv^{(1)}}{dr} - \frac{v^{(1)}}{r}\right)^2 + \left(\frac{dw^{(1)}}{dr}\right)^2\right] + 2\alpha_2\left(\frac{dw^{(1)}}{dr}\right) + 2\rho v^{(1)^2}\right\}dr\right]$$

$$- \frac{\Omega^4}{4!}\left[\int_a^b \frac{1}{r}\left\{16\alpha_1\left[\left(\frac{dv^{(1)}}{dr} - \frac{v^{(1)}}{r}\right)\left(\frac{dv^{(3)}}{dr} - \frac{v^{(3)}}{r}\right) + \frac{dw^{(1)}}{dr}\frac{dw^{(3)}}{dr}\right] + 8\alpha_2\,\frac{dw^{(1)}}{dr}\frac{dw^{(3)}}{dr}\right)\right\}dr\right.$$

$$+ \int_a^b \frac{1}{r}\left\{96(\gamma_3+\gamma_4+\gamma_5)\left[\left(\frac{dv^{(1)}}{dr} - \frac{v^{(1)}}{r}\right)^2 + \left(\frac{dw^{(1)}}{dr}\right)^2\right]^2 + 48\gamma_6\left(\frac{dw^{(1)}}{dr}\right)^2\left[\left(\frac{dv^{(1)}}{dr} - \frac{v^{(1)}}{r}\right)^2\right.\right.$$

$$\left.\left.\left. + \left(\frac{dw^{(1)}}{dr}\right)^2\right] + 8\rho v^{(1)}v^{(3)}\right\}dr\right] \qquad (51)$$

where various velocity components have been obtained earlier.

(c) Volume Flux

The volume discharge Q per unit time through a cross-section perpendicular to the pipes is given by

$$Q = 2\pi \int_a^b rw\,dr .$$

On using (47) the above formula may be written as

$$Q = 2\pi\Omega \int_a^b rw^{(1)} dr + \frac{2\pi\Omega^3}{3!} \int_a^b rw^{(3)} dr + O(\Omega^5). \qquad (52)$$

If we substitute $w^{(1)}$, $w^{(3)}$, from first and third order solutions into (52) and perform the necessary integration, we get

$$Q = 2\pi\Omega \left[\frac{G}{16\mu}\left\{(b^4-a^4) - \frac{(b^2-a^2)^2}{\ln b/a}\right\} + U\left\{\frac{b^2}{2} - \frac{b^2-a^2}{4\ln b/a}\right\} \right]$$

$$+ \frac{2\pi\Omega^3}{3!} \frac{(\beta_2+\beta_3)}{\mu} \left[12N^2B_1 \frac{(b^2-a^2)}{b^2a^2}\left[1 - \frac{b^4-a^4}{4b^2a^2\ln b/a}\right] \right.$$

$$+ 6(B_1^3-4N^2B_o)\left[\ln\frac{b}{a} - \frac{(b^2-a^2)^2}{4b^2a^2\ln b/a}\right] - 3B_o^3\left[\frac{b^6-a^6}{3}\right.$$

$$\left. \left. - \frac{(b^4-a^4)(b^2-a^2)}{4\ln b/a}\right] + 18B_o^2B_1\left[\frac{b^4-a^4}{4} - \frac{(b^2-a^2)^2}{4\ln b/a}\right] \right]$$

$$+ O(\Omega^5) \qquad (53)$$

5. Discussion

It is now of some interest to discuss the effect of consideration of the elasticity of the fluid on the torque and the volume flow rate. We note that the first term in (50) is the contribution, to the torque, from the viscous Newtonian part of the equation while remaining terms in this equation are due to consideration of viscoelastic effects. If we consider $N>0$, i.e., the rotation of the inner cylinder to be larger than of the outer one, it then follows that the torque is further lowered in a viscoelastic fluid, in comparison to the linearly viscous fluid case, provided $(\beta_2+\beta_3)>0$. On the other hand, if $(\beta_2+\beta_3)<0$, then we have the reverse effect. Similarly the volume discharge rate in equation (53) consists of two parts and it is found that it is lowered, in comparison to the linearly viscous fluid case, if we assume $(\beta_2+\beta_3)>0$. However, for $(\beta_2+\beta_3)<0$ the volume discharge rate increases in comparison to the viscous Newtonian case. We thus note that, since the combination of $(\beta_2+\beta_3)$ only appears in both the expressions, any one of the experiment, either involving with the determination of volume flow rate or determination of torque, can determine the correct magnitude of $(\beta_2+\beta_3)$ for that kind of material.

In order to consider the effect of coupling between different kind of flows we recall that the constant B_o gives the contribution due the pressure gradient, the constant N due to rotational motion of the pipes and the constant B_1 represents the linearly additive contribution due to B_o and the relative translational velocities of the pipes. The constants B_o and B_1 thus represent the axial motion while N represents the angular motion. From equation (53) we find that presence of the terms involving N contributes to the reduction of the flow rate Q provided $(\beta_2+\beta_3)>0$. For such fluids, that is, for which $(\beta_2+\beta_3)>0$, the rotation of the pipe, therefore, reduces the axial flow rate. Similarly from equation (50) we note that the presence of B_o and B_1 terms in the torque M, is to reduce further the value of M provided $(\beta_2+\beta_3)>0$. In both of the above cases the results are, however, reversed for $(\beta_2+\beta_3)<0$. As a final remark we point out that several important flows are a special case of the present problem. For example, when we set in the above solutions U=0 we have spiral Poiseuille flow, when we set $B_o=0$ we have spiral Couette flow, when se set $U=B_o=0$ we have circular Couette flow, when we set N=0, U=0, we have rectilinear Poiseuille flow and when we set N=0, $B_o=0$, we have Pochettino flow.

Acknowledgement

The work reported in this paper has been supported by Grant No. A7728 of N.S.E.R.C. of Canada, and by the Ministry of Colleges & Universities, Toronto, Ontario, Canada.

Nomenclature

a	radius of the inner pipe
b	radius of the outer pipe
(e_r, e_θ, e_z)	unit vectors in (r, θ, z) direction
p	indeterminate pressure
$\mathbf{A}_1, \mathbf{A}_2 .. \mathbf{A}_n$	Rivlin-Ericksen tensors of different orders
B_o	$\frac{G}{2\mu}$
C	arbitrary constant
G	pressure gradient
M	Torque permit length
N	$(1-\lambda)a^2b^2/(b^2-a^2$, a constant
Q	volume flux
$\mathbf{S}_1, \mathbf{S}_2, \mathbf{S}_3, \mathbf{S}_4$	extra stress tensors of first, second, third and fourth order
$\mathbf{T}$	stress tensor
$\bar{U} = U\Omega$	translational velocity of outer pipe relative to inner pipe
$\mathbf{V} = (u, v, w)$	velocity vector
μ	viscosity coefficient
α, α_2	material constants occurring in the stress components of second order approximation
$\beta_1, \beta_2, \beta_3$	material constants occurring in the stress components of third order approximation
$\lambda_1, \lambda_2 ... \lambda_8$	material constants occurring in the fourth order approximation
ψ	stream-function
Ω	angular velocity of the inner pipe
$\lambda\Omega$	angular velocity of the outer pipe

References

[1] Coleman, B.D. & Noll, W. J. Appl. Physics 30, 1508 (1959).

[2] Dierckes, A.C. and Schowalter, W.R. I & E.C. Fundamentals 5 263 (1966).

[3] Fosdick, R.L. & Kao, B.G. Rheol. Acta. 19 675 (1980).

[4] Fredrickson, A.G. Chem. Engng. Sci. 11 252 (1960).

[5] Joseph, D.D. & Fosdick, R.L. Arch. Ratl. Mech. Anal. 49 321 (1973).

[6] Rajagopal, K.R., Kaloni, P.N. & Tao, L. J. Math. Physical Sciences. (In press), 1989.

[7] Rivlin, R.S. J. Ratl. Mech. Anal. 5 179 (1956).

[8] Rivlin, R.S. & Ericksen, J.L. J. Ratl. Mech. Anal. 4 329 (1955).

[9] Savins, J.G. & Wallick, G.C. A.I.Ch.E.J. 12 357 (1966).

[10] Saut, J.C. & Joseph, D.D. Arch. Ration Mech. Anal. 81 53 (1981).

[11] Truesdell, C. & Noll, W. Non-linear Field Theories of Mechanics. Flugge's Handbuch der Physik III/3, Springer, Berlin 1965.

A.SIGINER

On the parallel and orthogonal superposition of pure and oscillatory shear flows of simple fluids

I. INTRODUCTION

Superposed oscillatory and steady shear flows of viscoelastic liquids are important in rheology and in a host of other fields and applications. This type of flow has always been an acid test to pass for anybody's favorite constitutive structure. A great many popular differential and single integral type constitutive equations fail to make even qualitative predictions in such flows, in some cases pointing at the opposite trend indicated by the experiments. This article is devoted to some nearly viscometric flows of this type, those driven simultaneously by a quasi-periodic pressure gradient oscillating about a non-zero mean and by a quasi-periodic forcing from the boundary made up of longitudinal and transversal boundary oscillations. A review of past work is made and new closed form solutions for mass flow enhancement effects due to the nonlinear dependance of the stress field on the deformations is given based on hierarchy equations. Of interest in either phenomena is the possibility of increasing the mass transport with the same power input as for the steady case or in any case increasing the discharge with less power input than would be required for the enhanced discharge under steady conditions.

A case is made for the use of hierarchy equations which are not very popular, in particular with experimental rheologists because of the rather large number of parameters embedded in the constitutive structure if fluids of order larger than two is considered. In view of the failure of most differential and single integral type structures to make quantitative and in many cases even qualitative predictions I argue for the promise of the generality and possible superiority of hierarchy equations for viscoelastic fluids with memory. I take the view that it is better to look for an equation as universal as possible, valid for as many motions of a limited class of fluids than to search for an equation which applies to a restricted class of motions of a large class of fluids.

The problem of the superposed oscillatory and pure shear flow is solved via a regular perturbation method for a fluid characterized by a multiple integral representation with strain history kernels. An algorithm based on established methods of rheometry and closed form solutions, at the lowest significant order, of the nearly viscometric motions considered in this paper is proposed. The procedure although long and tedious has the potential of determining the constitutive parameters of the fluid of order three.

II. REVIEW OF PAST WORK

II.1 Pulsating Gradient Driven Flow

This problem has been the subject of rather extensive investigations over the last two decades. The zero mean gradient flow has been studied in the context of different but related problems by Pipkin [34], Etter & Schowalter [14] and more recently by Randria & Bellet [36] who used the Rivlin-Ericksen fluid of complexity n and a three and four constant Oldroyd model respectively. Walters & Townsend [46], Barnes & al. [1,2], Townsend [45] explored the effects of pressure gradient oscillations around a non-zero mean in circular rigid pipes followed by the work of Edwards & al. [13], Davies & al. [12], Böhme & Nonn [6], Manero & Walters [27], Sundstrom & Kaufman [43] and Phan-Thien [29,30,31].

The emphasis on all these investigations is on differential models with the exception of Phan-Thien [31]. Available experimental data [2,43] cannot be predicted quantitatively by the three or four constant Oldroyd, Goddard-Miller, fourth order fluid, Maxwell, co-rotational and nonaffine network models and Wagner model used by the authors mentioned above. Experimental data for polyacrylamide solutions of different concentrations [2], with the amplitude of the fluctuation smaller than the mean gradient and thus pertaining to oscillatory shear superposed on a dominant Poiseuille flow, shows that the enhancement, defined as the ratio of the pressure gradient oscillation induced change in mass transport to the steady discharge driven by the mean gradient, increases with frequency for the limited range of frequencies for which data is available. Enhancement also shows an increase with the amplitude of the sinusoidal oscillation at fixed frequency and mean gradient and an increase and a decrease respectively in the range of relatively small and relatively high pressure gradients at fixed amplitude and frequency displaying a resonance effect, a sharp increase followed by a decrease, for a window in the range of intermediate pressure gradients. Sundstrom & Kaufman's [43] data with weakly viscoelastic polymeric solutions pertains to superposed equally dominant oscillatory and pure shear and is at best inconclusive. They claim however that the enhancement effects decrease with increasing frequency at fixed mean pressure gradient and amplitude based on analytical results with the inelastic Ellis model the use of which could be only partially justifiable if the period of the oscillation is much greater than the natural response time of the fluid. On the other hand Edwards et al. [13] predict that the enhancement is a constant value independent of the mean pressure gradient and frequency for the limited range of small frequencies for which experimental data is available. Phan-Thien analyzed the problem for sinusoidal as well as more general pressure gradient noises with constant mean values in a statistical sense using a generalized Maxwell model with strain-rate dependent memory kernel [30] and in a follow-up work using the nonaffine network model of Johnson-Siegelman and the Wagner model [31]. Frequency dependence of the enhancement data is not even qualitatively predicted by any of the constitutive structures mentioned above with the possible exception of the generalized Maxwell model.

II. 2 Interaction of Boundary Vibration and Pure Shear.

The effect of the presence of higher harmonics in oscillatory flows of rheologically complex fluids has been investigated by Goldstein & Schowalter [15] with a view towards using the results of the nonlinear analysis as a rheometrical tool. Although the presence of higher harmonics in oscillatory testing of non-Newtonian fluids had been already demonstrated by then, they are the first to predict possible flow enhancement effects together with Townsend [45]. The Bird-Carreau model used by Goldstein & Schowalter has serious shortcomings. It does not reduce to the equations of linear viscoelasticity for small strains in oscillatory shear and does not predict a transition from linear to nonlinear viscoelasticity at a finite strain no matter how high the strain rate is. Townsend [45] in his numerical, finite difference study of the Poiseuille flow of a four constant Oldroyd model in a round, rigid and straight tube predicted that oscillatory shear imposed from the boundary through the longitudinal oscillations of the pipe wall would yield large increases in the flow rate. Later Manero and co-workers [25,26] experimentally investigated the effect of longitudinal oscillations of the pipe wall on the Poiseuille flow in straight tubes. They determined that the change in the volumetric flow rate depends on the frequency and amplitude of the oscillation and on the constant pressure gradient in their experiments with aqueous solutions of polyacrylamide at various concentrations. This change in mass transport is an increasing function of both the frequency and amplitude at a fixed value of the pressure gradient and is a monotonically decreasing function of the pressure gradient at fixed frequency and amplitude. Maximum enhancement occurs at very small values of the pressure gradient with a magnitude up to ten times larger than the discharge corresponding to the constant pressure gradient driven flow. Their experiments correspond to the parallel superposition of a dominant, primary oscillatory shear onto a secondary pure shear. Earlier work for the superposed, parallel steady and oscillatory shear flows of fluids described by a generalized Rouse model was conducted experimentally by Booij [7]. Zahorski [48], Jones & Walters [17] and Bernstein [4] investigated the behaviour of simple fluids with superimposed proportional stretch histories, fluids of the integral type and K-BKZ fluids respectively. The last author in collaboration with Fosdick [5] also derived relationships between the storage modulus and the first normal stress difference for parallel shearing of the K-BKZ fluids. Simmons [42] is probably the first researcher to consider transversal shearing followed by Tanner & Williams [44] who investigated theoretically and experimentally the orthogonal shearing of K-BKZ fluids in the small gap between concentric cylinders with the inner and outer cylinders axially oscillating and steadily rotating respectively with a view towards showing that the K-BKZ constitutive equation is the best suited to describe nearly viscometric flows with rather inconclusive results. Their experiments fall into the category of primary steady shear perturbed by orthogonally superposed small strain oscillatory shear.

In an effort to make the data consistent with a theoretical framework Kazakia & Rivlin [21,22] investigated the effect of sinusoidal boundary vibration, either longitudinal or transversal, on the Poiseuille flow, between parallel plates and in straight round tubes, of a slightly non-Newtonian liquid in the sense of Langlois & Rivlin [23] and also its effect on the steady shear of the Rivlin-Ericksen fluids. Parallel plates are simultaneously and synchronously vibrated in their own plane in a direction either parallel or orthogonal to the constant pressure gradient driven flow and pipe wall is subjected to longitudinal and rotational vibrations. They show that inertia is the main driving mechanism and that the fluid has to be shear thinning for an increase in discharge to occur with both fluids although their analysis cannot account for the magnitude of the increase in volumetric flow rate. This increase in mass transport is proportional to the square of the frequency of the boundary oscillation for small frequencies and to the square root of the frequency for large frequencies in the case of the liquid slightly non-Newtonian in the sense defined in [23]. Their analysis together with that of Böhme & Nonn [6] who used the fluid of grade four in their investigation is open to the criticism leveled by recent stability studies which shed serious doubt on the modelling of time dependent flows with constitutive structures represented either by fluids of complexity n or of grade n, Renardy [37], Joseph [18]. Fluids of grade n arise as asymptotic expansions of the stress response functional of the simple fluid of the memory type with or without retarding the history of the strain $\underset{\sim}{G}(\underset{\sim}{X},\tau)$ of the particle $\underset{\sim}{X}$ defined in terms of the relative right

Cauchy-Green strain tensor $\underset{\sim}{C}_t(\underset{\sim}{X},\tau)$

$$\underset{\sim}{G}(t-s) = \underset{\sim}{C}_t(t-s)-\underset{\sim}{1} \quad , \quad s=t-\tau \quad , \quad \tau\leq t \tag{1}$$

If the history is retarded, i.e. the same motion is allowed to proceed at a much slower pace, by introducing a stretched time scale ϵs, $\epsilon < 1$ the retardation theorem of Coleman & Noll [8] allows the expansion of the history into a series pivoted about s = 0 assuming that the history is analytic or at least N times differentiable

$$\underset{\sim}{G}(t-\epsilon s) = \sum_{n=0}^{N} \frac{s^n}{n!} \frac{\partial^n}{\partial s^n} \underset{\sim}{G}(t-\epsilon s)|_{s=0} + o(\epsilon^N)$$

$$\underset{\sim}{G}(t-\epsilon s) = \sum_{n=0}^{N} \frac{s^n}{n!} (-\epsilon)^n \underset{\sim}{A}_n(t) + o(\epsilon^N) \tag{2}$$

where the nth order Rivlin-Ericksen tensor $\underset{\sim}{A}_n(t) = \underset{\sim}{A}_n\ [\underset{\sim}{u}(\underset{\sim}{x},t)]$ is defined as

$$\underset{\sim}{A}_n(t) = \frac{\partial^n}{\partial s^n} \underset{\sim}{C}_t(t-s)|_{s=0} \tag{3}$$

These considerations lead to an asymptotic expansion of the stress response functional

$$\underset{\sim}{\theta}\overset{\infty}{\underset{s=0}{[\underset{\sim}{G}(t-\epsilon s)]}} = \sum_{n=1}^{\infty} \underset{\sim n}{f} [\underset{\sim 1}{A},\ldots,\underset{\sim n}{A}] \qquad (4)$$

with $\underset{\sim n}{f}$ linear in $\underset{\sim n}{A}(t)$. Retarded motions may be considered as slowly varying slow motions, i.e. for which partial time derivatives are small. Rivlin [38] pursuing this entirely different approach to slow flows derived the same asymptotic expansion to the response functional by reducing the displacements and accelerations of a particle within a given time by postulating

$$\underset{\sim}{U}(\underset{\sim}{x},t) = \epsilon\, \underset{\sim}{U}^{*}(\underset{\sim}{x},t) \quad , \quad \dot{\underset{\sim}{U}}(\underset{\sim}{x},t) = \epsilon^{2}\, \dot{\underset{\sim}{U}}^{*}(\underset{\sim}{x},t) \; , \; \ldots$$

On the other hand Rivlin-Ericksen fluids of complexity n arise in the expansion of the response functional of the simple fluid when it is assumed that only values of $\underset{\sim t}{C}(s)$ very close to s=0 have any significance thus leading for incompressible simple fluids to representations of the type

$$\underset{\sim}{T} = -p\underset{\sim}{1} + \underset{\sim}{f}(\underset{\sim 1}{A},\ldots,\underset{\sim n}{A}) \quad , \quad n = 2,\ldots,\infty$$

with $\underset{\sim}{T}$ and p the total stress and the mechanical pressure respectively.

Although fluids of complexity n and of grade n arise from different considerations and they lead to different results for unsteady motions they coincide for steady flows. In particular rapidly varying small disturbances cannot be modelled by these constitutive structures by virtue of the inherent limitations of the constitutive assumptions. But they do have the same linearized forms for unsteady motions and the results of the stability studies apply to both. It is shown in [18,37] that these fluids are inadmissible either as exact or approximate models for the class of unsteady motions because they are unstable in the rest state to small disturbances in the sense of the spectral problem of the linearized theory when the effect of these disturbances is described by a linearized problem on the rest state. There is also computational evidence to suggest these theoretical results. If deformation rates, i.e. velocity gradients are unsteady retarded motion expansions may diverge, Larson [24]. A stagnant fluid which is prone to spontaneous motion without the application of any forcing either from the boundary or in the form of body forces cannot be expected to lead to physically meaningful results in the study of unsteady flows.

The analysis of [21,22] was extended by Phan-Thien [32] to the case of random longitudinal vibration of the boundary. For the results of these investigations to be meaningful discounting the stability considerations mentioned above the primary shear flow must be sufficiently slow for the slow flow approximation to be valid and the strains and strain rates driven by the oscillatory forcing from the boundary must remain small thereby restricting the boundary oscillation to small amplitudes and frequencies. Phan-Thien [33] further investigated the predictions of the generalized Newtonian, power law and single integral models with either strain or strain rate history type kernels. He concludes that essential features of the phenomena can be qualitatively described by these models and comments that with a proper choice of the viscosity function it may be possible to quantitatively predict the flow enhancement using the generalized Newtonian or equivalently the power law model. But the power law model points to an infinite enhancement in the limit of infinitesimal pressure gradients and it cannot explain resonance effects observed for certain values of the parameters.

III. CONSTITUTIVE STRUCTURE

The flows studied in this article fall into the category of nearly viscometric flows. Exact, universal equations for the description of viscometric flows are known, for instance the CEF (Criminale, Ericksen, Filbey) and K-BKZ (Kaye, Berstein, Kearsley, Zapas) models. Flows with large shear rate variations occur often in practice. But except for viscometric flows universal representations for stress-strain relations to characterize a given class of fluids are not known and may be unknowable. The only way we know how to overcome this towering difficulty is by considering either a restricted class of motions or a restricted class of fluids. I take the view that it is better to look for equations capable to describe as large a class of motions as possible for a limited class of fluids. It is by no means certain that universality in the sense of the Navier-Stokes equations can ever be attained. But it is important to realize that searching for an equation which applies to a restricted class of motions or a more general one characterizing a limited class of fluids are not mutually exclusive and the interaction would lead ultimately to canonical equations as universal as possible. For instance one may look for small strain perturbations of rigid body rotations or of viscometric flows. Conceptually a framework for a universal equation has been laid out by Noll [28] and his simple fluid theory has received wide acceptance. The response functional of an incompressible simple fluid may be expressed as

$$\underset{\sim}{S} = \underset{\sim}{\theta}\Big[\overset{\infty}{\underset{s=0}{}} \underset{\sim}{G}(\underset{\sim}{X},s)\Big] = \underset{\sim}{T} + p\underset{\sim}{1} \quad ; \quad s=t-\tau \tag{5}$$

where $\underset{\sim}{T}$ and $\underset{\sim}{p}$ represent the total stress and the isotropic stress respectively. Green & Rivlin [16] developed a uniform approximation to the response functional of a simple fluid based on the Stone-Weierstrass theorem.

$$\underset{\sim}{\theta}\Big[\overset{\infty}{\underset{s=0}{}} \underset{\sim}{G}(\underset{\sim}{X},s)\Big] = \int_0^\infty \underset{\sim}{K}_1(s)\underset{\sim}{G}(s)ds + \int_0^\infty\int_0^\infty \underset{\sim}{K}_2(s_1,s_2)\underset{\sim}{G}(s_1)\underset{\sim}{G}(s_2)ds_1ds_2$$

$$+ \int_0^\infty\int_0^\infty\int_0^\infty \underset{\sim}{K}_3(s_1,s_2,s_3)\underset{\sim}{G}(s_1)\underset{\sim}{G}(s_2)\underset{\sim}{G}(s_3)ds_1ds_2ds_3 \quad + \ldots \tag{6}$$

The integrands are tensor polynomials with even order kernels $\underset{\sim}{K}_i$ which ultimately define the material functions characterizing the fluid. Such an approximation is valid in a suitable function space and when the response functional is continuous with respect to a continuity measure appropriate to that space. For instance, Coleman and Noll [8,9] use a Hilbert space with a rapidly decaying weighted fading memory norm. Fading memory implies that the stress at the present time is determined to a large extent by the recent deformation history and is only weakly dependent on strains which occurred in the more distant past. Theories of fading memory may be framed on both normable and non-normable spaces with different topologies in which equation (6) has to live. The assumed topology specifies how the stress is more dependent on the recent history than on strains removed from the recent past. The domain of the response functional $\underset{\sim}{\theta}$ is defined by the particular topology assumed, that is the class of admissible deformation histories is restricted by the topology. The response functional $\underset{\sim}{\theta}$ is continuous in the assumed topology and consequently is only weakly dependent on strains experienced by the material in the not so recent past. The range of the functional $\underset{\sim}{\theta}$ is the collection of stresses generated by the domain of $\underset{\sim}{\theta}$ with the assumed topology. To obtain mathematically manageable forms of (6) $\underset{\sim}{\theta}$ is linearized by assuming functional differentiability at some deformation history $\underset{\sim}{G}_o$.

$$\underset{\sim}{\theta}[\underset{s=0}{\overset{\infty}{}}\underset{\sim}{G}(\underset{\sim}{X},s)] = \underset{\sim}{\theta}[\underset{\sim}{G}_o] + \delta\underset{\sim}{\theta}[\underset{\sim}{G}_o|\underset{\sim}{G}_{oo}] + \delta^2\underset{\sim}{\theta}[\underset{\sim}{G}_o|\underset{\sim}{G}_{oo}, \underset{\sim}{G}_{oo}] + O(|\underset{\sim}{G}_{oo}|^3)$$

$$\underset{\sim}{G}(\underset{\sim}{X},s) = \underset{\sim}{G}_o(\underset{\sim}{X},s) + \underset{\sim}{G}_{oo}(\underset{\sim}{X},s) \tag{7}$$

$\delta\underset{\sim}{\theta}$, $\delta^2\underset{\sim}{\theta}$ represent functional derivatives at $\underset{\sim}{G}_o$ linear and bilinear in $\underset{\sim}{G}_{oo}$ respectively. Theories of fading memory were formulated along the lines described above, by Wang [47], Coleman and Mizel [10] and more recently by Saut & Joseph [39]. Coleman & Noll frame equation (6) in a Hilbert space with a weighted Hilbert norm. The domain of $\underset{\sim}{\theta}$ in their formulation admits a large class of deformation histories with possibly non-smooth elements. In Wang's and Saut & Joseph's treatments spaces with different topologies are introduced to further restrict the domain space of $\underset{\sim}{\theta}$. For instance in Wang's theory of fading memory equation (6) makes sense in a non-normable Fréchet space with the elements of the dual, defined as the collection of all the possible linearized forms of $\underset{\sim}{\theta}$ under the assumed topology, given by Gateaux derivatives.

Restricting the domain of $\underset{\sim}{\theta}$ results in enlarging the dual. Important practical consequences stem from these topological considerations although it may be argued that real fluids know nothing of spaces and associated topologies. But any constitutive equation makes sense in a space with properties defined by the topology imposed on it and the dynamics of any motion is predicted in the context of the space in which the constitutive equation is defined and has to live. Hopefully the predicted dynamics will agree with experimental findings. In this context Saut & Joseph [39] show that the solution of the dynamical equations based on Coleman & Noll theory of fading memory may allow shocks whereas their theory smooths discontinuous solutions.

If $\underset{\sim}{\theta}$ is Fréchet differentiable at $\underset{\sim}{G}_o$ functional derivatives in (7) may be assumed to be in integral form. Although Riesz theorem justifies the representation of the first Fréchet derivative as a single integral with the integrand linear in $\underset{\sim}{G}_{oo}$ there are no representation theorems for higher orders and the representation of the second and third order Fréchet derivatives as double and triple integrals bilinear and trilinear respectively in $\underset{\sim}{G}_{oo}$ is merely a partially justifiable constitutive hypothesis. Canonical forms of the stress for small strain perturbations of steady, rigid body rotations were given by Joseph [19] up to and including second order. Small amplitude perturbations of steady viscometric flows were investigated by Pipkin & Owen [35] who gave canonical forms for the first Fréchet stress represented by the first functional derivative $\delta\underset{\sim}{\theta}$ in (7) and consistency relationships between elements of $\delta\underset{\sim}{\theta}$ and the viscometric functions. They determine that thirteen elements of $\delta\underset{\sim}{\theta}$ are non zero due to isotropy, symmetry and incompressibility considerations. The later work of Zahorski [48,49] is particularly relevant and important. He investigated complex flows with superposed proportional stretch histories. Nearly viscometric flows are a special case of this much larger class. He derives canonical forms for $\delta\underset{\sim}{\theta}$ with the same number of constitutive functions as in [35] and considers both parallel and orthogonal shearing oscillations. Two observations are in order. Application of $\delta\underset{\sim}{\theta}$ is limited only to very small deviations from viscometric behaviour and canonical forms for $\delta^2\underset{\sim}{\theta}$ may be needed if appreciable deviations from viscometric behaviour need to be described taking into account the nonlinear interaction of stress perturbation and the history of the deformation perturbations. Secondly knowledge of the kernel functions in $\delta\underset{\sim}{\theta}$ and $\delta^2\underset{\sim}{\theta}$ is essential. Rheologists, experimental and theoretical alike, shy away from the application of equations like (7) because of the rather large number of constitutive functions involved. I take the point of view that constitutive equations should not be too specialized. An arbitrary perturbation of a viscometric flow can be modelled in a general sense only by (7). In the absence of canonical expressions for $\delta^2\underset{\sim}{\theta}$ when $\underset{\sim}{G}_o$ is a viscometric flow we take $\underset{\sim}{G}_o = \underset{\sim}{0}$ as the rest state and use an extension of an algorithm developed by Joseph [20] to perturb the base state by imposing a small mean pressure gradient and oscillations of small amplitude around this mean gradient together with small amplitude boundary oscillations both in the longitudinal and transversal directions on a fluid of the multiple integral type. Under the requirement of isotropy the fluid of integral type of order three is obtained from (6)

$$\underset{\sim}{\theta}_3[\underset{\sim}{G}(\underset{\sim}{X},s)] = \sum_1^3 \underset{\sim}{S}_n \tag{8}$$

$$\underset{\sim}{S}_1 = \int_0^\infty \zeta(s)\underset{\sim}{G}(s)ds$$

$$\underset{\sim}{S}_2 = \sum_1^2 \underset{\sim}{S}_{2i} \quad ; \quad \underset{\sim}{S}_3 = \sum_1^4 \underset{\sim}{S}_{3i}$$

$$\underset{\sim}{S}_{21} = \int_0^\infty \int_0^\infty \beta_{21}(s_1,s_2)\underset{\sim}{G}(s_1)\underset{\sim}{G}(s_2)ds_1 ds_2$$

$$\underset{\sim}{S}_{22} = \int_0^\infty \int_0^\infty \beta_{22}(s_1,s_2)[tr\underset{\sim}{G}(s_1)]\underset{\sim}{G}(s_2)ds_1 ds_2$$

$$\underset{\sim}{S}_{31} = \int_0^\infty \int_0^\infty \int_0^\infty \beta_{31}(s_1,s_2,s_3)\underset{\sim}{G}(s_1)\underset{\sim}{G}(s_2)\underset{\sim}{G}(s_3)ds_1 ds_2 ds_3$$

$$\underset{\sim}{S}_{32} = \int_0^\infty \int_0^\infty \int_0^\infty \beta_{32}(s_1,s_2,s_3)[tr\underset{\sim}{G}(s_1)]\underset{\sim}{G}(s_2)\underset{\sim}{G}(s_3)ds_1 ds_2 ds_3$$

$$\underset{\sim}{S}_{33} = \int_0^\infty \int_0^\infty \int_0^\infty \beta_{33}(s_1,s_2,s_3)[tr\underset{\sim}{G}(s_1)][tr\underset{\sim}{G}(s_2)]\underset{\sim}{G}(s_3)ds_1 ds_2 ds_3$$

$$\underset{\sim}{S}_{34} = \int_0^\infty \int_0^\infty \int_0^\infty \beta_{34}(s_1,s_2,s_3)tr[\underset{\sim}{G}(s_1)\underset{\sim}{G}(s_2)]\underset{\sim}{G}(s_3)ds_1 ds_2 ds_3$$

the history of the strain perturbation $\underset{\sim}{G}_{oo}$ in [7] is expanded in a series

$$\underset{\sim}{G}_{oo} = \epsilon^n \underset{\sim}{G}_n \quad , \quad \epsilon < 1 \quad , \quad tr\underset{\sim}{G}_{oo} = \epsilon^2 tr\underset{\sim}{G}_2 + O(\epsilon^3)$$

and (8) is rewritten as

$$\underset{\sim}{\theta}_3[\underset{\sim}{G}(\underset{\sim}{X},s)] = \sum_1^3 \epsilon^n \underset{\sim}{S}^{(n)} = \delta\underset{\sim}{\theta} + \delta^2\underset{\sim}{\theta} + \delta^3\underset{\sim}{\theta} \tag{9}$$

where the Fréchet derivatives at various orders can be expressed in terms of the first Rivlin-Ericksen tensor and new kernel functions obtained through partial differentiation

$$\underset{\sim}{S}^{(1)} = \int_0^\infty G(s)\ \underset{\sim}{A}_1{}^{(1)}\ (t-s)ds \tag{10}$$

$$\underset{\sim}{S}^{(2)} = \int_0^\infty G(s)\ \underset{\sim}{A}^{(1)}(t-s_1)\ \underset{\sim}{A}_1{}^{(2)}\ (t-s)\ ds + \int_0^\infty G(s)\ \underset{\sim}{L}_1(t-s)\ ds$$

$$+\int_0^\infty\int_0^\infty \gamma(s_1,s_2)\, \underset{\sim}{A}_1^{(1)}(t-s_2)\, ds_1 ds_2 \tag{11}$$

$$\underset{\sim}{S}^{(3)} = \int_0^\infty G(s)\underset{\sim}{A}_1^{(3)}ds + \int_0^\infty G(s)[\underset{\sim}{L}_2 + 1/2\underset{\sim}{L}_3 + \underset{\sim}{L}_4]ds$$

$$+\int_0^\infty\int_0^\infty \gamma(s_1,s_2)\, [\underset{\sim}{A}_1^{(1)}(s_1)\, \underset{\sim}{A}_1^{(2)}(s_2) + \underset{\sim}{A}_1^{(2)}(s_1)\, \underset{\sim}{A}_1^{(1)}(s_2)]\, ds_1 ds_2$$

$$+\int_0^\infty\int_0^\infty \gamma(s_1,s_2)[\underset{\sim}{A}_1^{(1)}(s_1)\, \underset{\sim}{L}_1(s_2) + \underset{\sim}{L}_2(s_1)\, \underset{\sim}{A}_1^{(1)}(s_2)]\, ds_1 ds_2$$

$$+\int_0^\infty\int_0^\infty 2\alpha(s_1,s_2)\, \underset{\sim}{U},_j^{(1)}(s_1)\, \underset{\sim}{\xi}^*,_j(s_1)\, \underset{\sim}{A}_1^{(1)}(s_2)\, ds_1 ds_2$$

$$+\int_0^\infty\int_0^\infty\int_0^\infty \sigma_1(s_1,s_2,s_3)\, \underset{\sim}{A}_1^{(1)}(s_1)\, \underset{\sim}{A}_1^{(1)}(s_2)\, \underset{\sim}{A}_1^{(1)}(s_3)\, ds_1 ds_2 ds_3$$

$$+\int_0^\infty\int_0^\infty\int_0^\infty \sigma_4(s_1,s_2,s_3))\mathrm{tr}[\underset{\sim}{A}_1^{(1)}(s_1)\, \underset{\sim}{A}_1^{(1)}(s_2)]\underset{\sim}{A}_1^{(1)}(s_3)\, ds_1 ds_2 ds_3 \tag{12}$$

with the following definitions

$$\underset{\sim}{\xi}^* = \int_t^\tau \underset{\sim}{U}^{(1)}(\underset{\sim}{X},\tau')d\tau' \quad , \quad t > \tau$$

$$\underset{\sim}{\xi}^{**} = \int_t^\tau \underset{\sim}{U}^{(2)}(\underset{\sim}{X},\tau')d\tau' \quad , \quad \underset{\sim}{L}_\xi = (\underset{\sim}{\xi}^{**} - \frac{d\xi}{d\epsilon} \bullet \nabla\underset{\sim}{\xi}^* - \frac{1}{2}\,\underset{\sim}{\xi}^*)\bullet\nabla\underset{\sim}{\xi}^*$$

$$\underset{\sim}{L}_j = \underset{\sim}{\xi}^*\bullet\nabla\underset{\sim}{A}_1^{(j)} + \underset{\sim}{A}_1^{(j)}\bullet\nabla\underset{\sim}{\xi}^* + (\underset{\sim}{A}_1^{(j)}\bullet\nabla\underset{\sim}{\xi}^*)^T \quad ; \quad j=1,2$$

$$\underset{\sim}{L}_3 = \underset{\sim}{\xi}^*\bullet\nabla\underset{\sim}{L}_1 + \underset{\sim}{L}_1\bullet\nabla\underset{\sim}{\xi}^* + (\underset{\sim}{L}_1\bullet\nabla\underset{\sim}{\xi}^*)^T$$

$$\underset{\sim}{L}_4 = \underset{\sim}{L}_\xi\bullet\nabla\underset{\sim}{A}_1^{(1)} + \underset{\sim}{A}_1^{(1)}\bullet\nabla\underset{\sim}{L}_\xi + (\underset{\sim}{A}_1^{(1)}\bullet\nabla\underset{\sim}{L}_\xi)^T$$

The notation $n!(.)^{(n)}$ refers to the nth order partial derivative with respect to ϵ evaluated at $\epsilon = 0$ and

$$\underset{\sim}{A}_1^{(n)}(s) = \underset{\sim}{A}_1[\underset{\sim}{U}^{(n)}(\underset{\sim}{X},t-s)] \tag{13}$$

We expand the velocity and pressure fields in power series

$$\underset{\sim}{U}(\underset{\sim}{X},t;\epsilon) = \epsilon^n \underset{\sim}{U}^{(n)}(\underset{\sim}{X},t) \quad , \quad \Phi(\underset{\sim}{X},t;\epsilon) = \epsilon^n \Phi^{(n)}(\underset{\sim}{X},t) \tag{14}$$

The algorithm is good for small pressure gradients and small amplitudes. Although they are small, they are not necessarily of the same order of magnitude and the algorithm places no restrictions on the frequency thereby allowing the simulation of rapidly varying oscillatory flow superposed on a dominant steady Poiseuille flow and vice-versa.

IV. MATHEMATICAL FORMULATION

Flow between parallel plates is considered

$$\rho \frac{D\underset{\sim}{U}}{Dt} = - \nabla\Phi + \nabla\bullet\underset{\sim}{\theta} \quad , \quad \Phi = p+\rho g y \quad , \quad \nabla\bullet\underset{\sim}{U} = 0 \tag{15}$$

$$p,_x = -\epsilon(P+\lambda_k \sin\omega_k t) \quad , \quad (P,\lambda_k) > 0 \quad , \quad \epsilon < 1 \tag{16}$$

$$\underset{\sim}{U}(x,\pm d,z,t) = \epsilon(\underset{\sim}{e}_x\lambda_{nx}\sin\omega_{nx}t + \underset{\sim}{e}_z\lambda_{rz}\sin\omega_{rz}t) \tag{17}$$

$$k = 1,\ldots, K \; ; \; n = 1,\ldots, N \; ; \; r = 1,\ldots, R$$

IV. 1. First Order Solution

The application of (9,13,14) to (15,16,17) yields

$$\rho\underset{\sim}{U},_t^{(1)} = - \nabla\Phi^{(1)} + \nabla\bullet\underset{\sim}{S}^{(1)} \quad , \quad \nabla\bullet\underset{\sim}{U}^{(1)} = 0 \tag{18}$$

$$p,_x^{(1)} = - (P + \lambda_k \sin\omega_k t) \tag{19}$$

$$\underset{\sim}{U}^{(1)}(x,\pm d,z,t) = \underset{\sim}{e}_x \lambda_{nx}\sin\omega_{nx}t + \underset{\sim}{e}_z \lambda_{rz}\sin\omega_{rz}t \tag{20}$$

$\underset{\sim}{S}^{(1)}$ is the linear viscoelasticity stress-strain relationship. The first order velocity field is of the following form

$$\underset{\sim}{U}^{(1)} = U^{(1)} \underset{\sim}{e}_x + W^{(1)}\underset{\sim}{e}_z \tag{21}$$

and the linearly viscoelastic solution is recovered

$$U^{(1)}(y,t) = U_{1m} + 2 \,\mathrm{Re}\sum_k^K A_k \, e^{i\omega_k t} + 2 \,\mathrm{Re}\sum_j^N U_j \, e^{i\omega_{jx} t} \tag{22}$$

$$W^{(1)}(y,t) = 2 \,\mathrm{Re}\sum_j^R W_j e^{i\omega_{jz} t} \tag{23}$$

$$A_k(y) = \frac{\lambda_k}{2\rho\omega_k} \left(\frac{\mathrm{Ch}\Lambda_k y}{\mathrm{Ch}\Lambda_k d} - 1\right) \quad , \quad U_{1m}(y) = \frac{P}{2\mu} (d^2-y^2) \tag{24}$$

$$a_j(y) = \frac{\lambda_{jk}}{2i} \frac{\mathrm{Ch}\Lambda_{jk} y}{\mathrm{Ch}\Lambda_{jk} d} \quad , \quad k = x,z; \; a= U,W \tag{25}$$

$$\Lambda^2 = \frac{i\rho\omega}{\int_0^\infty G(s)e^{-i\omega s}ds} = \frac{i\rho\omega}{\eta^*} \qquad (26)$$

It is well known that for th linear viscoelastic model to be stable in the linearized sense the near relaxation modulus G(s) must have a positive representation rapidly decaying in time. There is no change in the Newtonian volumetric flow rate at this order. Symmetries of the problem require the velocity field at the second order to be zero with a pressure field oscillating with harmonics of the fundamental frequencies in (19) and (20) superposed on a mean field. Details are of no interest here and can be found in Siginer [40,41].

IV. 2. Third Order Solution

The field equations and boundary conditions at the third order are obtained form (9,12,13,14) and (15,16,17)

$$\rho \underset{\sim}{U}_{,t}^{(3)} + \rho[\underset{\sim}{U}^{(1)}\bullet \nabla\underset{\sim}{U}^{(2)} + \underset{\sim}{U}^{(2)}\bullet \nabla\underset{\sim}{U}^{(1)}] = -\nabla\Phi^{(3)} + \nabla\bullet\underset{\sim}{S}^{(3)} \qquad (27)$$

$$\nabla\bullet\underset{\sim}{U}^{(3)} = 0 \quad , \quad \underset{\sim}{U}^{(3)}(x,\pm d,z,t) = 0 \qquad (28)$$

The divergence of the Fréchet stress at this order is given by

$$\nabla\bullet\underset{\sim}{S}^{(3)} = \int_0^\infty G(s)\nabla^2\underset{\sim}{U}^{(3)}(t-s)ds + 2\int_0^\infty\int_0^\infty(\gamma+\alpha)[\underset{\sim}{e}_x M_{1,y} + \underset{\sim}{e}_z M_{2,y})ds_1 ds_2$$

$$+ \int_0^\infty\int_0^\infty\int_0^\infty(\sigma_1+2\sigma_4)\ [\underset{\sim}{e}_x(P_1+P_2)_{,y} + \underset{\sim}{e}_z(P_3+P_4)_{,y}]ds_1 ds_2 ds_3 \qquad (29)$$

M_i, P_j are obtained by evaluating the integrands in (12)

$$M^*(s_1) = U_{,y}^{(1)}(t-s_1)\ \xi^*_{x,y}(s_1) + W_{,y}^{(1)}(t-s_1)\ \xi^*_{z,y}(s_1)$$

$$M_2 = W_{,y}^{(1)}(t-s_2)\ M^*(s_1) \quad , \quad M_1 = U_{,y}^{(1)}(t-s_2)\ M^*(s_1)$$

$$P_1 = U_{,y}^{(1)}(t-s_1)\ U_{,y}^{(1)}(t-s_2)\ U_{,y}^{(1)}(t-s_3)$$

$$P_2 = U_{,y}^{(1)}(t-s_1)\ W_{,y}^{(1)}(t-s_2)\ W_{,y}^{(1)}(t-s_3)$$

$$P_3 = U_{,y}^{(1)}(t-s_1)\ U_{,y}^{(1)}(t-s_2)\ W_{,y}^{(1)}(t-s_3)$$

$$P_4 = W_{,y}^{(1)}(t-s_1)\ W_{,y}^{(1)}(t-s_2)\ W_{,y}^{(1)}(t-s_3)$$

ξ^* has been already defined after (12)

$$\xi^* = [-sU_{1m} + 2\mathrm{Re}[U_j a(\omega_{jx},s)]]\underset{\sim}{e}_x + 2\mathrm{Re}[W_j a(\omega_{jz},s)]\underset{\sim}{e}_z$$

$$a(\omega_{jn},s) = \frac{e^{i\omega_{jn}t}}{i\omega_{jn}} (e^{-i\omega_{jn}s} - 1) \quad ; \quad n = x,z \; ; \; (j,n) \text{ no sum}$$

The velocity field is of the following form

$$\underset{\sim}{U}^{(3)}(y,t) = U^{(3)} \underset{\sim}{e}_x + W^{(3)}\underset{\sim}{e}_z$$

$$U^{(3)} = U_{3m} + [U_{m_1..m_K n_1..n_N r_1..r_R} e^{i(m_\alpha\omega_\alpha + n_\gamma \omega_{\gamma x} + r_\kappa\omega_{\kappa z})t} + \text{Conj.}] \quad (30)$$

$$W^{(3)} = W_{k_1..k_K p_1..p_N s_1..s_R} e^{i(k_\alpha\omega_\alpha + p_\gamma\omega_{\gamma x} + s_\kappa\omega_{\kappa z})t} + \text{Conj.} \quad (31)$$

Summation over repeated indices is implied in (30) and (31). The range of the indices $m_\alpha, n_\gamma, r_\kappa, k_\alpha, p_\gamma$ and s_κ for α,γ or κ fixed is $(0,\pm1,\pm2,\pm3)$.

$$1 \le \alpha \le K \quad , \quad 1 \le \gamma \le N \quad , \quad 1 \le \kappa \le R$$

The longitudinal velocity field is composed of a mean component and components which oscillate with harmonics of the fundamental frequencies and their combinations. For instance if there are two waves each in the pressure gradient and on the boundary in orthogonal directions, i.e. $K = N = R = 2$, flow components which oscillate with frequencies equal to

$$(2\omega_1-\omega_2) \; , \; (\omega_{1x}-2\omega_{2z}) \; , \; (\omega_{1z}-2\omega_{2x}) \; , \; (\omega_{1x} - \omega_{1z} - \omega_{2z})$$

will exist among others. If frequencies are such that

$$\omega_2 = 2\omega_1 \; , \; \omega_{1x} = 2\omega_{2z}, \ldots$$

time independent flows will take place. In other words the interaction of the components of the oscillatory shear imposed from the boundary and the pressure gradient may add to the mean flow field if the frequencies satisfy certain relationships. The number of these relationships depend on the number of superposed waves.

To make expressions simpler we consider the case $K = N = R = 1$ and keep only those terms in (30) and (31) which may contribute to the mean velocity field. Writing $\omega_{1x} = \omega_x$, $\omega_{1z} = \omega_z$

$$U^{(3)}(y,t) = U_{3m}(y) + \Big[U_{2(-1)0}\, e^{i(2\omega_1-\omega_x)t} + U_{1(-2)0}\, e^{i(\omega_1-2\omega_x)t}$$

$$+ U_{01(-2)} e^{i(\omega_x - 2\omega_z)t} + \text{Conj.}] \tag{32}$$

$$W^{(3)}(y,t) = W_{10(-1)}(y) e^{i(\omega_1-\omega_z)t} + W_{01(-1)}(y) e^{i(\omega_x -\omega_z)t}$$

$$+W_{0(-2)}(y) e^{i(\omega_z -2\omega_x)t} + \text{Conj.} \tag{33}$$

The mean longitudinal flow $U_{3m}(y)$ is made up of a steady component generated by the constant pressure gradient plus components due to the interaction of the oscillatory shear, forced from the boundary and through the pressure gradient oscillations, with the pure shear.

$$- \mu U_{3m}(y) = \left(\frac{P}{\mu}\right)^3 \frac{y^4}{4} \psi_o + \frac{P}{4\mu} \sum_j^3 \psi_j \phi_j + C \tag{34}$$

$$\psi_j = 2\int_0^\infty\int_0^\infty (\gamma+\alpha)\Psi_j s_1 ds_1 ds_2 + \int_0^\infty\int_0^\infty\int_0^\infty (\sigma_1+2\sigma_4) \Psi_j^* ds_1 ds_2 ds_3 \tag{35}$$

$$\Psi_o = 1 \quad , \quad \Psi_o^* = -1$$

$$\Psi_1^* = -2 [\cos\omega_1(s_3-s_1) + \cos\omega_1(s_2-s_1) + \cos\omega_1(s_3-s_2)] \tag{36}$$

$$\Psi_1 = 2 \omega_1^{-1} [\sin\omega_1 s_1 + \sin\omega_1 s_2 + \omega_1 s_1 \cos\omega_1(s_1-s_2) + \sin\omega_1(s_1-s_2)] \tag{37}$$

$$\Psi_2 = \Psi_1 , \Psi_2^* = \Psi_1^* \quad \text{with } \omega_1 \to \omega_x \text{ in (36, 37)}$$

$$\Psi_3^* = 2 \cos\omega_z(s_1-s_2) \quad , \quad \Psi_3 = -2 (\omega_z)^{-1} \sin\omega_z s_1$$

$$\phi_1 = (\rho\omega_1)^{-2} \lambda_1^2 \phi_1^*(\omega_1) \ , \ \phi_2 = \lambda_x^2 \phi_1^* (\omega_x) \ , \ \phi_3 = \lambda_z^2 \phi_1^* (\omega_z)$$

$$\phi_1^*(\omega) = \left|\frac{\Lambda}{\mathrm{Ch}\Lambda d}\right|^2 \Big(y \Big[\frac{\mathrm{Sh}(\Lambda + \bar{\Lambda})y}{(\Lambda + \bar{\Lambda}} - \frac{\mathrm{Sh} (\Lambda - \bar{\Lambda})y}{(\Lambda - \bar{\Lambda})}\Big]$$

$$+ \frac{\mathrm{Ch} (\Lambda - \bar{\Lambda})y}{(\Lambda - \bar{\Lambda})^2} - \frac{\mathrm{Ch} (\Lambda + \bar{\Lambda})y}{(\Lambda + \bar{\Lambda})^2}\Big)$$

The additional terms in (32) may generate a mean flow if $2\omega_1 = \omega_x$ or $\omega_1 = 2\omega_z$. For instance we find for the last relationship

$$U_{01(-2)} = \psi_4(\omega_x, \omega_z) \phi_4(\omega_x, -\omega_z) + C \tag{38}$$

$$\psi_4(\omega_x, \omega_z) = \int_0^\infty\int_0^\infty\int_0^\infty (\sigma_1+2\sigma_4) \Psi_4^* (\omega_x, \omega_z) ds_1 ds_2 ds_3$$

$$+ 2\int_0^\infty\int_0^\infty (\gamma+\alpha)\ \Psi_4(\omega_x,\ \omega_z)ds_1ds_2$$

$$\Psi_4^*(\omega,\hat{\omega}) = e^{-i\omega s_3}\ e^{-i\hat{\omega}(s_1-s_2)}$$

$$\Psi_4(\omega,\ \hat{\omega}) = e^{-i\omega s_2}\ (e^{-2i\hat{\omega}s_1} - e^{-i\hat{\omega}s_1})\ (i\hat{\omega})^{-1}$$

$$\phi_4(\omega_x,\ \omega_z) = \frac{\lambda_x\Lambda_x(\lambda_z\bar{\Lambda}_z)^2}{32i\mu(\text{Ch}\Lambda_x d)\ (\text{Ch}\bar{\Lambda}_z d)^2}\left[\frac{\text{Ch}(\Lambda_x+\bar{\Lambda}_z)y}{\Lambda_x+2\bar{\Lambda}_z}\right.$$

$$\left.+\frac{\text{Ch}(\Lambda_x-2\bar{\Lambda}_z)y}{\Lambda_x-2\bar{\Lambda}_z} - \frac{2\text{Ch}\Lambda_x y}{\Lambda_x}\right]$$

It is interesting that the components $U_{2(-1)0}$, $U_{1(-2)0}$ and $U_{01(-2)}$ in (32) give steady velocity components independent of the steady shear, i.e. constant pressure gradient, if $2\omega_1 = \omega_x$, $\omega_1 = 2\omega_x$ or $\omega_x = 2\omega_z$ respectively as the expression (38) for $U_{01(-2)}$ clearly shows. The implication is that a steady flow can be generated in the absence of pure shear if for instance two orthogonal vibrations satisfying $\omega_x = 2\omega_z$ are imposed on the boundary or if the pressure gradient oscillates about a zero mean with frequency ω_1 and the boundary oscillates longitudinally with frequency ω_{1x} and they satisfy either $2\omega_1 = \omega_x$ or $\omega_1 = 2\omega_x$.

The orthogonal field (33) which is equivalent to secondary flows in circular pipes can generate a steady transversal flow if $\omega_1 = \omega_z$ or $\omega_x = \omega_z$ or $\omega_z = 2\omega_x$. The first two of these and the last are mean pressure gradient dependent and independent respectively. For instance if the longitutinal pressure gradient in a circular pipe oscillates around a non-zero mean with frequency ω_1 and a rotational vibration of frequency ω_2 is imposed on the pipe wall a mean secondary flow dependent on the non-zero mean longitudinal pressure gradient will take place if $\omega_1 = \omega_z$. On the other hand if the pipe wall vibrates simultaneously in the longitudinal and transversal directions with frequencies ω_{1x} and ω_{1z} a pressure gradient dependent and independent secondary flow will take place if $\omega_x = \omega_z$ and $\omega_z = 2\omega_x$ respectively. We find

$$W_{01(-1)} = \frac{P}{4\mu^2} \psi_5(-\omega_x,\omega_z)\ \phi_5(\omega_x,\omega_z) + C$$

$$W_{0(-2)1} = \psi_6(-\omega_x,\omega_z)\ \phi_6(-\omega_x,\omega_z) + C$$

when $\omega_x = \omega_z$ and $\omega_z = 2\omega_x$ respectively.

$$\psi_j(\omega,\hat{\omega}) = \int_0^\infty\int_0^\infty\int_0^\infty (\sigma_1+2\sigma_4)\ \Psi_j^*(\omega,\hat{\omega})ds_1ds_2ds_3$$

$$+ (-1)^j\ 2\int_0^\infty\int_0^\infty(\gamma+\alpha)\Psi_j(\omega,\hat{\omega})ds_1ds_2 \quad ; \quad j = 5,6$$

$$\Psi_5^*(\omega,\hat{\omega}) = e^{-i\hat{\omega}s_3}(e^{-i\omega s_2} + e^{-i\omega s_1})$$

$$\Psi_5(\omega,\hat{\omega}) = s_1\ e^{-i\omega s_1}\ e^{-i\hat{\omega}s_2} - \frac{e^{-i\hat{\omega}s_2}}{i\omega}(e^{-i\omega s_1}-1)$$

$$\Psi_6(\omega,\hat{\omega}) = \Psi_4(\hat{\omega},\omega) \quad , \quad \Psi_6^*(\omega,\hat{\omega}) = \Psi_4^*(\hat{\omega},\omega)$$

$$\phi_5(\omega_x,\omega_z) = \frac{\lambda_x\lambda_z\bar{\Lambda}_x\Lambda_z}{Ch\bar{\Lambda}_xd)(Ch\Lambda_zd)}\Big(y[\frac{Sh(\bar{\Lambda}_x+\Lambda_z)y}{\bar{\Lambda}_x+\Lambda_z} - \frac{Sh(\bar{\Lambda}_x-\Lambda_z)y}{\bar{\Lambda}_x-\Lambda_z}]$$

$$+ \frac{Ch(\bar{\Lambda}_x-\Lambda_2)y}{(\bar{\Lambda}_x-\Lambda_z)^2} - \frac{Ch(\bar{\Lambda}_x+\Lambda_z)y}{(\bar{\Lambda}_x+\Lambda_z)^2}\Big)$$

$$\phi_6(\omega_x,\omega_z) = \phi_4(\omega_z,\omega_x)$$

We will consider the mean volumetric flow in the longitudinal direction in the absence of any special frequency relationships. Integrating (34) we obtain

$$Q_m^{(3)} = \frac{2}{5}\frac{P^3d^5}{\mu^4}\psi_o - \frac{P}{2\mu}\sum_j^3 \psi_j\hat{\phi}_j = Q_o^{(3)} + Q_1^{(3)} \tag{39}$$

where ψ_j are defined in (35) and $\hat{\phi}_j$ are given by

$$\hat{\phi}_1 = (\rho\omega_1)^{-2}\ \lambda_1^2\ \phi_1^{**}\ (\omega_1)\ ,\ \hat{\phi}_2 = \lambda_x^2\ \phi_1^{**}(\omega_x)\ ,\ \hat{\phi}_3 = \lambda_z^2\ \phi_1^{**}\ (\omega_z) \tag{40}$$

$$\phi_1^{**}(\omega) = |\frac{\Lambda}{Ch\Lambda d}|^2\Big(2d[\frac{Chmd}{m^2} + \frac{cosnd}{n^2}] + [d^2 - \frac{2}{n^2}]\ \frac{sinnd}{n^2} - [d^2 + \frac{2}{m^2}]\ \frac{Shmd}{m}\Big) \tag{41}$$

where $m = 2\mathrm{Re}\Lambda$ and $n = 2\mathrm{Im}\Lambda$ and $Q_o^{(3)}$ and $Q_1^{(3)}$ denote the frequency independent and dependent parts respectively of the discharge at this order.

V. Asymptotic Behaviour

V.1. Small Frequencies

For very small frequencies we obtain from (41) and (35)

$$\phi_1^{**}(\omega) \sim -\frac{2}{3} d^5 \left(\frac{\rho\omega}{|\eta^*|}\right)^2 , \quad \omega < 1$$

$$\psi_j(\omega) \sim 6\psi_o , \quad j = 1,2 \quad ; \quad \psi_3(\omega) \sim 2\psi_o , \quad \omega < 1$$

and with the definition of a vibratory Reynolds number Re_n, the frequency dependent part of the volumetric flow rate $Q_1^{(3)}$ becomes

$$Q_1^{(3)} \sim \frac{2Pd^3\psi_o}{|\eta^*|^2} \left[\left(\frac{Re_1}{\rho\omega}\right)^2 + Re_x^2 - \frac{1}{3} Re_z^2\right] \quad ; \quad Re_n = \frac{d\lambda_n\omega_n}{\gamma} \tag{42}$$

Coleman & Markovitz [11] have shown, at least in principle, how the coefficients of the fluid of grade n can be related to the kernels of the integral fluids. The constitutive constants of the fluid of grade n and the kernels in (11) and (12) satisfy the following moment relationships

$$\beta_2 = -\int_0^\infty\int_0^\infty \gamma(s_1,s_2)s_1 ds_1 ds_2 \tag{43}$$

$$\beta_4 = -\int_0^\infty\int_0^\infty \alpha(s_1,s_2)s_1 ds_1 ds_2 \tag{44}$$

$$\beta_5 = \int_0^\infty\int_0^\infty\int_0^\infty \sigma_4(s_1,s_2,s_3) ds_1 ds_2 ds_3 \tag{45}$$

$$\beta_6 = \int_0^\infty\int_0^\infty\int_0^\infty \sigma_1(s_1,s_2,s_3) ds_1 ds_2 ds_3 \tag{46}$$

$$\beta_3 = \beta_4 + \beta_5 + \frac{\beta_6}{2} \tag{47}$$

in the case of the fluid of grade three,

$$\underset{\sim}{S} = \sum_i^3 \underset{\sim}{S}_i \quad , \quad \underset{\sim}{S}_1 = \mu\underset{\sim}{A}_1 \quad , \quad \underset{\sim}{S}_2 = \alpha_1\underset{\sim}{A}_2 + \alpha_2\underset{\sim}{A}_1^2$$

$$\underset{\sim}{S}_3 = \beta_1\underset{\sim}{A}_3 + \beta_2(\underset{\sim}{A}_2\underset{\sim}{A}_1 + \underset{\sim}{A}_1\underset{\sim}{A}_2) + \beta_3(\mathrm{tr}\underset{\sim}{A}_2)\underset{\sim}{A}_1$$

Comparison of (35) and (43,...,47) shows

$$\psi_o = - 2(\beta_2+\beta_3)$$

The combination $(\beta_2+\beta_3)$ is negative if the liquid is shear thinning. Then (42) shows that the contribution of the pressure gradient oscillation tends to a finite, positive value in the limit of zero frequency whereas the contributions of the longitudinal and transversal boundary vibrations tend to zero. It is also clear that the enhancement through boundary vibration is an inertial phenomena. In both cases the fluid has to be shear thinning for an increase in the volumetric flow rate to occur. Also transversal boundary vibration acts to decrease the longitudinal mean flow at least for small frequencies.

V. 2 Large Frequencies

To determine the behaviour at rapid deformation rates assumptions need to be made about ϕ adn ψ in (39). Without postulating explicit expressions we write, based on experimental evidence

$$\eta^* = \eta' - i\eta'' \quad , \quad \eta' \sim O(\omega^{x_1}) \quad , \quad \eta'' \sim O(\omega^{x_2}) \ , \ x_1 \le 0, \ x_2 < 0 \ , \ \omega \to \infty$$

The possibility of the existence of a second Newtonian plateau η'_∞ is represented by $x_1 = 0$. The asymptotic behaviour of ϕ depends on $2\Lambda = m + in$. We obtain

$$m^2 = \frac{2\rho\omega}{|\eta^*|^2} (|\eta^*|-\eta') \quad , \quad n^2 = - \frac{2\rho\omega}{|\eta^*|^2} (|\eta^*|+\eta'') \tag{48}$$

Defining $n \sim O(\omega^{x_3})$, $m \sim O(\omega^{x_4})$ we find that $(48)_2$ implies $n \to \infty$ always as $\omega \to \infty$ regardless $\eta'/\eta'' \lesseqgtr 1$. But although $m \to \infty$ if $\eta'/\eta'' > 1$ it may also tend to a finite limit when $\eta'/\eta'' < 1$. Among the five possible combinations of the signs of (x_3,x_4) in view of these restrictions the most interesting is $(x_3,x_4) > 0$, $x_3 = x_4$, in which case

$$\phi(\omega) \sim \frac{|\Lambda|^2}{\Lambda+\bar{\Lambda}} \quad , \quad \omega \to \infty$$

Assuming $\psi_j \sim O(\omega^b)$, $b \leq 0$; $Q_1^{(3)} \sim O(\omega^c)$, $c = b - \frac{1}{2}(3+x_1)$.

Unless $b = 0$, i.e. ψ_j tend to a constant as $\omega \to \infty$ or η' tends to zero much faster than ψ, $Q_1^{(3)}$ will almost certainly approach zero at large frequencies.

VI. Intermediate Frequency Behaviour

Explicit expressions for the constitutive functions need to be introduced to determine the enhancement at moderate frequencies. Rapidly decaying representations compatible with the moment theory may be assumed

$$\gamma = \alpha_2 \sum_i^N C_{1i}\, k_i^2\, e^{-k_i(s_1+s_2)} \quad ; \quad \sum_i^N C_{1i} = 1 \tag{49}$$

$$\alpha = -\beta_4 \sum_i^N C_{2i}\, \rho_i^3\, e^{-\rho_i(s_1+s_2)} \quad ; \quad \sum_i^N C_{2i} = 1$$

$$\sigma_1 = \beta_6 \sum_i^N C_{3i}\, m_i^3\, e^{-m_i(s_1+s_2+s_3)} \quad ; \quad \sum_i^N C_{3i} = 1$$

$$\sigma_4 = \beta_5 \sum_i^N C_{4i}\, n_i^3\, e^{-n_i(s_1+s_2+s_3)} \quad ; \quad \sum_i^N C_{4i} = 1$$

$$G(s) = \frac{\mu - \eta'_\infty}{\theta^k \Gamma(k)}\, s^{k-1}\, e^{-s/\theta} + \delta(s)\eta'_\infty \tag{50}$$

$$\theta = \frac{\alpha_1}{k(\eta'_\infty - \mu)} > 0 \quad , \quad \alpha_1 < 0 \quad , \quad 0 < k \leq 1 \quad , \quad \eta'_\infty < \mu$$

δ and Γ are the Dirac and Gamma functions respectively. The representation (50) for the shear relaxation modulud G(s) implies

$$\eta'_\infty \sim O(\omega^{-k}) \sim \eta'' \quad , \quad \frac{\eta''}{\eta'} \sim \tan\left(k\frac{\pi}{2}\right)$$

If $\eta'_\infty \neq 0$, $Q_1^{(3)} \sim O(\omega^c)$, $c = -\frac{1}{2}(3-k)$ and if $\eta'_\infty = 0$, $c = -\frac{3}{2}$.

The smaller the power index k the faster $Q_1^{(3)}$ tends to zero as $\omega \to \infty$. The constitutive constants $\mu, \alpha_1, \alpha_2, k_1, k_2, \theta, k, (\beta_2+\beta_3)$ can be determined using established methods of rheometry. Working with the lowest orders in the representations of α, σ_1 and σ_4 a parametric study of the enhancement can be conducted by assigning values to the unknowns $\rho_1, m_1, n_1, \beta_4, \beta_5, \beta_6$. Such studies are reported in Siginer [40,41].

VII. A PROCEDURE FOR THE SEQUENTIAL DETERMINATION OF THE CONSTITUTIVE PARAMETERS

A program which combines experiments with closed form analytic solutions of the type obtained in this article and established methods of viscometry can be devised to determine the constitutive constants of a particular fluid. For instance the first two Rivlin-Ericksen constants may be determined through cone and plate rheometry and $(\beta_2+\beta_3)$ obtained from steady shear experiments. The coefficients C_i and relaxation times k_i^{-1} in the representation (49) of γ can be determined with enough accuracy up to and including second order from mean free surface deformation measurements in the motion driven by an oscillating rod in a large vat, Beavers [3], which in turn would determine β_2 in (43). As $(\beta_2+\beta_3)$ is already determined from steady shear experiments β_3 may be considered known. Then (47) puts a constraint on $\beta_4, \beta_5, \beta_6$. Working with the first terms in the series for $\alpha, \sigma_1, \sigma_4$ analytical form of $Q_1^{(3)}$ may be expressed as a function of the unknown parameters $\beta_4, \beta_5, \beta_6, \rho_1, m_1, n_1$ and possibly the power index k. This expression together with (47) defines a constrained maxima problem. If accurate enough experimental volumetric flow rate values can be obtained for different frequencies the unknown parameters can be determined in principle.

REFERENCES

[1] Barnes, H. A., Townsend, P. and Walters, K., Flow of Non-Newtonian Liquids Under a Varying Pressure Gradient, Nature, 224, 585-587, 1969.

[2] Barnes, H. A., Townsend, P. and Walters, K., On Pulsatile Flow of Non-Newtonian Liquids, Rheol. Acta, 10, 517-527, 1971.

[3] Beavers, G. S., The Free Surface on a Simple Fluid Between Cylinders Undergoing Torsional Oscillations, Part II: Experiments, Arch. Rat. Mech. Anal., 62, 4, 343-352, 1976.

[4] Bernstein, B., Small Shearing Oscillations Superposed on Large Steady Shear of the BKZ Fluid, Int. J. Non-linear Mech., 4, 183, 1969.

[5] Bernstein, B. and Fosdick, R. L., On Four Rheological Relations, Rheol. Acta, 9, 186, 1970.

[6] Böhme, G. and Nonn, G., Instationäre Rohrströmung Viskoelastischer Flüssigkeiten Mabnahmen zur Durchsatzsteigerung, Ingenieur - Archiv, 48, 35-49, 1979.
[7] Booij, H. C., Effect of Superimposed Steady Shear Flow on Dynamic Properties of Polymeric Fluids, Ph.D. Thesis, Leiden, 1970.
[8] Coleman, B. D. and Noll, W., An Approximation Theorem for Functionals with Applications in Continuum Mechanics, Arch. Rat. Mech. Anal., 6, 355-370, 1960.
[9] Coleman, B.D. and Noll, W., Foundations of Linear Viscoelasticity, Rev. Mod. Physics, 33, 239-249, 1961.
[10] Coleman, B. D. and Mizel, V., On the General Theory of Fading Memory, Arch. Rat. Mech. Anal., 29, 18-31, 1968.
[11] Coleman, B. D. and Markovitz, H., Normal Stress Effects in Second Order Fluids, J. Appl. Phys., 35, 1-9, 1964.
[12] Davies, J.M., Bhumiratana, S. and Bird, R.B., Elastic and Inertial Effects in Pulsatile Flow of Polymeric Liquids in Circular Tubes, J. Non-Newt. Fluid Mech., 3, 237-259, 1978.
[13] Edwards, M.F., Nellist, D.A. and Wilkinson, W.R., Pulsating Flow of Non-Newtonian Fluids in Pipes, Chem. Eng. Sc., 27, 545-553, 1972.
[14] Etter, I. and Schowalter, W.R., Unsteady Flow of an Oldroyd Fluid in a Circular Tube, Trans. Soc. Rheol., 9, 351-369, 1965.
[15] Goldstein, C. and Schowalter, W.R., Nonlinear Effects in the Unsteady Flow of Viscoelastic Fluids, Rheol. Acta, 12, 253-262, 1973.
[16] Green, A.E. and Rivlin, R.S., The Mechanics of Non-linear Materials with Memory, Part I, Arch. Rat. Mech. Anal., 1, 1-21, 1957.
[17] Jones, T.E.R. and Walters K., The Behaviour of Materials under Combined Steady and Oscillatory Shear, J. Phys. A. Gen. Phys., 4, 85, 1971.
[18] Joseph, D.D., Instability of the Rest State of Fluids of Arbitrary Grade Greater than One. Arch. Rat. Mech. Anal., 75, 251-256. 1981.
[19] Joseph, D.D., Rotating Simple Fluids, Arch. Rat. Mech. Anal., 66, 311-344, 1977.
[20] Joseph, D.D., Stability of Fluid Motions II, Springer-Verlag, New York, 1976.
[21] Kazakia, J.Y. and Rivlin, R.S., The Influence of Vibration on Poiseuille Flow of a Non-Newtonian Fluid, I., Rheol. Acta, 17, 210-226, 1978.
[22] Kazakia, J.Y., and Rivlin, R.S., The Influence of Vibration on Poiseuille Flow of a Non-Newtonian Fluid, II., Rheol. Acta, 18, 244-255, 1979.
[23] Langlois, W.E. and Rivlin, R.S., Steady Flow of Slightly Viscoelastic Fluids, Tech. Rep. No. DA-4725/3, Division of Appl. Math., Brown Univ., Providence, R.I.
[24] Larson, R.G., Constitutive Equations for Polymer Melts and Solutions, Butterworths, 1988.
[25] Manero, O. and Mena, B., An Interesting Effect in Non-Newtonian Flow in Oscillating Pipes, Rheol. Acta, 16, 573-576, 1977.
[26] Manero, O., Mena, B. and Valenzuela, R., Further Developments on Non-Newtonian Flow in Oscillating Pipes, Rheol. Acta, 17, 693-697, 1978.
[27] Manero, O. and Walters, K., On Elastic Effects in Unsteady Pipe Flows, Rheol. Acta, 19, 277-284, 1980.
[28] Noll, W., A Mathematical Theory of the Mechanical Behaviour of Continous Media, Arch. Rat. Mech. Anal., 2, 197, 1958.
[29] Phan-Thien, N., On Pulsatile Flow of Polymeric Fluids, J. Non-Newt. Fluid Mech., 4, 167-176, 1978.

[30] Phan-Thien, N., Flow Enhancement Mechanisms of a Pulsating Flow of Non-Newtonian Liquids, Rheol. Acta, 19, 285-290, 1980.
[31] Phan-Thien, N., On a Pulsating Flow of Polymeric Fluids: Strain Dependent Memory Kernels, J. Rheol., 25, 293-314, 1981.
[32] Phan-Thien, N., The Effects of Random Longitudinal Vibration on Pipe Flow of a Non-Newtonian Liquid, Rheol. Acta, 19, 539-547, 1980.
[33] Phan-Thien, N., The Influence of Random Longitudinal Vibration on Channel and Pipe Flows of a Slightly Non-Newtonian Liquid, J. Appl. Mech., 48, 661-664, 1981.
[34] Pipkin, A.C., Alternating Flow of Non-Newtonian Fluids in Tubes of Arbitrary Cross-section, Arch. Rat. Mech. Anal., V. 15, 1, 1-13, 1964.
[35] Pipkin, A.C. and Owen, D.R., Nearly Viscometric Flows, Physics of Fluids, 10, 836-843, 1967.
[36] Randria, P. and Bellet, D., Theoretical and Experimental Studies on Unsteady Flows of Polymer Solutions, Proc. Int. Conference on Viscoelasticity of Polymeric Liquids, Alp d'Huez, France, Jan. 1986.
[37] Renardy, M., On the Domain Space for Constitutive Laws in Linear Viscoelasticity, Arch. Rat. Mech. Anal., 85, 21-26, 1984.
[38] Rivlin, R.S., Viscoelastic Fluids, in Research Frontiers in Fluid Dynamics, New York 1965.
[39] Saut, J.C. and Joseph, D.D., Fading Memory, Arch. Rat. Mech. Anal., 81, 53-95, 1982.
[40] Siginer, A., On the Pulsating Pressure Gradient Driven Flow of Simple Fluids, forthcoming.
[41] Siginer, A., Effect of Boundary Vibration on Poiseuille Flow of an Elastico-viscous Liquid, forthcoming.
[42] Simmons, J.M., Dynamic Modulus of Polyisobutylene Solutions in Superposed Steady Shear Flow, Rheol. Acta, 7, 184, 1968.
[43] Sundstrom, D.W. and Kaufman, A., Pulsating Flow of Polymer Solutions, Ind. Eng. Chem. Proc. Des. Dev., 16, 320-325, 1977.
[44] Tanner, R.I. and Williams, G., On the Orthogonal Superposition of Simple Shearing and Small-strain Oscillatory Motions, Rheol. Acta, 10, 528-538, 1971.
[45] Townsend, P., Numerical Solutions of Some Unsteady Flows of Elastico-viscous Liquids, Rheol. Acta, 12, 13-18, 1973.
[46] Walters, K. and Townsend, P., The Flow of Viscous and Elastico-viscous Liquids in Straight Pipes under a Varying Pressure Gradient, Proc. 5th Int. Congress on Rheology, 4, 471-483, Kyoto, 1968.
[47] Wang, C.C., The Principle of Fading Memory, Arch. Rat. Mech. Anal., 18, 343-366, 1965.
[48] Zahorski, S., Motions with Superposed Proportional Stretch Histories as Applied to Combined Steady and Oscillatory Flows of Simple Fluids, Arch. Mech. Stos., 25, 575, 1973.
[49] Zahorski, S., Flows with Proportional Stretch History, Arch. Mech. Stos., 24, 681, 1972.

A. Siginer

Department of Mechanical Engineering
Auburn University
Auburn, AL 36849 USA

K. WALTERS, A.Q. BHATTI AND N. MORI

The influence of polymer conformation on the rheological properties of aqueous polymer solutions

1. INTRODUCTION

There are both esoteric and practical reasons to study the correlation between polymer conformation and rheology. From the purist standpoint, there is the hope that the resulting rheology can be simulated by one of the increasing number of available microrheological theories. From the practical standpoint, there is the growing importance of biopolymers like xanthan gum and synthetic polymers like polyacrylamide in such widely differing applications as the food industries and oil extraction. In the former, xanthan gum is widely employed as a thickening or gelling agent [cf. 1], while in the latter both xanthan gum and polyacrylamide are of potential use in Enhanced Oil Recovery [cf. 2]. Particularly interesting and somewhat unexpected features emerge from the present study, which help to illustrate, in passing, that the meaning of the term 'highly elastic liquid' can be misleading, unless the particular context in mind is made clear. For interest, we therefore pose the question: Are aqueous solutions of polyacrylamide more elastic than comparable solutions of xanthan gum? Of course, we have to define what we mean by 'comparable', i.e. we have to compare 'like with like' and in this respect the present work differs from similar detailed studies by a number of workers in the field [1, 3-5]. Our intention is to choose the concentrations of the two polymers such that the resulting shear viscosities are very similar over a reasonably wide shear-rate range. We can then compare and contrast their response under other rheometrical flow conditions.

The present experiments have involved the xanthan gum Keltrol F (supplied by Kelco) and the polyacrylamide Magnaflox E10 (supplied by Allied Colloids).By a process of trial and error we have found that near room temperature 3%, 2% and 0.9% aqueous solutions of xanthan gum have a similar shear-viscosity response to 2%, 1.5% and 0.75% aqueous solutions of polyacrylamide, respectively, and we have investigated the rheological properties of these six solutions at one temperature (20°C). Detailed results will only be given for the most concentrated solutions, but the general conclusions reached are applicable to all the chosen concentrations.

The conformation of the xanthan gum molecule has been well covered by Rochefort and Middleman [1] and Lim et al [3]. It is usually considered to be 'semi rigid'. In contrast, polyacrylamide is generally regarded as being 'very flexible' and it is largely this difference which gives rise to the substantially different rheological response.

We remark that the concentrations employed in the present series of experiments place the xanthan solutions in the liquid-crystaline category (see [6] for an excellent review of the rheology of rod-like polymers in the liquid-crystaline state).

2. BASIC RHEOMETRY

Consider a steady simple shear flow represented by Cartesian velocity components

$$v_x = \dot{\gamma}y, \quad v_y = v_z = 0, \tag{1}$$

where $\dot{\gamma}$ is the constant shear rate. The corresponding stress distribution for a non-Newtonian elastic liquid can be written in the form [7,8]

$$\begin{aligned} &\sigma_{xy} = \sigma = \dot{\gamma}\eta(\dot{\gamma}), \\ &\sigma_{xx} - \sigma_{yy} = N_1(\dot{\gamma}), \quad \sigma_{xx} - \sigma_{zz} = N_2(\dot{\gamma}), \end{aligned} \tag{2}$$

where σ_{ik} is the stress tensor, η is the apparent (or shear) viscosity and N_1 and N_2 are the first and second normal-stress differences, respectively. In conventional rheometry, it is customary to limit attention to η and N_1.

In a uniaxial extensional flow represented by the Cartesian velocity components:

$$v_x = \dot{\epsilon}x, \quad v_y = -\frac{\dot{\epsilon}y}{2}, \quad v_z = -\frac{\dot{\epsilon}z}{2}, \tag{3}$$

where $\dot{\epsilon}$ is a constant strain rate, the corresponding stress distribution can be written [7,8]

$$\begin{aligned} &\sigma_{xx} - \sigma_{yy} = \sigma_{xx} - \sigma_{zz} = \dot{\epsilon}\eta_E(\dot{\epsilon}), \\ &\sigma_{ik} = 0 \text{ for } i \neq k, \end{aligned} \tag{4}$$

where η_E is called the extensional viscosity.

Finally, in an oscillatory shear flow given by

$$v_x = \alpha\omega y \cos \omega t, \quad v_y = v_z = 0, \tag{5}$$

where α is a *small* amplitude, the relevant stress is σ_{xy} given by [7,8]

$$\sigma_{xy} = \alpha\omega[\eta'\cos\omega t + \frac{G'}{\omega}\sin\omega t], \tag{6}$$

where η' is the dynamic viscosity and G' the dynamic rigidity. The complex viscosity η^* is defined by

$$\eta^* = \eta' - i\,\frac{G'}{\omega} \tag{7}$$

For nonlinear materials such as polymer solutions, the response will only be linear if the amplitude is very small; otherwise nonlinear effects must be expected and have to be accommodated (see §5).

In the following sections, we shall refer to all the rheometrical functions defined above, except for the second normal stress difference N_2. The experiments were performed on a Weissenberg Rheogoniometer and a Controlled Stress Rheometer (both manufactured by Carrimed U.K.)

3. STEADY SHEAR AND EXTENSIONAL BEHAVIOUR

Published shear results already exist for the 2% polyacrylamide solution and the 3% xanthan solution [8]. The experiments have been repeated in the present study and the data are reproduced in Fig 1. Note that the viscosity of the xanthan solution is still increasing at very low shear rates (and is higher than the polyacrylamide solution in this region) but the existence of a *linear*-viscoelastic response (see § 5) is convincing evidence of the absence of a yield stress for this solution (see also 3).

Figure 1 clearly shows that the first normal stress difference is much higher for the polyacrylamide solution. Since N_1 is (rightly) viewed as a measure of the viscoelastic response of the polymer solution, it is evident that, in this deformation mode at least, the polyacrylamide solution is more elastic than its xanthan gum counterpart.

The two solutions have also been investigated in extensional rheometers and the relevant data are reproduced in Fig 2 [cf. 8]. Here again, as expected, η_E for the solution of the flexible polymer is significantly higher than that for the solution of the semi-rigid

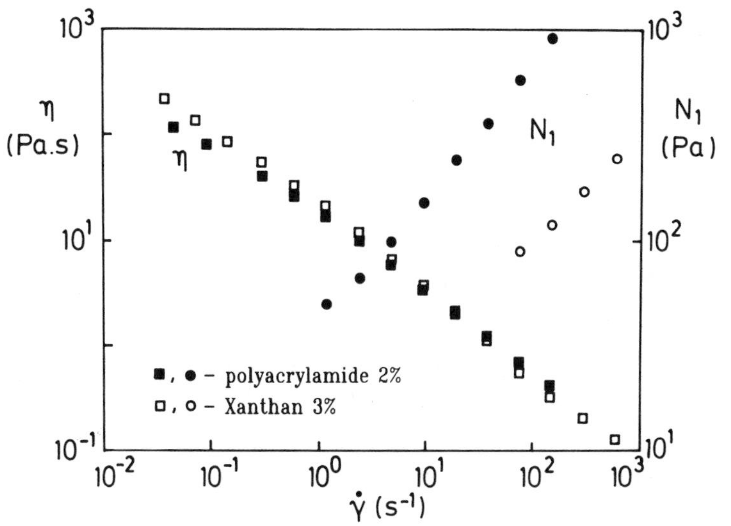

Fig 1. Steady shear data for a 2% polyacrylamide solution and a 3% xanthan gum solution (see also 8).

xanthan gum. Indeed, the behaviour of the two solutions is *qualitatively* different, the polyacrylamide solution being tension thickening (with η_E increasing with strain rate $\dot{\epsilon}$) while the xanthan solution is tension thinning (with η_E decreasing with $\dot{\epsilon}$). Yet again the flexible polyacrylamide is showing the features of a highly elastic liquid and the xanthan solution is relegated to being only 'weakly elastic' [cf. 8].

The significant differences in the *rheometrical* response of the two solutions discussed in this section are exaggerated in flow through complex geometries, where any extravagant viscoelastic response is invariably associated with the polyacrylamide solution [cf. 2].

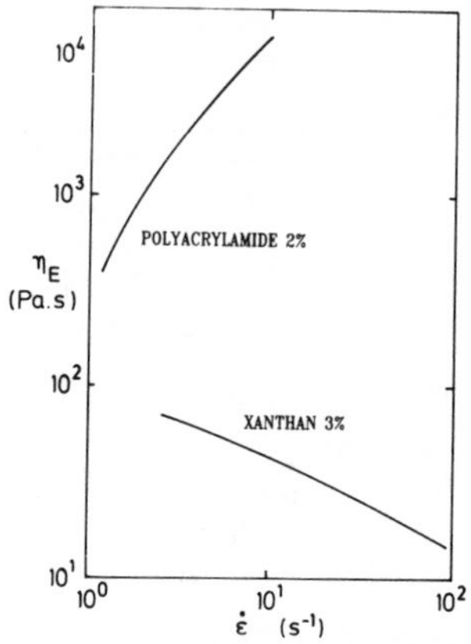

Fig 2. Extensional viscosity behaviour for a 2% polyacrylamide solution and a 3% xanthan gum solution (see also 8).

4. OSCILLATORY SHEAR FLOW

(i) Effect of pre shearing

It is well known that, unlike polyacrylamide solutions, xanthan gum solutions suffer from the effect of 'pre shearing', which has to be taken into account if rheometrical tests are to have objective meaning [cf. 1]. We can best illustrate the effect by reference to Fig 3.

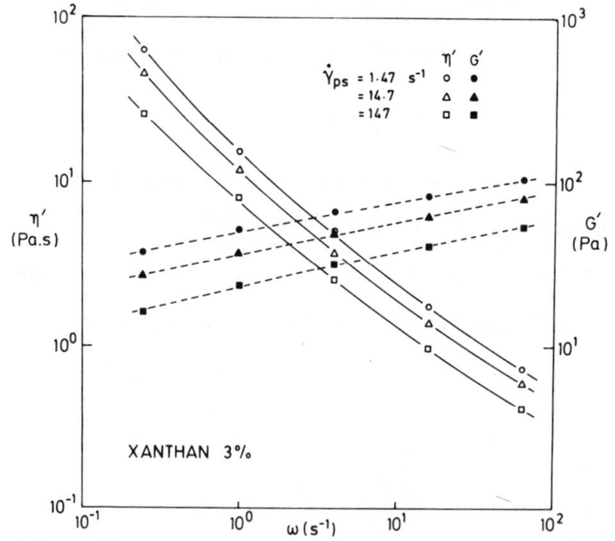

Fig 3. The effect of pre shearing on the dynamic properties of a 3% xanthan gum solution. $\dot{\gamma}_{ps}$ is the pre shearing rate.

Before each oscillatory test, the sample was exposed to a steady shear $\dot{\gamma}_{ps}$ for a period of 60s. It is clear that both η' and G' drop significantly as the pre shear rate is increased. Figure 4 considers the problem in more detail for three prescribed frequencies. $|\eta^*|$ is the modulus of the complex viscosity given by

$$|\eta^*| = \left[(\eta')^2 + \left(\frac{G'}{\omega}\right)^2 \right]^{\frac{1}{2}} . \tag{8}$$

It is interesting that all the curves have a maximum at a frequency near $1s^{-1}$. If the test sample is left to recover from its low values at high $\dot{\gamma}_{ps}$, periods of up to a day can be involved [cf. 1]. However, gentle shearing at a frequency near $1s^{-1}$ for a period of less

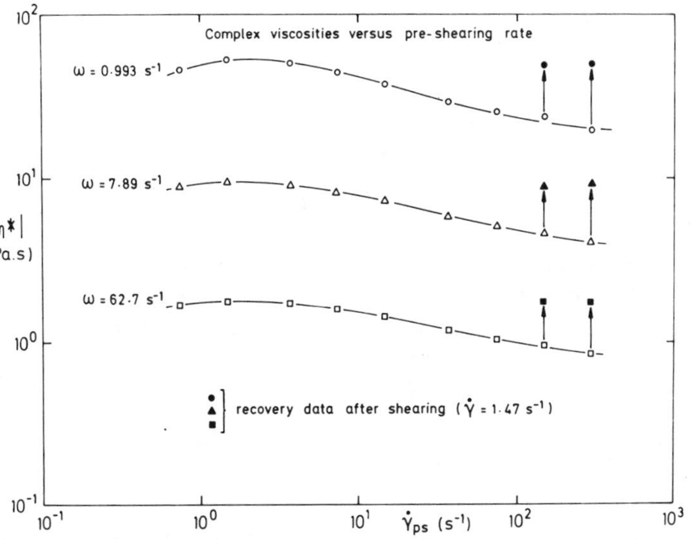

Fig 4. $|\eta^*|$ as a function of $\dot{\gamma}_{ps}$ for three frequencies. 3% aqueous solution of xanthan gum.

than 60s had the same effect! Now, the process of loading a rheometer usually involves an element of shearing and it was therefore thought prudent to formally pre shear at a low shear rate for 60s before each oscillatory test in order to obtain consistent and repeatable results.

We are not in a position to comment on the precise reason for the maximum in $|\eta^*|$ after gentle pre shearing and we have yet to determine how the observed maximum in the curves varies with temperature, salinity or polymer concentration.

(ii) Oscillatory testing

Dynamic testing on the two polymer solutions has lead to some provocative findings (see Fig 5). Although there is no fundamental reason to expect overlapping η' behaviour, this was in fact observed for the polyacrylamide and xanthan solutions. Of far more interest in the present context is the dramatic reversal of roles so far as G' is concerned. The gel-like

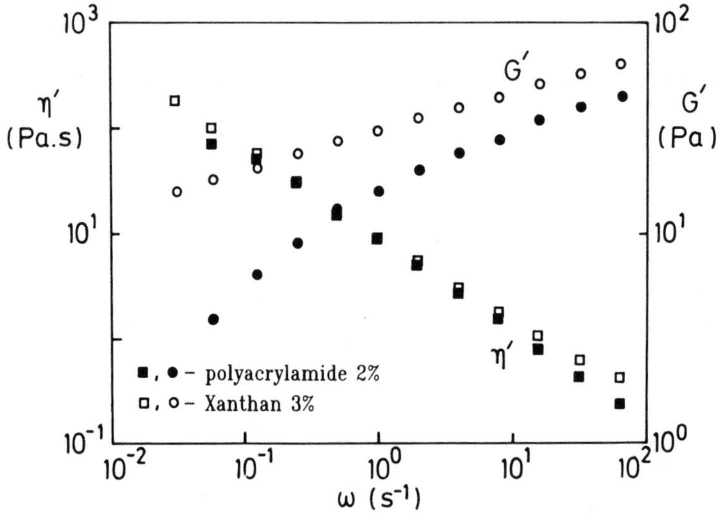

Fig 5. Dynamic data for a 2% aqueous solution of polyacrylamide and a 3% aqueous solution of xanthan gum.

response for the xanthan solution is associated with significantly higher values of G' than for the comparable polyacrylamide solution, so that, *in this deformation mode,* the xanthan gum solution is more elastic!

To the theoretician, the above conclusion would be unexpected in view of the continuum-mechanics requirements [8]:

$$\left.\begin{aligned} \eta'(\omega)\Big|_{\omega\to 0} &= \eta(\dot\gamma)\Big|_{\dot\gamma\to 0}, \\ \frac{G'(\omega)}{\omega^2}\bigg|_{\omega\to 0} &= \frac{N_1(\dot\gamma)}{2\dot\gamma^2}\bigg|_{\dot\gamma\to 0}. \end{aligned}\right\} \qquad (9)$$

The second relationship would certainly lead one to expect a 1.1 correspondence between N_1 and G'. This is invariably found to be so when just one polymer type is being studied, but we are here discussing two *different* polymers and, across this divide, conventional expectations are not realized.

There is of course the problem of reconciling the data of Figs 1 and 5 with the limiting relations (9). This is attempted in Figs 6-9.

The first of the relations in (9) is clearly not violated, with $\eta' \leq \eta$ over the available $\omega/\dot{\gamma}$ ranges. The expected plateau regions in η and η' are not reached at low shear rates/frequencies in the present experiments.

The second of the relations in (9) is also not violated by the data, although, once again, the anticipated plateau regions in $N_1/2\dot{\gamma}^2$ and G'/ω^2 at very low $\dot{\gamma}/\omega$ have not been reached in the present experiments.

The Cox-Merz [9] rule has often been found to apply for polymeric liquids. In this rule, $|\eta^*|$ as a function of frequency ω is predicted to be a similar function to η plotted as a function of $\dot{\gamma}$. Figures 10 and 11 show that the polyacrylamide solution obeys the Cox-Merz rule, but the xanthan solution does not [cf. 4].

Figures 12 and 13 further illustrate the change of roles for the polyacrylamide and xanthan solutions, this time for two lower concentrations. With evidence from even lower concentrations, not reproduced here, we are in a position to conclude that, near the rest state, the xanthan solutions are more elastic than their polyacrylamide counterparts, but that, in steady shear and extensional flow, the polyacrylamide solutions are certainly more elastic. Loosely, we may say that the xanthan solutions behave more like colloidal systems, with a strong rest structure which is readily broken down by shear. The polyacrylamide solutions behave like typical polymeric liquids, with a strong viscoelastic response in steady shear and extensional flow.

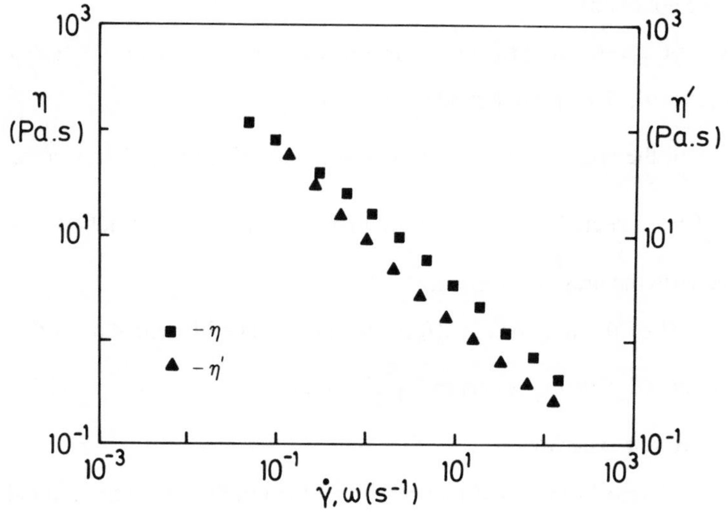

Fig 6. η and η′ data for the 2% polyacrylamide solution.

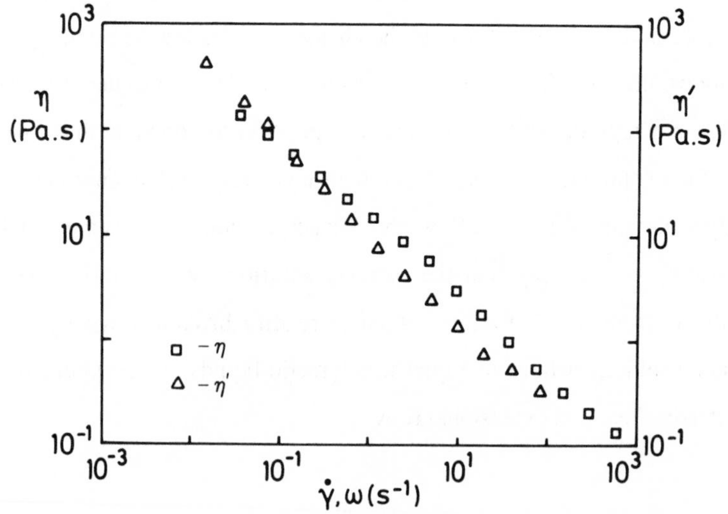

Fig 7. η and η′ data for the 3% xanthan solution.

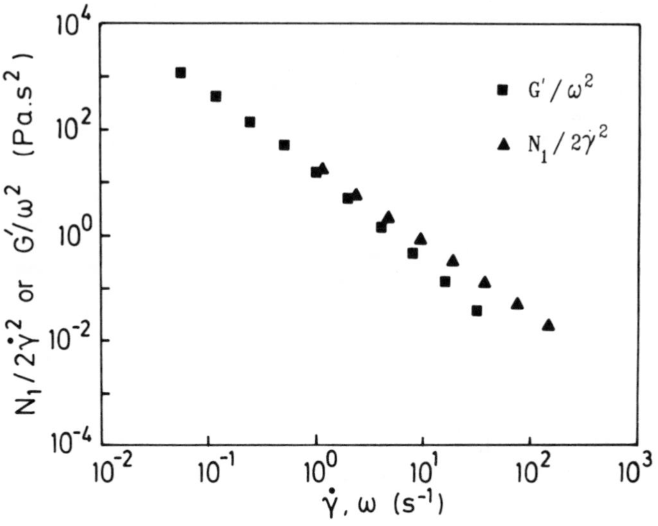

Fig 8. G' and N_1 data for the 2% polyacrylamide solution.

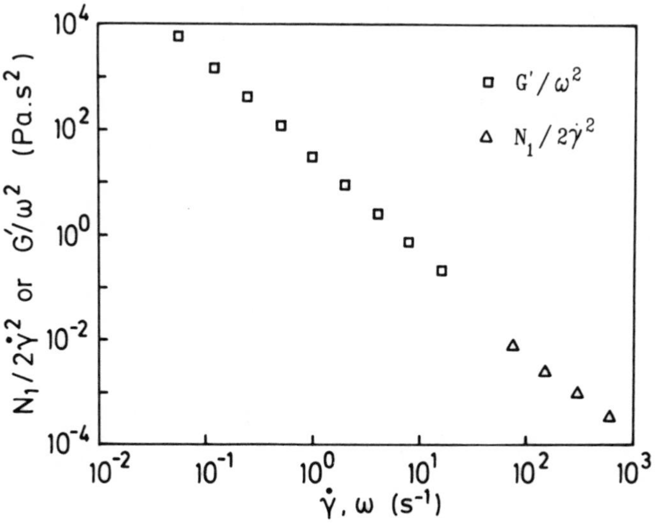

Fig 9. G' and N_1 data for the 3% xanthan solution.

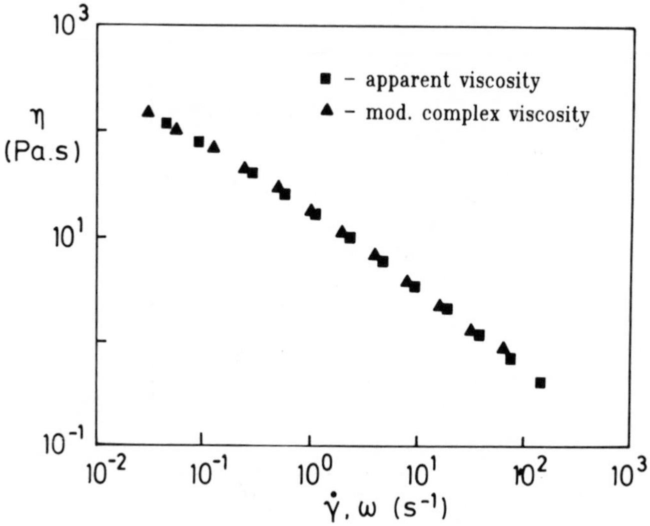

Fig 10. Application of the Cox-Merz rule to the 2% polyacrylamide solution.

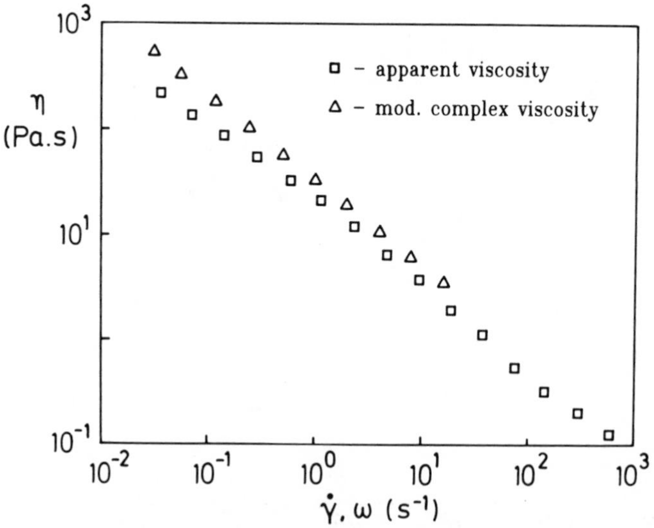

Fig 11. Application of the Cox-Merz rule to the 3% xanthan solution.

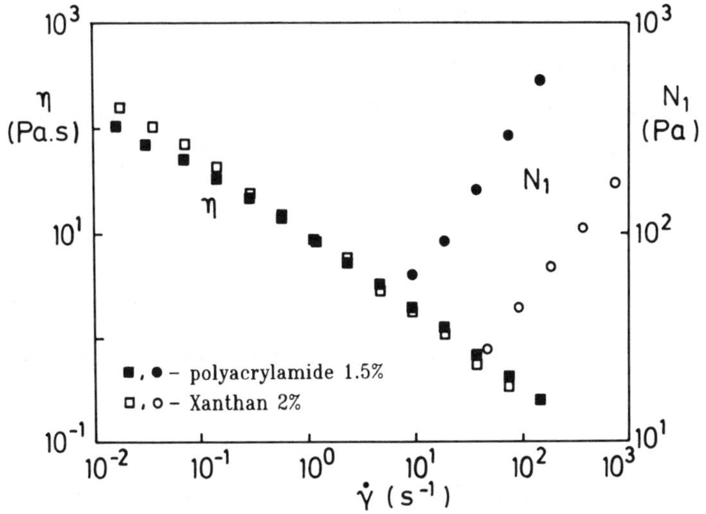

Fig 12. Steady shear data for a 2% xanthan solution and a 1.5% polyacrylamide solution.

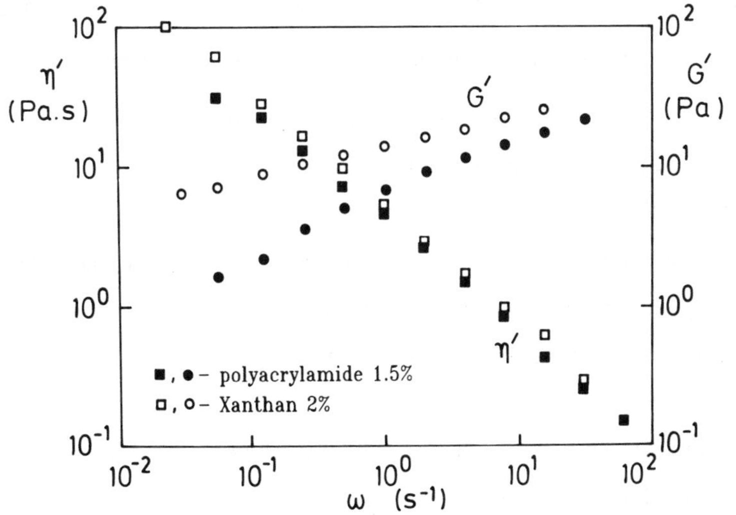

Fig 13. Dynamic data for a 2% xanthan solution and a 1.5% polyacrylamide solution.

5. NON LINEAR BEHAVIOUR IN OSCILLATORY SHEAR FLOW

It is of interest to consider *large strain* oscillatory shear flow for two reasons. First, it is important to confirm that conventional dynamic testing is carried out in the linear regime. Secondly, the non linear effects themselves are able to tell us something about the rheology of the solutions.

Figures 14 and 15 contain representative dynamic data for one frequency. The behaviour is typical of that expected for elastic liquids, with a linear region followed by decreasing values of both η' and G' as the strain is increased, the fall in G' being greater than the fall in η' [cf. 7].

In Figs 14 and 15, the fall in both η' and G' is more pronounced for the xanthan solution. In the light of the earlier discussion, this is not unexpected.

Finally, we show results for a *combined* steady and oscillatory shear flow in which the velocity field is given by [7]

$$v_x = \gamma y + \alpha\omega y\cos\omega t, \quad v_y = v_z = 0, \tag{10}$$

and the dynamic properties are now functions of γ as well as ω. Figures 16 and 17 contain the relevant data. The polyacrylamide results are in line with expectation [cf. 7], but the η' results for the 3% xanthan solution are unusual with $\eta'(\gamma,\omega) > \eta'(0,\omega)$ for some frequencies and shear rates. This may be another manifestation of the structure build up and breakdown discussed in relation to Figs 3 and 4.

6. CONCLUSIONS

(i) The polymer concentration of aqueous solutions of polyacrylamide and xanthan gum can be chosen to yield very similar shear viscosity behaviour.

(ii) The dynamic properties of xanthan solutions can be significantly affected by pre shearing.

(iii) The xanthan solutions are highly elastic near the rest state with a gel-like structure, but this structure is easily broken down by shear.

(iv) The polyacrylamide solutions behave as typical polymeric liquids in shear flow and extensional flow with a strong viscoelastic response.

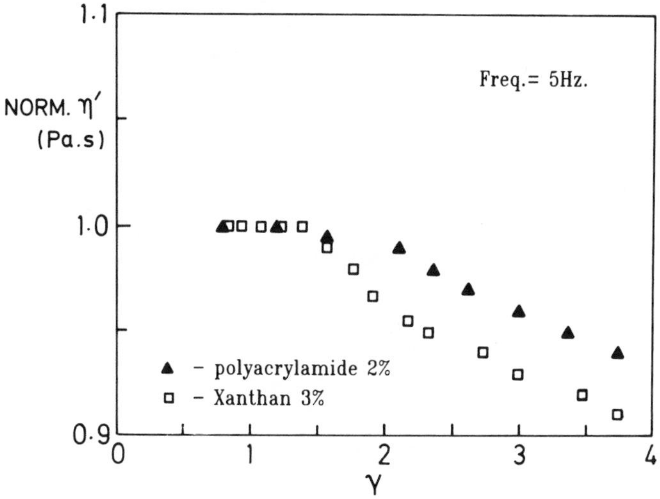

Fig 14. Dynamic viscosity data normalized with respect to its vanishingly small strain value for ω=5Hz. γ is strain.

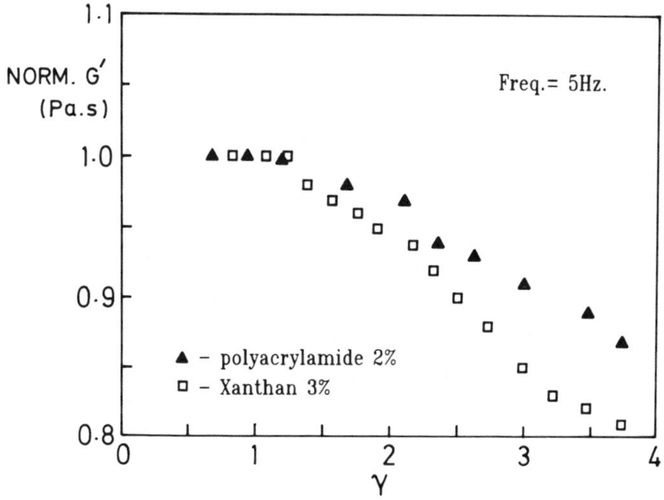

Fig 15. Dynamic rigidity data normalized with respect to its vanishingly small strain value for ω=5Hz. γ is strain.

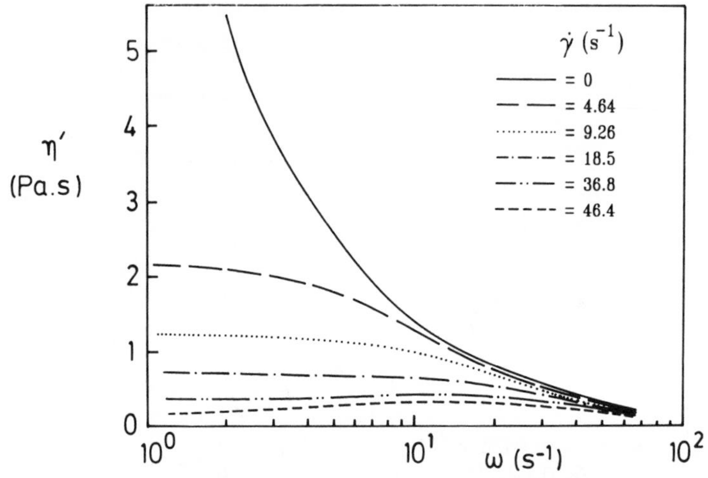

Fig 16. Combined steady and oscillatory shear η′ data for the 2% polyacrylamide solution.

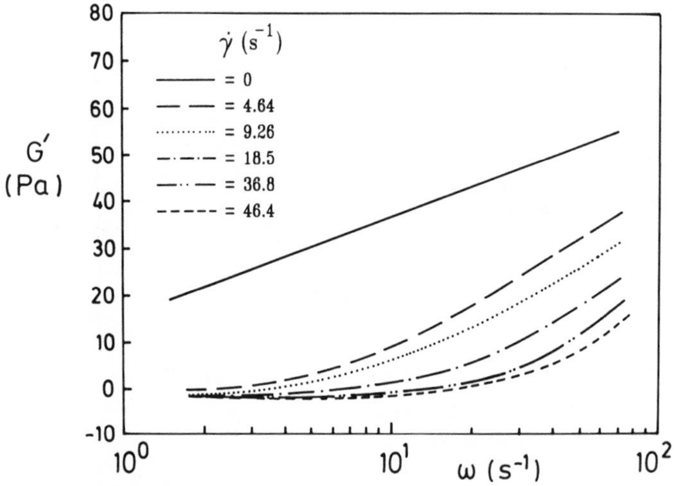

Fig 17. Combined steady and oscillatory shear G′ data for the 2% polyacrylamide solution.

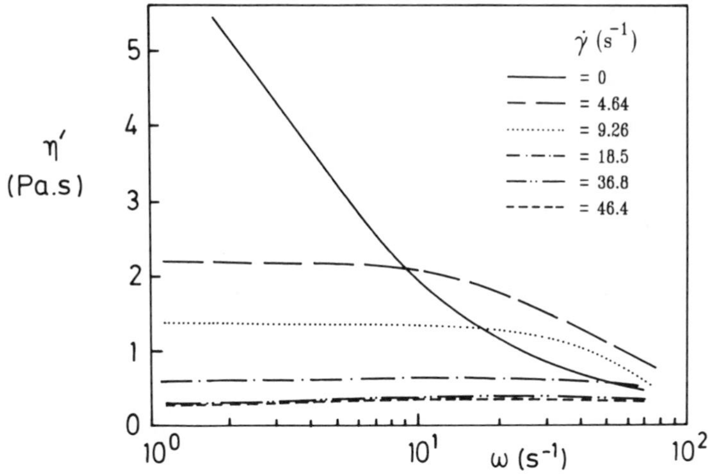

Fig 18. Combined steady and oscillatory shear η′ data for the 3% xanthan gum solution.

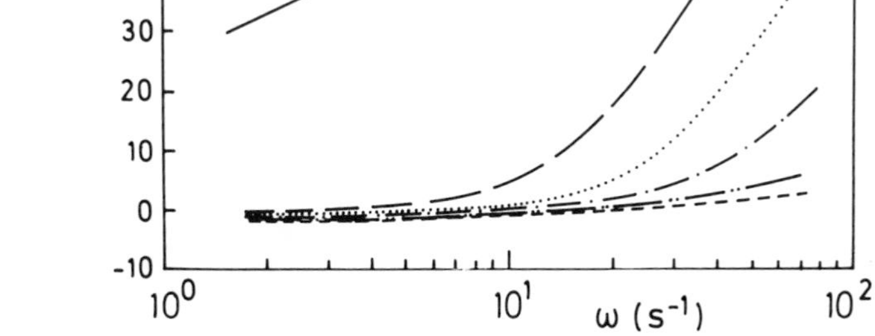

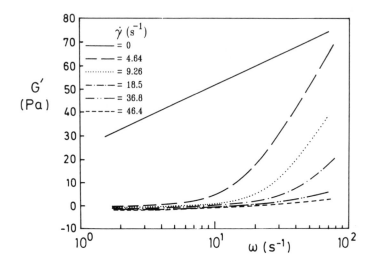

Fig 19. Combined steady and oscillatory shear G′ data for the 3% xanthan gum solution.

(v) It is dangerous to expect a 1:1 correspondence between N_1 and G' if the exercise is carried out across polymer types.

(vi) There is no simple answer to the question: "What is meant by the expression 'highly elastic liquids'?" without a reference to the particular deformation mode in mind.

ACKNOWLEDGMENTS

We have benefitted from discussions and assistance from D M Binding, M C Couch, R E Evans, D M Jones and W M Jones.

REFERENCES

1. W E Rochefort and S Middleman, Journal of Rheology 31 (1987) 337-369.
2. D M Jones and K Walters. To appear in Rheologica Acta.
3. T Lim, J T Uhl and R K Prud'homme, Journal of Rheology 28 (1984) 367-379.
4. K C Tam and C Tiu, Journal of Rheology 33 (1989) 257-280.
5. A Ait-Kadi, P J Carreau and G Chauveteau, Journal of Rheology 31 (1987) 537-561.
6. K F Wissbrun, Journal of Rheology 25 (1981) 619-662.
7. K Walters, 'Rheometry', Chapman and Hall, 1975.
8. H A Barnes, J F Hutton and K Walters, 'An Introduction to Rheology', Elsevier 1989.
9. W P Cox and E H Merz, Journal Polymer Sci. 28 (1958) 619-622.

Professor K Walters
Department of Mathematics
University College of Wales
Aberystwyth, UK.

Mr A Q Bhatti
Department of Mathematics
University College of Wales
Aberystwyth, UK.

Dr N Mori
Department of Mechanical Engineering
Nara University
Nara, Japan.

P. SCHÜMMER, H.G. THELEN AND C. BECKER

Rheological problems in slender bodies-jets

1. Introduction

Fluid jets play an important role in technical applications such as fibre spinning, ink–jet printing or extrusion of polymers. For reasons of process design the prediction of jet behaviour is of great interest. An important point in this field of investigations is the controlled break–up of liquid jets. Research into this topic was begun by Lord Rayleigh [23], who used the method of linear perturbation analysis. Modern publications deal with the non–linear problem by means of perturbation analysis of higher order [6,7] or by numerical methods [21,27]. The liquid jets analysed so far mostly consist of inviscid or linear–viscous fluids.

This lecture deals with the controlled break–up of viscoelastic liquid jets. In addition to the effects of surface tension, inertia and viscous friction, elastic forces play an important role and a special feature appears. The growing droplets are connected by filaments which hinder the break–up process considerably or even make it impossible. This phenomenon is due to the resistance of viscoelastic liquids to extensional deformations [24].

Green and Naghdi [14,17,18,19] proposed a one–dimensional continuum theory for slender bodies allowing an economical calculation, which we intended to use for jet break–up of a viscoelastic liquid. Another possibility of deriving a set of one-dimensional equations to describe the motion of slender bodies is also given by Green et al. [16,17,18]. Roughly spoken the position vector to a material point of the jet is expanded in a McLaurin series about the jet axis. Then, taking into account the slenderness of the jet and thus retaining only terms of first order, the resulting approximation of the position vector is introduced into the three–dimensional equations. By integration over a cross–section perpendicular to the jet axis a system of one–dimensional equations of motion can be derived. Either of the two methods furnishes essentially the same equations which can be completely identified under certain assumptions [18].

In what regards the constitutive equation we started with the usual, three–dimensional Jeffreys model to find the one–dimensional form by the above mentioned

procedure of integration over the jet cross–section. Bechtel et al. developed a one-dimensional Jeffreys law from indifference principle by a direct approach but dealt with other forms of a jet. Bousfield et al. [9] calculated the same geometry but in this special case without the influence of inertia. On the other hand they took into account an initial stress which is realistic for other special problems.
In this lecture we would like to demonstrate the good agreement between the theoretical predictions and experimental observation.

2. The equation of motion of a slender body in Green's continuum theory

Due to the slenderness of liquid jets, Green and Naghdi [14,17,18,19] simplified the general three–dimensional balance of momentum to a one–dimensional equation, considering the influence of the radial distribution of the field–functions. In several papers, Bogy [3–8] has shown the efficiency of Green's theory with reference to Newtonian jet break–up.
Hence, the "equation of motion" for an axisymmetric, horizontal, incompressible liquid jet is

$$T = \frac{1}{We} X + Y \qquad (1)$$

where T designates "the dimensionless inertia forces"

$$T = v_t + vv_z - \frac{1}{2} R R_z (v_{zt} + vv_{zz} - \frac{1}{2} v_z^2 - \frac{1}{8} R^2(v_{zzt} + vv_{zzz}) \qquad (2)$$

X represents the dimensionless forces of the free surface

$$X = \frac{R_z}{R^2(1+R_z^2)^{1/2}} + \frac{R_z R_{zz}}{R(1+R_z^2)^{1/2}} + \frac{R_{zzz}}{(1+R_z^2)^{3/2}}$$

$$- \frac{3 R_z R_{zz}^2}{(1+R_z^2)^{5/2}} \qquad (3)$$

and Y stands for forces of the internal friction

$$Y = (n_z^* - q_z^* + p_{zz}^*)/R^2 \;. \qquad (4)$$

n^*, q^* and p^* are to be fixed by an appropriate constitutive assumption. Equations (1)–(4) are dimensionless with reference to the radius a of the initially undisturbed jet, the mean jet velocity $\bar{v}$, the mass density ρ and the surface tension σ of the fluid "We" stands for the Weber number

$$We = \rho \frac{a \ \bar{v}^2}{\sigma} \quad . \qquad (5)$$

The subscripts indicate partial differentiation with respect to time or to the local co-ordinate.

R and v are the shape and velocity functions, both dependent on time and location. The forces n^* and q^* consider the axial and radial and the azimuthal stresses respectively which are integrated over the cross-section. p^* stands for the cross-sectional integral of the moment of the shear stress. Depending on the constitutive equation further dimensionless numbers will appear in Y. For the inviscid fluid with Y=0, the We-number is the only parameter. For a Newtonian liquid the Re-number and for a viscoelastic liquid additional dimensionless rheological parameters appear.

3. Boundary and initial conditions of the jet problem

There are two fundamentally different approaches to the simulation of jet break-up.

The first is to consider the spatial instability of time harmonic disturbances of a semi–infinite jet. The second is to look at the problem as a temporal instability of a spatially harmonic disturbed infinite jet. The latter procedure is approximately equivalent to the use of a frame of reference moving along the jet axis with the mean velocity of the jet. Because of the spatial periodicity of this problem only one wavelength has to be considered in the simulation and calculations can be done much more economically as in the first case, which however is the more realistic approach. It corresponds to the use of a co–ordinate system fixed at the nozzle from which the jet emanates. Nevertheless, in our investigations we chose the second method because computation is much more costly with the first approach. Another advantage is, that boundary conditions can be specified in a relatively simple way. For high Weber–numbers, i.e. high jet velocities, the results calculated in the moving co–ordinate system are in good agreement with those in the fixed frame of reference [20].

Taking the picture of the moving co–ordinate system it can be stated that the equations describing the dynamics of jet break–up are the same as in the fixed frame of reference because the motion is uniform. Nevertheless, the definition of the dimensionless numbers is no more valid because the mean jet velocity $\bar{v}$ cannot be employed in these definitions in the moving system. One has to define a new reference velocity

$$\bar{v}' = \sqrt{\frac{\sigma}{\rho a}} \quad . \tag{6}$$

Introducing this into the dimensionless numbers yields:

$$We' = 1 \tag{7}$$

$$Re' = \rho\frac{\bar{v}' a}{\eta_0} = \frac{\sqrt{\sigma\rho a}}{\eta_0} = \frac{1}{Oh} \tag{8}$$

$$De' = t_0 \frac{\bar{v}'}{a} = \frac{t_0}{a}\sqrt{\frac{\sigma\rho a}{\eta_0}} \tag{9}$$

with the Ohnesorge–number (Oh) and the Deborah–number (De) which will be introduced later.

Considering these quantities in the fixed co–ordinate system one gets

$$Oh = \frac{\sqrt{We}}{Re} \qquad \text{and} \qquad De' = \frac{De}{\sqrt{We}} \quad . \tag{10}$$

The shape function in the moving system is

$$R' = R \quad . \tag{11}$$

De and De' respectively can be interpreted as a dimensionless time and so

$$t' = \frac{t}{\sqrt{We}} \tag{12}$$

$$v' = v\sqrt{We} \tag{13}$$

and

$$F' = F \cdot We \tag{14}$$

where F' and F are dimensionless forces.

So, if one uses different Weber–numbers calculating jet break–up in the moving co–ordinate system they only represent a stretching factor for all quantities containing the dimension of time.

Now we have to look at the boundary and initial conditions valid for the infinite jet problem. The surface of the liquid jet is described by a dynamic and a kinematic condition. The dynamic condition of vanishing shear stress at the surface is already implied in equation (3). The kinematic condition requires that the temporal gradient of the shape function dR/dt equals the radial component of the velocity. The application of this condition leads to

$$R^2_t + (v\, R^2)_z = 0 \quad . \tag{15}$$

The jet shape is initially disturbed by

$$R(z,t=0) = 1 + \delta^1(0) \cos(2\pi z/\lambda) \quad , \tag{16}$$

$\delta^1(0)$ is the dimensionless amplitude of the shape disturbance and λ the dimensionless wavelength. The periodical form of the problem permits the solution of the governing equations in the range of half a wavelength

$$0 \leq z \leq \lambda/2 \tag{17}$$

The initial conditions for the velocity v and the internal forces n^* , q^* and p^* will be derived from a linear stability analysis [25].

4. Constitutive equation of an upper–convected Jeffreys fluid

The approach to the constitutive equation cited below is orientated at a formalism introduced by Caulk [10] for elliptical jets of a Newtonian fluid. We start with the equation for an upper–convected Jeffreys fluid (or Oldroyd–B fluid)

$$\underline{\underline{S}}^* + t_1 \overset{\nabla}{\underline{\underline{S}}^*} = 2\eta_0(\underline{\underline{D}} + t_2 \overset{\nabla}{\underline{\underline{D}}}) \tag{18}$$

with

$$\underline{\underline{S}} = -p\underline{\underline{I}} + \underline{\underline{S}}^* \tag{19}$$

and the operation $\overset{\nabla}{\underline{\underline{T}}}$ of upper–convected time derivative of a tensor $\underline{\underline{T}}$

$$\overset{\nabla}{\underline{\underline{T}}} = \dot{\underline{\underline{T}}} - (\nabla\underline{v})^T \cdot \underline{\underline{T}} - \underline{\underline{T}} \cdot \nabla\underline{v} \tag{20}$$

In contrast to, for example, the corotational Maxwell fluid [1], equation (18) describes the resistance of viscoelastic liquids to extensional deformations. The enlarged stability of viscoelastic jets is connected with this feature [24].
The internal stresses of the Jeffreys fluid can be composed of a Maxwell and a

Newtonian part:

$$\underline{\underline{S}}^* = \underline{\underline{S}}_M + \underline{\underline{S}}_N \,. \qquad (21)$$

with

$$\underline{\underline{S}}_N = 2\eta_N \underline{\underline{D}} \quad ; \quad \underline{\underline{S}}_M + t_1 \overset{\nabla}{\underline{\underline{S}}} = 2\eta_M \underline{\underline{D}} \qquad (22)$$

The formulation with these two partial stresses should be used for an effective numerical procedure.

The relations of the different parameters from the Jeffreys, Maxwell and the Newtonian model are:

$$\eta_0 = \eta_M + \eta_N \quad ; \quad t_2 = t_1 \frac{\eta_N}{\eta_M + \eta_N} \qquad (23)$$

where η_M is the partial viscosity which belongs to the relaxation time t_1.

A set of dimensionless parameters will appear in connection with eqs. (23), namely

$$Re_M = \rho \frac{\bar{v}\,a}{\eta_M} \;, \quad Re_N = \rho \frac{\bar{v}\,a}{\eta_N}$$

$$De_2 = De_1 \frac{Re_M + Re_N}{Re_N} \,. \qquad (24)$$

These can be related to the usually employed dimensionless numbers

$$Re = \rho \frac{\bar{v}a}{\eta} \quad ; \quad De = t_1 \frac{\bar{v}}{a} \quad ; \quad \lambda = \frac{t_2}{t_1} \qquad (25)$$

by

$$1/Re = 1/Re_M + 1/Re_N \qquad (26)$$

and

$$De = De_1 \quad ; \quad \lambda = \frac{De_2}{De_1} \tag{27}$$

The consequent treatment of the three–dimensional form of the Maxwell constitutive equation leads to the one–dimensional equivalent [28]:

$$n^*_M + De_1(n^*_{Mt} + vn^*_{Mz} - v_z n^*_M) = \frac{2}{Re} Re^2 v_z \tag{28}$$

$$\bar{q}_M + De_1(\bar{q}_{Mt} + v\bar{q}_{Mz} + 2v_z\bar{q}_M) = \frac{1}{Re_M} R^2(\frac{1}{8} R R_z v_{zz} - v_z) \tag{29}$$

$$p^*_M + De_1(p^*_{Mt} + v\, p^*_{Mz} + v_z p^*_M) = -\frac{1}{8\, Re_M} R^4 v_{zz} \tag{30}$$

for the resultants from the Maxwell part of the stress, where the transformed force

$$\bar{q}_M = q^*_M - R_z\, p^*_M \, . \tag{31}$$

has been used to decouple the equations for n^*, $\bar{q}$ and p^*.

The one–dimensional resultants from the Newtonian part of the stress can be inserted directly into the term of the internal forces. One gets finally:

$$Y = \frac{1}{Re_N}\left[\frac{6R_z v_z}{R} + v_{zz}(3 - \frac{3}{2} R_z^2 - \frac{1}{2} RR_{zz}) - RR_z v_{zzz} - \frac{R^2}{8} v_{zzzz}\right]$$

$$+ \frac{1}{R^2}\left[n_{M_z} - q_{M_z} + p_{M_{zz}} - \frac{R_z}{R} p_{M_z} - (\frac{R_{zz}}{R} - \frac{R_z^2}{R})p_M\right] \tag{32}$$

5. Numerical solution

By approximation of the spatial and time derivatives in eqs. (1–3,15,28–30,32) by finite differences, a numerical solution of these equations is available. The finite

differences proposed by Shine et al. [26] are useful for this purpose. One gets a set of five pentdiagonal equations, which have to be solved by an iterative procedure. We use the balance of mass eq.(15) to calculate the shape–function R(z,t) from the initial values. The equation of motion then determines the velocity function.
The constitutive equations of the Maxwell–type (eqs.28–30) are used to calculate the non–linear part of the inner forces n_M, q_M, p_M.
This results in an algorithm in which the field equations are solved for each small incremental time step. The computation starts in the first time level using the solution of the linear analysis, which is a modification of Middleman's [22] and Goren and Gottlieb's [13] analysis. A perturbation of the radius as given by eq.(7) is impressed on the jet. We introduce the disturbed quantities

$$\delta = R - 1 \quad , \quad w = v - 1 \quad , \quad n = n^* \quad ,$$
$$q = \bar{q} \quad , \quad p = p^* \tag{33}$$

and make the following ansatz for these functions:

$$f = f_0 \, e^{\alpha t - ikz} \tag{34}$$

where f is δ, w, n, q and p respectively. k is the dimensionless wave–number of the periodic disturbance

$$k = \frac{2\pi}{\lambda} \tag{35}$$

and α is the temporal growth rate of the disturbance.
With (33) and (34) we get the dispersion relation

$$\alpha^2 + \frac{1}{Re}\,\frac{24+k^2}{8+k^2}\,k^2\,\frac{1+De_2\alpha}{1+De_1\alpha}\,\alpha = \frac{4}{We}\,\frac{(1-k^2)k^2}{8+k^2} \tag{36}$$

from the linearized equations.

Moreover, with given δ_0 and k we get

$$w_0 = -2 \frac{\alpha}{k} \delta_0 i \tag{37}$$

$$n_0 = -\frac{4}{Re} \delta_0 \alpha \frac{1+De_2\alpha}{1+De_1\alpha} \tag{38}$$

$$q_0 = \frac{2}{Re} \delta_0 \frac{1+De_2\alpha}{1+De_1\alpha} \tag{39}$$

$$p_0 = -\frac{1}{4Re} \delta_0 \alpha k \frac{1+De_2\alpha}{1+De_1\alpha} i \tag{40}$$

for the amplitudes and

$$w(z,0) = w_0 \sin kz, \tag{41}$$

$$n^*(z,0) = n_0 \cos kz, \tag{42}$$

$$\bar{q}(z,0) = q_0 \cos kz, \tag{43}$$

$$p^*(z,0) = p_0 \sin kz \quad . \tag{44}$$

With the assumed disturbance of the shape function it suffices to consider only half a wavelength and appropriate boundary conditions have to be prescribed for the numerical solution of the problem. Because of the symmetry these conditions are

$$R(z) = R(-z) \tag{45}$$

$$v(z) = -v(-z) \tag{46}$$

$$n^*(z) = n^*(-z) \tag{47}$$

$$\bar{q}(z) = \bar{q}(-z) \tag{48}$$

and

$$p^*(z) = -p^*(-z) \tag{49}$$

for $z=0$ and $z=\lambda/2$ respectively.

6. Results

Of special interest is the influence of elasticity on jet break–up. The variation of parameters illustrates that the formation of a long living filament is mainly influenced by the amount of inertia. The behaviour of a Maxwell fluid with Re = 37.4, $De_1 = 1$, $De_2 = 0$ hardly differs from that of a Newtonian with $De_1 = 0$. It is only through the reduction of the inertial forces that filament formation occurs. For a jet with 20 times the viscosity of water (Re = 1.87) the formation of the filaments is easily observed. This underscores the experimental result we got [24]: the flow in the filament region can be seen as a homogeneous extensional flow. This kind of flow is only possible if the condition of negligible inertia is fulfilled. The development of the velocity function as well as the inner forces underscore the elongational flow situation in the thin filaments. We have shown all this in detail elsewhere [25]. The uniaxial extension of the filament region is confirmed by two additional functions. The differential equation for the filament radius of a Maxwell fluid [24] has an asymptotic solution of the form

$$R/R_k = e^{-\frac{1}{3}\frac{t}{De_1}} \quad , \tag{50}$$

where R_k is a suitable constant. Figure 1 shows that the development of the radius at the neck $R(\lambda/2)$ approaches this asymptotic solution. This result is similar to the one of Bousfield et al. [9] who calculated the break–up of a liquid jet of a Maxwell fluid without inertia and shear stress.

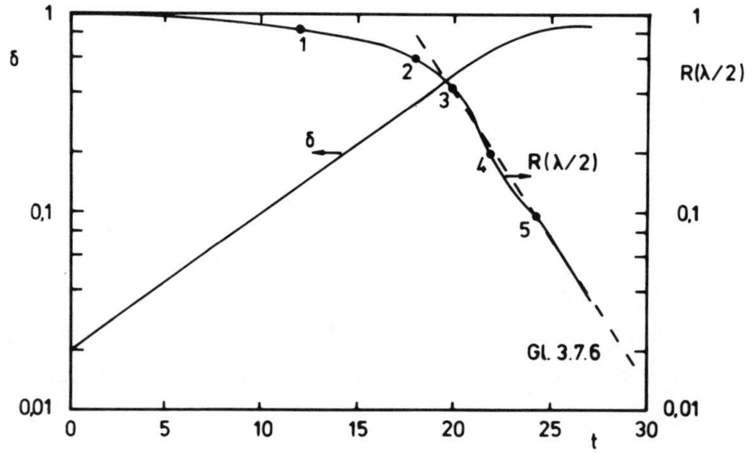

Fig.1 Temporal growth of the disturbance $\delta(t)$ and development of the neck $R(\lambda/2)$; dashed line: eq. (50)

If we look at a perturbation measure defined by

$$\delta(t) = \frac{1}{2} \{R\ (0,t) - R(\lambda/2,t)\} \quad , \qquad (51)$$

the following can be stated: In the beginning the growth of the perturbation is exponentially in time, corresponding to the results of the linear analysis. The appearance of the viscoelastic filaments dampens the growth of the perturbation remarkably (Fig. 1). The introduction of a retardation time in the constitutive equation leads to no recognizable influence on the development of the shape function $R(z,t)$. But it has remarkable influence on the numerical procedure. As shown in figure 2 for a very viscoelastic Maxwell–fluid the development of the perturbation δ as well as the filament radius $R(\lambda/2)$ tends to oscillate in the subsequent non–linear

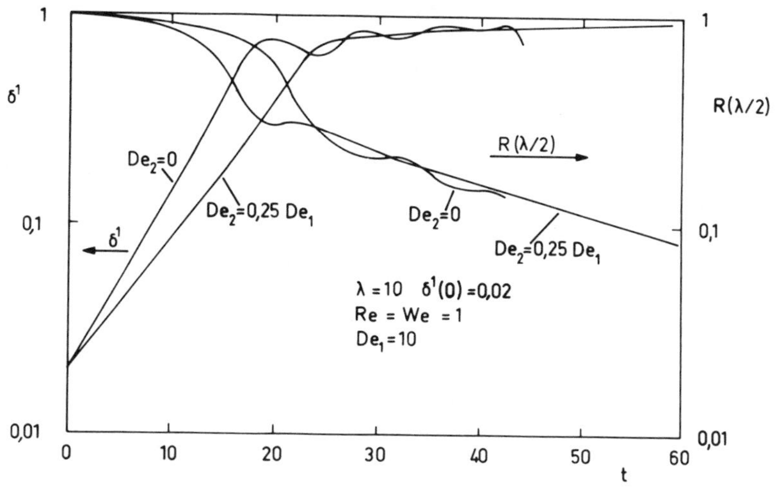

Fig.2 Influence of the retardation time on the calculation of jet break–up

region. Introducing a Jeffreys–fluid with $De_2=0{,}25De_1$ leads to a stable, monotonous numerical solution. Moreover, the algorithm marches further on in time compared to the Maxwell calculation. Furthermore, the calculation is done in only a third of the time.

We want to compare our numerical results with the real behaviour of a viscoelastic fluid jet. The photographs of the jet show the jet shape in the fixed laboratory system. So, in a first step, the results of the numerical calculation, which stem from a moving system, have to be transformed into a fixed frame of reference. This can be done by the transformation

$$t = x \text{ and } z = x - T \qquad (52)$$

(non–dimensional by $\bar{v}$ and a), which transforms the results of the calculation R(z,t) to the shape function in the fixed frame R*(x,T) [11]. In figure 3 a comparison is given of a jet of 5% polyvinylalcohol in water, disturbed periodically by a vibrating nozzle, to the calculated shape function. The dimensionless parameters of the expe–

rimentally realized jet were

$$\mathrm{We} = 180 \qquad \mathrm{Re} = 24$$

$$\mathrm{De}_1 = 40 \qquad \mathrm{De}_2 = 10.$$

Only the amplitude of the initial disturbance $\delta^1(0) = 0.1$ is fitted to the experiment. The elasticity parameter De_1 is adjusted with measurements we worked out in another connection [24] while De_2 is estimated from such data.

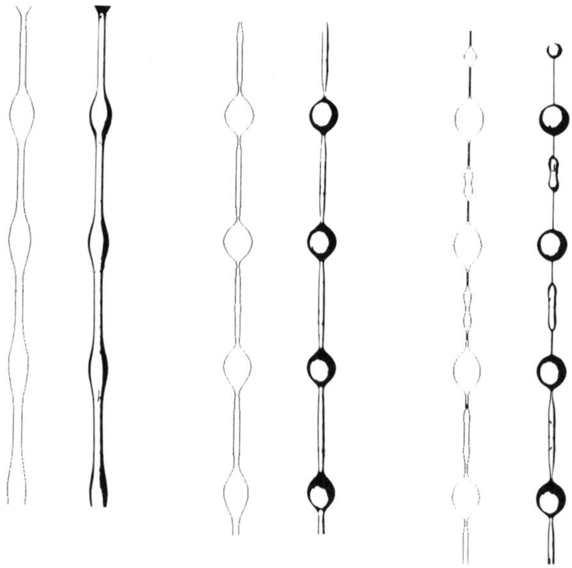

Fig.3 Comparison of a calculated and an experimentally realized break–up of a viscoelastic jet. For illustration purposes the jet is divided into three parts.

The numerically calculated shape of the jet agrees very well with that of the photograph. The upper part of figure 3 shows the zone of the jet where the visco–elastic filaments appear. Due to the influence of inertia the extensional flow in the filaments is disturbed after a while as shown in the middle and the filaments tend to form satellite droplets. Now the viscoelastic nature of the fluid gains importance again. As can be seen in the lower part of figure 3 main and satellite drops are connected with a thin filament. The shape of the satellite drop changes through the forces of surface tension from a dumbbell to a sphere.

Summary

Certain continuous materials can be considered as structured or directed. They may be structured either in a physical or a geometrical sense. The present paper deals with jets of viscoelastic liquids which are an example for the latter case. Their structure is based on the slenderness of the body, i.e. the fact that the change of field properties in axial direction is strong compared to the change in radial direction.

In considering structured continua certain effects can often be neglected. In the present case for instance, all effects higher than linear order are neglected in the radial velocity distribution. Thus, the theory utilized here is not able to describe for example the fully developed parabolic velocity profile of a liquid emanating from the nozzle. Green et al. used this simplification and obtained a one–dimensional balance of momentum and mass respectively by integration over a cross–section of the slender body.

We also employed this method and transformed the upper–convected Jeffreys constitutive equation into one dimension. With this set of equations we investigated viscoelastic jet break–up numerically using a finite difference technique.

For this purpose another simplification has been made by considering the problem as a temporal instability of spatially harmonic disturbances of an infinite column of liquid. In spite of the simplifications the theory employed here yields quite good results in the numerical simulation of the disintegration of a jet. These are so encouraging that related problems should be considered on the basis of this or similar methods.

Acknowledgement

We are grateful to the "Deutsche Forschungsgemeinschaft" for supporting this study.

Nomenclature:

a	– radius of the undisturbed jet
De	– Deborah–number
k	– wave number
n^*	– axial and radial stresses integrated over the jet cross–section
p^*	– moment of shear stress integrated over the jet cross–section
q^*	– azimuthal stresses integrated over the jet cross–section
R	– shape function
Re	– Reynolds–number
$\underline{\underline{S}}$	– Cauchy stress tensor
$\underline{\underline{S}}^*$	– extra–stress tensor
t	– time
t_1	– relaxation time
t_2	– retardation time
T	– dimensionless inertia forces
v	– velocity function
$\bar{v}$	– mean jet velocity
We	– Weber–number
X	– dimensionless forces of the free surface
Y	– dimensionless forces of the inner friction
z	– axial co–ordinate
$\delta^1(0)$	– dimensionless amplitude of the shape disturbance
$\delta(t)$	– perturbation measure
λ	– wavelength of the shape disturbance
η	– viscosity
ρ	– mass density
σ	– surface tension

References

[1] Bechtel, S.E. , Forest, M.G. , Bogy, D.B. , J. Non–Newt. Fluid Mech. 21 (1986) 21–273

[2] Bechtel, S.E. , Lin, K.J. , Forest, M.G. , J. Non–Newt. Fluid Mech 26 (1987) 1–41

[3] Bogy, D.B. , Phys. Fluids 21 (1978) 190

[4] Bogy, D.B. , J. App. Mech. 45 (1978) 469

[5] Bogy, D.B. , Ann. Rev. Fluid Mech. 11 (1979) 207

[6] Bogy, D.B. , Phys. Fluids 22 (1979) 224

[7] Bogy, D.B. , IBM J. Res. Develop 23 (1979)

[8] Bogy, D.B. , Shine, S.J. , Talke, F.E. , J. Comp. Phys. 38, 3(1980) 294–326

[9] Bousfield, D.W. , Keunings, R. , Marucci, G. , Denn, M.M. , J. Non-Newt. Fluid Mech 21 (1986) 79–97

[10] Caulk, D.A. , Ph.D. Thesis , Uni. Calf. Berkeley (1976)

[11] Chaudhary, K.C. , Redekopp, L.G. , J. Fluid Mech. 96 (1980) 257–274

[12] Crochet, M.J. , Keunings, R. , J. Non–Newt. Fluid Mech. 10 (1982) 339–356

[13] Goren, S.L. , Gottlieb, M. , J. Fluid Mech. (1982) 120–245

[14] Green, A.E. , Laws, N. , Proc. R. Soc. Ser. A 283 (1966) 145–155

[15] Green, A.E. , Laws, N. , Int. J. Eng. Sci. 6 (1968) 317–328

[16] Green, A.E. , Laws, N. , Naghdi, P.M. , Proc. Camb. Phil. Soc. 64, (1968), 895–913

[17] Green, A.E. , Naghdi, P.M. , Int. J. Solids Structure 6 (1970) 209–244

[18] Green, A.E. , Naghdi, P.M. , Wenner, M.L. , Proc. R. Soc. London A 337 (1974) 451–507

[19] Green, A.E. , Int. J. Eng. Sci. 14 (1976) 49

[20] Keller, J.B. , Rubinow, S.J. and Tu, Y.O. , Phys. Fluids, 16 (1973) 2052

[21] Keunings, R.J. , Comp. Phys. 62 (1986) 199

[22] Middleman, S. , Chem. Eng. Sci (1965) 20–1037

[23] Lord Rayleigh, Proc. London Math. Soc. 10 (1878) 4

[24] Schümmer, P. , Tebel, K.H. , Proceedings 2nd World Congress of Chem. Engineering Montreal Canada 1981

Germ. Chem. Eng. 5 (1982) 209

[25] Schümmer, P. , Thelen, H.G. , Rheol. Acta 27 (1988) 39–43

[26] Shine, S.J. , Bogy, D.B. , Talke, F.E. , Jour. Comp. Phys. 38, 3(1980) 294–326

[27] Shokoohi, F. , Ph. D. Thesis Columbia Univ. New York (1976)

[28] Thelen, H.G. , Dissertation RWTH Aachen 1988

Author's adress:

P. Schümmer, H.-G. Thelen, C. Becker
RWTH Aachen
Institut für Verfahrenstechnik
Turmstraße 46
D-5100 Aachen

D.F. JAMES

Extensional viscosity and its measurement

1. The Character of Extensional Viscosity

It is well established that resistance to extensional motion is an important fluid property in some applications. These applications include polymer processing, porous media flows, coating flows, and turbulent boundary layer flows, to name some dominant ones. While these applications are well known, less known is the difficulty of measuring this property. Consequently the goal of this article is to review the nature of this elusive fluid property and then examine the progress made in measuring it.

The resistance to extensional motion is usually expressed as a viscosity (analogous to resistance to deformation in shear) and is variously known as the extensional viscosity, tensile viscosity, Trouton viscosity and the elongational viscosity. The first seems to be the most common choice in the literature and consequently will be used here. As for notation, the symbols $\bar{\eta}$ and η_E are the two most frequently used and the latter is selected here because it has the blessing of the Society of Rheology [5].

The property of interest, then, is the extensional viscosity η_E, defined by

$$\eta_E = \frac{\tau_{11} - \tau_{22}}{\dot{\varepsilon}},$$

where τ_{11} is the normal stress in the direction of extension, τ_{22} is the normal stress in a perpendicular direction and $\dot{\varepsilon}$ is the rate of extension of the material. Unlike shear viscosity which involves only a single stress component, η_E requires a stress difference because the total normal stress in the 1 direction includes the pressure, which cannot be determined independently and which can be eliminated only by involving a second stress.

When this standard definition is given in rheology texts, such as those by Bird et al. [2], Tanner [24], and Petrie [21], there is no explicit statement as to what η_E depends on. Since stress in a general fluid is a functional of the history of deformation, the normal stress difference $\tau_{11} - \tau_{22}$ has the same dependence and hence so does η_E. To use the formalism of Tanner (page 133 of [24]),

$$\eta_E = (\dot{\varepsilon})^{-1} \mathop{\mathcal{F}}_{t'=-\infty}^{t'=t} [C(t')],$$

where F is a functional, C is the strain tensor, t is the present time and t′ is a past time. This formalism is much too general and abstract to be of practical use, but the equation is written here to emphasize the point that η_E depends non-simply on strain history and is not generally a function of a single variable such as strain rate. That η_E depends on strain history is a natural consequence of physics at the molecular level. Whether the polymeric liquid is a melt or a solution, resistance to stretching partly depends on how much the polymer chains have already been stretched, i.e. on the history of past stretching.

To reduce the dependence on strain history, measurements of η_E should ideally be carried out while the material is stretched at a constant rate, i.e., while the extensional rate $\dot{\varepsilon}$ is constant. After sufficient time, it may be expected that stresses reach terminal values and then η_E is a function of $\dot{\varepsilon}$ only. In shear flow, a steady state is normally achieved and thus the shear viscosity is a function of shear rate only. In such flows, polymer molecules are subjected to a combination of shear and extension and, given enough time, the time-averaged conformations of the chains do not change. While an equilibrium is normally reached in shear, there is not necessarily an equivalent state in extension. Such a state requires that average molecular conformations do not change with time; with tension constantly tending to elongate the chains, a state of constant conformation is not necessarily reached.

For melts, steady state has sometimes been achieved (an example is presented in Section 3), but cases of steady state are not usual and should be treated as exceptions. For polymer solutions, it appears that steady state values of η_E have never been attained. However, even if these values were measured, they would have limited practical use because the strains in industrial flows are modest and much less than the values expected to be necessary to achieve a steady state.

Since steady-state values of η_E are rarely achieved experimentally, obtainable measurements of η_E are transient and are sometimes referred to in that way, e.g., in [19,22]. Since η_E generally depends on strain history, the adjective 'transient' is unnecessary and η_E instead should be understood to be a function of strain history, in the same way that shear viscosity is generally understood to be a function of shear rate. Adjectives should be used for special cases, such as when steady state is achieved. The adjective 'steady-state', however, is not the most appropriate for this condition because it suggests a state which is desirable and routinely achieved, as it is in shear flows. An adjective like 'ultimate' seems preferable because it implies that the associated η_E values are rare and achieved only after long times (or large strains). And to distinguish ultimate values from transient ones, a different symbol, such as η_E^{∞}, could be used. This ultimate value is related to η_E^+ in start-up flow. That is, when $\dot{\varepsilon} = \dot{\varepsilon}_0 H(t)$, where H(t) is the Heaviside step function, then η_E is η_E^+ [5] and

$$\eta_E^+ = \text{fctn}\ (\dot{\varepsilon}_0, t).$$

If normal stresses reach steady values after sufficient time, then $\eta_E^+ \rightarrow \eta_E^\infty = \text{fctn}\ (\dot{\varepsilon}_0)$. Many authors ignore strain history and write $\eta_E = \text{fctn}\ (\dot{\varepsilon}_0)$, but only η_E^∞ can properly be a function of $\dot{\varepsilon}_0$.

2. Constitutive Equations

Before examining measurement techniques, it might be useful to continue in a theoretical vein and to briefly relate the foregoing to constitutive relations.

A popular constitutive equation is the second-order model, and some comment is in order about the suitability of this and other retarded-motion models because they have been used in calculations of extensional flows. In these models, the stress depends on powers and derivatives of strain rate. This formulation cannot describe a stress which depends on strain or strain history and so is unsuitable for extensional motion. In the case of start-up flow, for example, the derivative terms are zero and the model predicts that stress depends only on strain rate. In reality, however, stress depends also on strain (or time). Therefore, although the constants in a retarded-motion model can be related to η_E^∞, the resulting constitutive equation is not valid while extensional stresses grow to their ultimate values.

Another popular group of constitutive equations for polymer solutions are the differential types, typically the Maxwell and Oldroyd B models. Both are sometimes thought to be unsuitable for extensional flows because both predict infinite stress for steady extensional flow when $\dot{\varepsilon}$ equals a critical value. The singularity arises because of terms like $(1 - \lambda\dot{\varepsilon})^{-1}$, where λ is the relaxation time; because of this singularity, constants in these constitutive equations cannot generally be related to η_E^∞. However, because steady state is never reached for solutions, differential models may still be useful for practical extensional flows of solutions and some melts. These models are physically based on deviations from zero-flow equilibrium and so should be most useful when strains are not large and at any strain rate.

3. Measurements for Melts

The most ligitimate measurements of η_E have been made for melts. Because melts are so viscous, they can be handled almost like solids and thus it is possible to grip a sample and extend it in a controlled manner. Extensional rheometers have been designed to generate the strain rate history $\dot{\varepsilon} = \dot{\varepsilon}_o \, H(t)$ and consequently it is possible to make proper measurements of $\eta_E^+ (\dot{\varepsilon}_o, t)$. Illustrative data for η_E^+ are presented in Figure 1 for two polystyrene samples [20]. These data were chosen because they demonstrate very different behaviour at long times. For the sample labelled I, the extensional viscosity

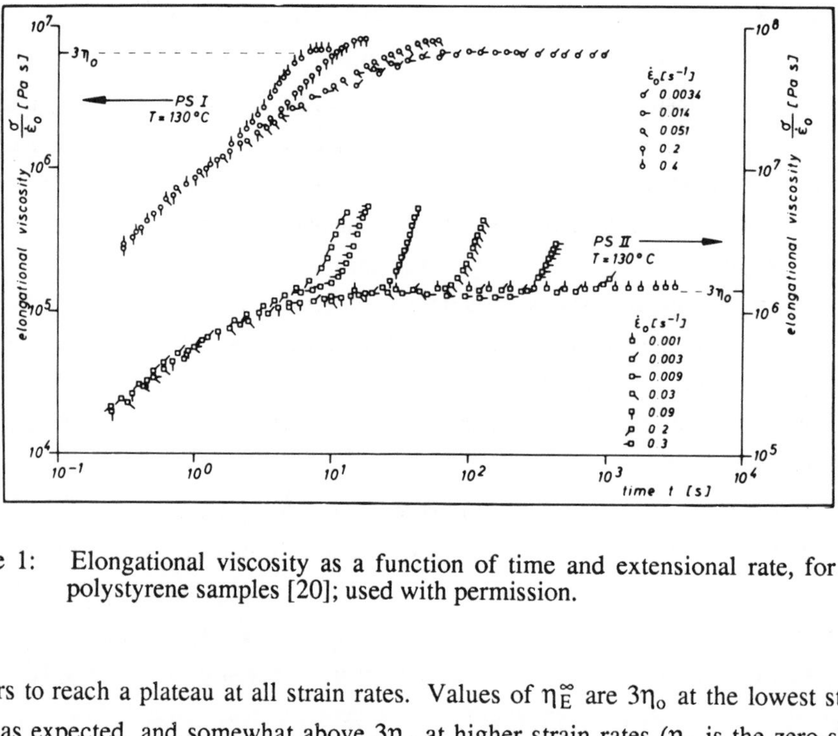

Figure 1: Elongational viscosity as a function of time and extensional rate, for two polystyrene samples [20]; used with permission.

appears to reach a plateau at all strain rates. Values of η_E^∞ are $3\eta_0$ at the lowest strain rates, as expected, and somewhat above $3\eta_0$ at higher strain rates (η_0 is the zero-shear viscosity). For the other sample, however, there is a plateau (equal to $3\eta_0$) only for the very lowest strain rates and, for the other strain rates, there is not even a suggestion of ultimate behaviour. In Figure 1 and in other cases where asymptotic behaviour has been obtained, the values of η_E^∞ are typically no more than 100% above the Newtonian value $3\eta_0$.

This is not surprising when one remembers that to, reach a plateau, molecular conformations must remain the same even as material points move apart. At low strain rates, thermal agitation keeps the chains randomly-oriented and the material isotropic – hence the observed Newtonian behaviour. At higher rates, the chains are extended somewhat and the material becomes anisotropic. One can imagine that chain stretching and orienting are balanced by thermal restoring forces such that average conformations remain constant and equilibrium is achieved. But it is difficult to envisage that such a balance is normal, particularly as strain rates increase. Hence it is not surprising to find that η_E^∞ values are the exception and not the rule in elongational measurements and that, when values are found, they are not far removed from Newtonian values.

4. Measurements for Mobile Liquids

Viscoelastic fluids with lower viscosities cannot be gripped and extended, and so other ways must be found to stretch the materials. A recent article by Gupta & Sridhar [8] reviews the techniques that have been developed for mobile fluids, in addition to techniques for melts. Prior reviews had been limited to extensional rheology for melts, and these are summarized in Table 8.1 of [8]. The review by Gupta & Sridhar is particularly useful in describing the measurement of flow variables: attainable ranges, recommended procedures, accuracy, sensitivity and stability. They show how each technique makes a reasonably accurate measurement of η_E. The present article will take a different tack and focus on the velocity fields in the extensional rheometers for mobile liquids. Of particular concern are the levels of shearing prior to and during the extensional motion, the accuracy of knowledge of the velocity and stress fields, and the variety and suitability of the straining histories. It will be seen that, despite fairly accurate measurements of η_E, values from different instruments cannot be compared because the strain rate histories (even if known accurately) vary considerably from instrument to instrument.

5. Fano Flow

In thinking of ways to measure the extensional viscosity of mobile liquids, one naturally considers exploiting the well-known Fano flow or tubeless siphon. This flow, shown in Figure 2, is a natural candidate for several reasons: the experiment is easily set up, the boundary is stress-free everywhere, the fluid is initially undeformed, and measurements of flow rate, tension and diameter are not hard to make. A number of experimenters have investigated this flow, but only a few will be referenced here. It has been found, for some fluids at least [13,16], that the rate of extension is constant over a major portion of the filament, which is highly advantageous. But, whatever the strain rate variation, the stress in the filament always increases in the flow direction because the tension is constant and the diameter decreases. Hence the relation between stress and strain rate history is hard to decipher from experimental data. The other obvious factors which complicate this technique are surface tension, gravity and inertia. They all affect the determination of stress, but they are reliably accounted for by a momentum balance along the filament. This balance is necessarily one-dimensional and so estimates of stress are accurate only above a certain height – roughly above where the slope is 20° from the vertical. Extensional rates are likewise uncertain near the bottom, and the initial stress and history of deformation are not known. For better calculations, a two-dimensional finite element technique would have to be employed. But such techniques require a constitutive equation, which is a stumbling block because an appropriate constitutive equation for extensional flows has not yet been found. Consequently, there appears to be no straightforward method for determining the dependence of stress on strain rate history.

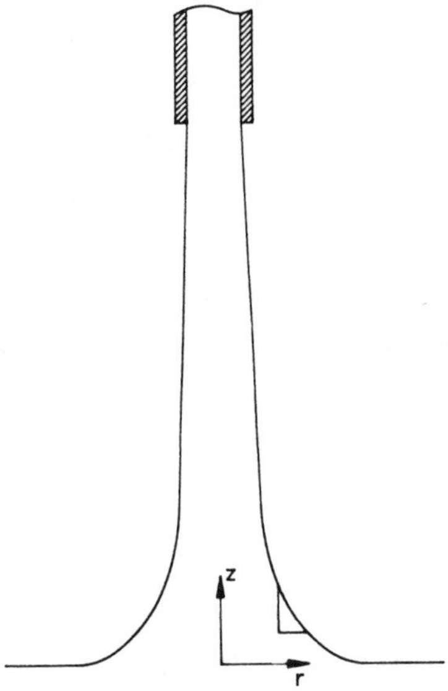

Figure 2: Fano flow or the tubeless siphon.

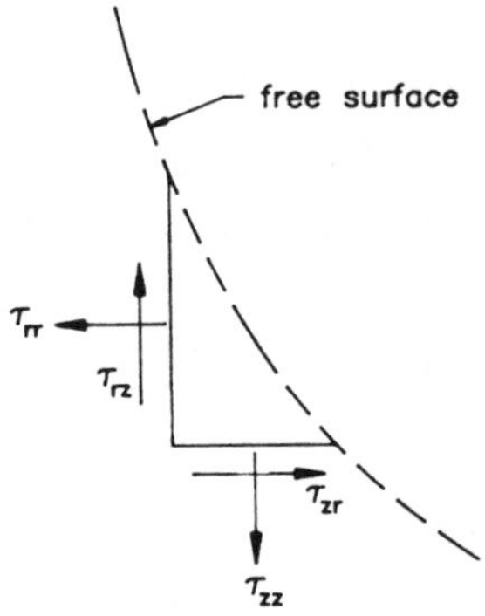

Figure 3: Cross section of a ring element adjacent to the free surface.

There is a further problem with Fano flow, one which is less known but more serious than those mentioned above, and the problem is considerable shearing in the flow. The flow has a stress-free boundary everywhere, but it is not shear-free, except perhaps in the upper nearly-cylindrical portion. Shear is generated because the streamlines are curved and the boundary has zero stress. Perhaps the easiest way to understand this is to consider a ring element of fluid adjacent to the free surface, as shown in Figure 3. The cross section of the element is triangular because the free surface is oblique. The normal and tangential stresses on the oblique face are taken to be zero. There are normal stresses on the orthogonal faces because the fluid is being extended. These stresses, particularly τ_{zz}, must be balanced by other stresses. Surface tension will not be a factor at this location, but gravity and acceleration can create stresses and these are in the same direction as τ_{zz}. These three stresses must be counterbalanced by an upward shear stress on the vertical surface. This shear stress creates shearing across the flow, with the axial velocity increasing inward. This is precisely the flow pattern which Matthys has observed [16]; using a photosensitive dye and pulsed lines of light, he exposed the shearing in the flow. From sequences of photographs he determined velocities and data are shown in Figure 4.

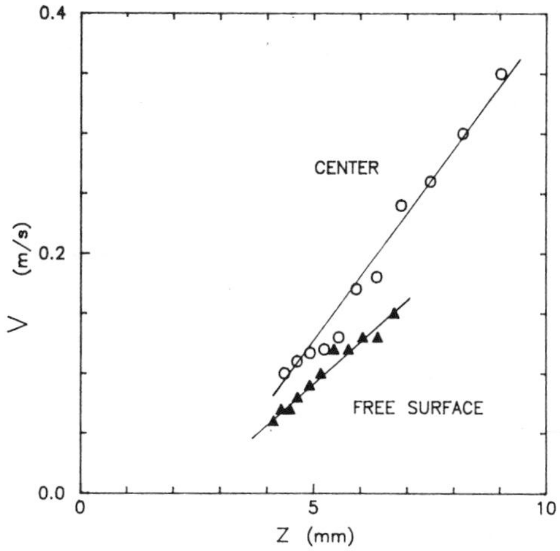

Figure 4: The centreline and free-surface axial velocities, as a function of height, in Fano flow [16]; used with permission.

This figure demonstrates that the difference in axial velocity between the centre-line and the free surface is at least 30%, and the difference does not diminish in the flow direction. Furthermore, rates of shear and extension can be estimated from his data and the former are larger by a factor of about 2. With significant shear present, deciphering extensional resistance from this flow will be extremely difficult.

Hence, despite its simplicity and stress-free boundary, Fano flow does not promise to be a good method of measuring extensional viscosity.

6. Falling Bob

A noteworthy new technique has been developed by Matta [15]. The technique consists of placing a drop of fluid between the end faces of two cylindrical bobs, one of which is stationary, and the other is allowed to fall under gravity to pull out the drop. The filaments formed in this way are found to be cylindrical, as the accompanying sketch, Figure 5, indicates. This device for mobile liquids is the closest analogue so far to the extensional rheometers used for melts. From high-speed photographs taken during the descent, the bob velocity and filament diameter are determined as functions of time, and consequently so is the rate of extension. The tension in the thread is calculated from the reduction in acceleration from the gravitational value, taking into account surface tension. Hence the tensile stress and then the extensional viscosity are found as functions of time. The method has the advantages that the fluid is initially undeformed, the motion is purely extensional, and measurements of stress and strain rate are accurate. The technique appears to be reliable because accurate measurements of η_E for several Newtonian fluids have been obtained.

The major drawback of the technique is that the strain rate history is particular to this method and depends on the material tested. The extensional rate is initially zero, rises rapidly to a value of $O(10^2)s^{-1}$, and then decreases inversely with time. During the typical 0.1 s duration of a run, $\dot{\varepsilon}$ falls to roughly one third of its peak value. The fall depends, of course, on the resistance offered by the thread and fluids with higher resistance have lower $\dot{\varepsilon}$ values overall. Consequently two viscoelastic fluids cannot be compared for extensional viscosity unless the viscosities are close; otherwise the strain histories are different and then a comparison is not meaningful.

Matta presents data for two polymer solutions, both with room-temperature viscosities around 2.5 Pa·s. Extensional viscosity does not correlate with extensional rate for either one, as expected, but for one, data with different bob weights are correlated by total strain. In the past, researchers have generally assumed that η_E is a function of $\dot{\varepsilon}$ and plotted their data in this fashion. It may come as a surprise, then, that it is more appropriate to plot η_E versus total fluid strain for some fluids. However, in many experiments it is difficult to plot results in this way because total strain cannot easily be determined from the flow field.

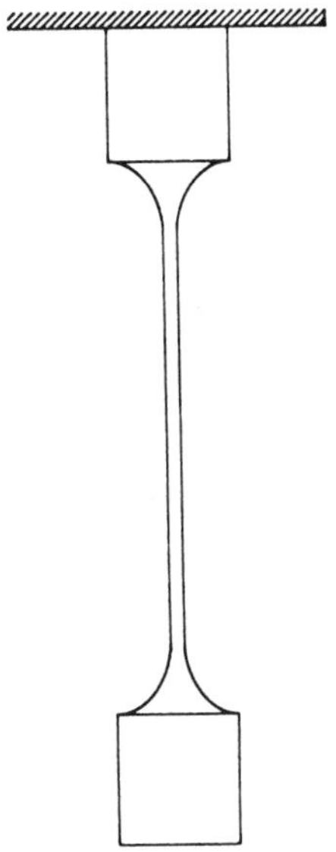

Figure 5: Matta's falling bob technique for measuring extensional viscosity [15].

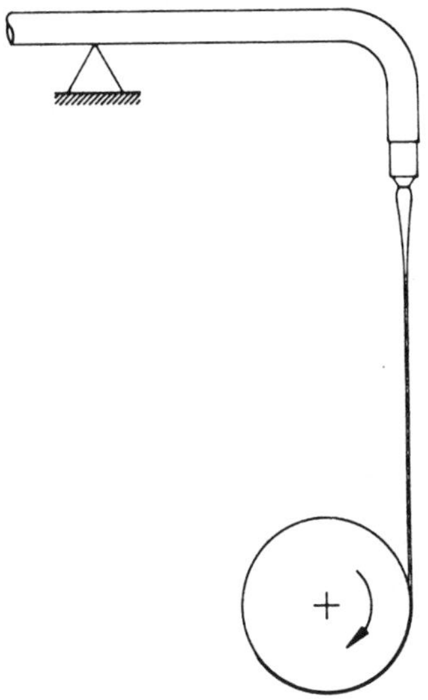

Figure 6: Fibre-spinning apparatus for measuring extensional viscosity.

7. Fibre-Spinning

The most developed method of measuring η_E for mobile liquids is by fibre spinning. Like the two prior methods, this one depends on drawing out a fluid filament. As shown in Figure 6, the fluid is drawn down from an orifice or tube by a rotating drum. A steady flow can be achieved if the fluid is spinnable, i.e., if its extensional viscosity is O(10) Pa·s and higher. As in Fano flow, the diameter along the filament is determined from images, and flow rate and filament tension are measured. Also like Fano flow, the extensional rate is often constant over most of the length of the filament, as shown in Chang & Denn [4] and Ferguson & Missaghi [6], for example. The only commercial extensional rheometer currently available is based on this design (and sold by Carri-Med). This instrument and laboratory devices like it are not hard to operate but such instruments are limited to stress levels above 10^2 Pa so that forces due to gravity, inertia and surface tension do not dominate those due to viscosity. In this regard, the recent paper by Secor et al. [22] usefully analyzes the experimental limits of fibre-spinning and determines the instrument's accuracy under various operating conditions.

The major rheological difficulty with the instrument is deformation of the fluid upstream. The fluid is sheared in the issuing tube and thus the macromolecules do not start in an undeformed state before being extended. Because the tube flow is laminar, the pre-shearing is well defined. However, the shear rate varies across the flow and thus fluid particles have variable deformation histories. Equally important, the shear rate is large – about an order of magnitude larger than the rate of extension in the filament. The maximum shear rate $\dot{\gamma}$ in the upstream tube is $8V_0/D_0$ where the subscript o indicates tube values. The rate of extension $\dot{\varepsilon}$ is V/L on average, where V is the filament velocity a distance L from the nozzle. Hence the ratio $\dot{\gamma}/\dot{\varepsilon}$ is about $8LD^2/D_0^3$, where D is the filament diameter at L, and D is much smaller than D_0. Based on typical values from Chang & Denn [4], the ratio is about 20, so that upstream shear is not negligible. The deformation history is complicated further because of the rearrangement of stresses just after the tube exit. The tangential stress decays to zero downstream, but the tensile stress is not relieved because of constant tension along the filament. Hence, before stresses start to rise due to stretching, there are initial stresses which are not likely uniform across the filament. Consequently the initial conditions for extensional motion are highly uncertain. Because of the initial tension and the high level of pre-shearing, values of η_E obtained with this device are likely to be higher than those obtained in other flows.

Shearing has been observed in drawn filaments, similar to what was found in Fano flow. Using small particles to monitor local velocities, Ferguson & Missaghi [6] detected a transverse variation in axial velocity and observed that the slower-moving particles were located on the periphery of the filament. Their data, Figure 7, suggest that velocities varied all along the filament and by as much as two to one. It is surprising that the velocity would not become more uniform far from the nozzle, but the filament length was only 12 times the nozzle diameter. The variation created by tube shear and then by surface curvature obviously persisted over this short length. Ferguson & Missaghi thought that the non-uniformity was caused by air drag and consequently arranged an air flow parallel to the filament and at the same speed. This removed air friction as a factor but the non-uniformity remained. It is likely that the effect is induced by the free-surface condition, as argued earlier in Section 5.

Similar to the fibre-spinning devices just described is the suction device of Sridhar & Gupta [23,3]. Instead of a rotating drum to draw out the filament, a suction tube is used. The fluid starts from a pressurized reservoir, flows through a short capillary tube,

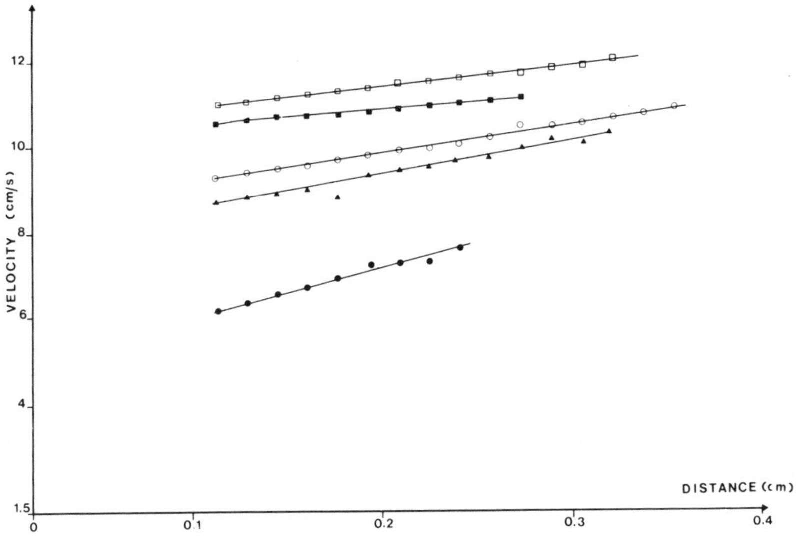

Figure 7: The variation in axial velocity along a drawn filament; the higher velocities are for streamlines near the axis. From Ferguson & Missaghi [6], used with permission.

and is pulled downward into a suction tube of smaller diameter. Tension in the filament is generally very small and is cleverly measured by changes in reservoir pressure. First, the driving pressure and flow rate are measured without suction, i.e., without tension in the filament, and then with suction the driving pressure is lowered to make the flow rate the same. The difference in driving pressures is shown to be equal to be the tensile stress [23]. By measuring this difference, the factors related to inertia, surface tension and gravity are all properly accounted for. What is also accounted for, but not previously recognized, is the extensional flow resistance in the flow entering the upstream capillary tube. This inclusion is important because the rate of extension in that converging flow is about equal to that in the stretched filament, based on an analysis of the flow data in [3]. Hence, it is possible to make accurate measurements of filament tension for low-viscosity fluids, particularly dilute aqueous solutions [3] but, like other drawn-filament techniques, the strain rate history is complex.

8. Sudden Contraction Flows

Since extensional motion is produced when the area of flow decreases, a common method of measuring η_E is to utilize flows in which the fluid passes through an orifice or tube contraction. For Newtonian fluids, the streamline pattern is virtually radial, except far from the opening where walls dictate the flow pattern. The same radial pattern is found for viscoelastic fluids at low flow rates, i.e., at deformation rates low enough that induced elastic stresses are small and do not alter the flow field. At higher flow rates, the induced stresses are large enough to cause flow reversal and a vortex ring forms around the opening on the upstream side. Even when the vortical motion is steady (it usually is not), determining extensional rates in the central flow is difficult. Some authors have observed that the flow near the opening is nearly radial and measured the angle of the confining cone, as Metzner & Metzner [17] did for η_E measurements of dilute aqueous solutions. They acknowledged the difficulty of identifying the correct angle and incorporated this uncertainty in their results.

Other authors have noted that sink flow is not a good approximation in general and that velocities in the central flow generally depend on the fluid and on the flow rate. Hence, aside from the difficulty in estimating rates of extension, there is little commonality in flow histories. The variation from wide-angle conical flow to wine-glass-stem flow is well illustrated in a series of photographs by Binding & Walters (their Figure 17 in [1]). Remarkably, these very different flow patterns are for one fluid. Hence η_E values for different flow rates correspond to different deformation histories, even for the same fluid.

The most reliable measurements of strain rates in these flows appear to have been made by Hasegawa & Iwaida [9]. They used laser Doppler anemometry to measure streamwise velocities in flows of dilute aqueous solutions of two high-molecular-weight polymers. As shown in Figure 8, the centre-line strain rate is virtually constant for much of the flow for the polyethylene oxide solution. But, for the solution of polyacrylamide, the strain rates are closer to those of sink flow. Because both polymers are linear and have about the same molecular weight (4×10^6), and because both solutions are reasonably dilute at 100 parts per million, one might expect similar flow patterns. But, as the figure shows, the strain histories are quite different and so it must be concluded that the molecular dynamics are different for these two polymers. This difference is consistent with the variation noted for these two polymers for pre-sheared extensional flow [12].

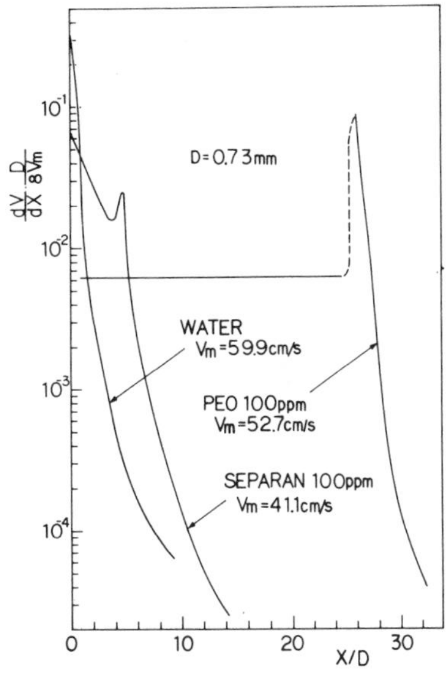

Figure 8: Comparison of dimensionless extensional rates along the centre-line of flow through an orifice of diameter 0.73 mm, as a function of dimensionless distance from the opening. Curves are based on LDA measurements by Hasegawa and Iwaida, [9], used with permission.

Axial tension in contraction flows must be inferred. No direct measurement can be made and stress is deduced from the additional driving pressure or from the thrust of the jet leaving the orifice. Some authors – for example, Binding & Walters [1] and Hasegawa & Fukutoni [10] – have assumed a power-law relation between $\tau_{rr} - \tau_{\theta\theta}$ and $\dot{\varepsilon}$ and thus related the normal stress difference to the extra driving pressure. Since stresses generally depend on strain as well as strain rate, (and the dependence is sometimes on strain only, as described earlier), this assumption is questionable. To use jet thrust to find the tensile stress, assumptions are needed about the stress field, particularly about the initial stress in the fluid, where the jet exits the orifice and rapidly expands. To shed some light on these two techniques, Hasegawa & colleagues [10,11] measured the tensile stress by both jet thrust and driving pressure and found that results by the two methods in essence agreed. This is a remarkable finding, and its explanation must await a more rigorous analysis. Sudden contraction flows may eventually prove to be a useful way of measuring η_E when techniques for measuring extensional stresses are improved.

9. Converging Channel Flow

A measurement technique based on converging channel flow has been developed by James and co-workers [14]. The shape of their channel is shown in Figure 9. The principle of the rheometer is that a viscoelastic fluid is pushed through the channel at high

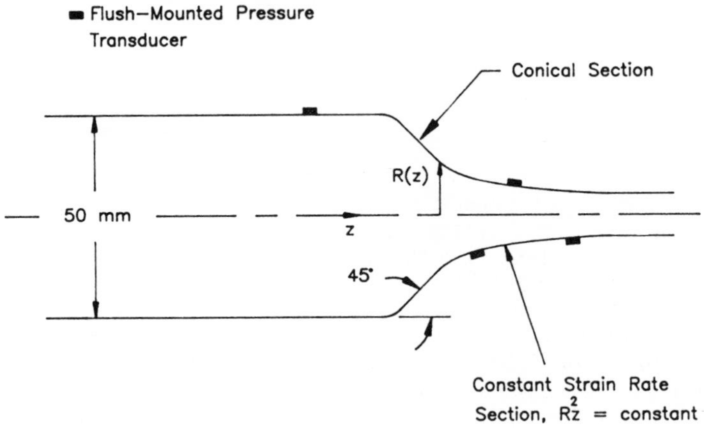

Figure 9: Converging channel extensional rheometer [14].

Reynolds numbers so that shearing occurs only in the thin boundary layer near the wall, leaving a shear-free extensional flow in the core. The channel starts as a large diameter tube where shear rates are low and the initial deformation rate is taken to be zero. Then it converges quickly and the smoothly converging section is designed to produce a constant rate of extension in the core flow. This shape produces constant $\dot{\varepsilon}$ motion exactly at extremely large Reynolds numbers and reasonably well at Reynolds numbers less than unity. For the moderate Reynolds numbers applicable to channel test conditions, the same shape produces nearly constant extensional rates, according to analytical and numerical results [14]. The evidence is only moderately compelling in the absence of confirming measurements by, for example, laser Doppler anemometry. If constant strain rates are achieved in the smoothly converging section, then the flow history approximates start-up flow, which is the desirable motion already achieved for melts.

Tensile stress in the central flow is monitored by flush-mounted pressure transducers along the channel, as shown in the Figure. Stresses developed in the core are transmitted across the boundary layer to the wall. To relate measured wall pressures to stresses in the core, the momentum equations are integrated with the stress components unspecified i.e., no constitutive equation is assumed. The integration along the centre line (r = 0) is

$$p_\infty = p(0,z) + \frac{1}{2}\rho V^2(0,z) - \tau_{zz}(0,z) \qquad (1)$$

where p_∞ is the upstream or driving pressure, V is the axial velocity, τ_{zz} is the normal stress in the z direction due to deformation, and z is an arbitrary location downstream. Integration in the radial direction at that location yields

$$p_w = p(0,z) - I - \int_0^R \frac{N_2}{r}\,dr - \tau_{rr}(0,z) \tag{2}$$

where p_w is the wall pressure at z, I is a known integral which depends on the velocity profile, N_2 is the second normal stress difference in shear, and R(z) is the local channel radius. When equations (1) and (2) are combined, then the pressure drop in the converging channel is

$$p_\infty - p_w = -(\tau_{zz} - \tau_{rr}) + \frac{1}{2}\rho\,V^2 - I - \int_0^R \frac{N_2}{r}\,dr \tag{3}$$

To find the normal stress difference in extension from the measured pressure drop, the last three terms on the right must be known. The first two of these depend on knowing the velocity field in the channel and a method for determining this field is described in [14]. The last term depends on N_2, which is extremely difficult to measure. This term has been dealt with by assuming that $N_2 = -\frac{1}{10}N_1$, which is true on average for polymer solutions. Since the term involving N_2 contributes from 10% to 50% of $\tau_{zz} - \tau_{rr}$, accuracy for η_E depends on measured values of N_2. But, since one significant figure still represents an achievement in this field, lack of data for N_2 is not crippling. If the last three terms in equation (3) can be calculated accurately, then the technique has the promise of measuring η_E as a function of strain, at the different transducer locations, as well as strain rate. A major deficiency of the technique is that reliability cannot be validated by measurements with Newtonian fluids. Since the instrument operates at Reynolds numbers of $O(10^2)$, stresses related to shear viscosity, including extensional stresses, are almost completely dominated by inertial stresses and are therefore undetectable.

There is an additional difficulty with this technique which has not been highlighted before, and that is the magnitude of pre-shear. The large inlet tube would seem to assure negligible deformation rates before extension, but this is not the case. If we use the formula developed earlier for the ratio of initial shear rate to overall extensional rate – the formula for $\dot{\gamma}/\dot{\varepsilon}$ in Section 7 – and if channel values of $L = 50$ mm, $D_0 = 50$ mm and $D = 5.7$ mm are substituted, then $\dot{\gamma}/\dot{\varepsilon}$ is about 0.4. This value is larger than expected (but small compared to values encountered in spin-line experiments), and needs to be lower to ignore pre-shearing. Since non-Newtonian effects in extension start at deformation rates one-half those in shear, an acceptable value for the ratio would be about 0.2. Such a value could be achieved by redesigning the channel.

10. Opposed Entry Tubes

Perhaps the most promising technique for the measurement of η_E is that developed by Fuller and co-workers [7,18]. As illustrated in Figure 10, fluid is drawn into opposing tube inlets to create a stagnation flow around the mid-plane. One inlet is stationary and the other is part of a balance arm to measure the force exerted by the entering fluid.

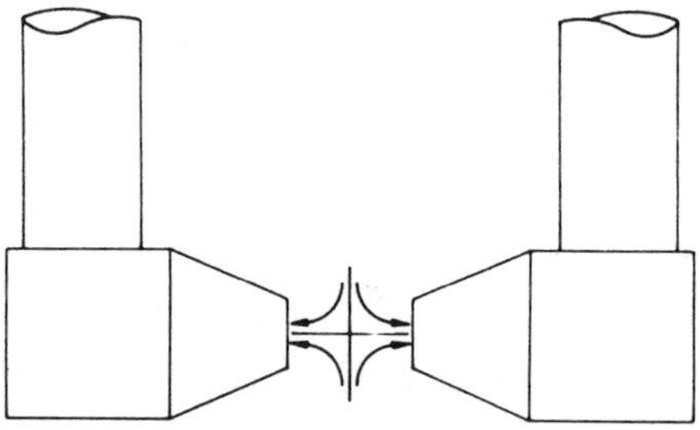

Figure 10: The stagnation flow created by opposed entry tubes, from Fuller et al. [7].

This force, divided by the tube area, is taken to be the tensile stress at the inlet. The extensional rate is approximated by the entering velocity divided by the distance from the mid-point to the tube entrance. Since a stagnation zone has been created, the velocity field must closely approximate a shear-free flow, i.e., the velocity increases linearly from the stagnation point and thus the extensional rate is constant. The only region excepted is that in the immediate vicinity of the inlet, particularly at the tube edge. But this region is small in comparison to the rest of the flow field, and so the formula for calculating extensional rate should be accurate.

Tests with variable gap distances for Newtonian fluids have shown that η_E values of $3\eta_o$ are achieved when the gap distance is equal to the tube diameter. Comparable tests have not been done, it appears, with viscoelastic fluids to check sensitivity to gap distance.

A rheological instrument based on this device is being developed by Rheometrics and is expected to be available in the summer of 1990. The instrument has obvious advantages. It is simple to operate and fluids of all viscosities can be tested; that is, they do not have to be spinnable and thus the extensional viscosity of dilute aqueous systems can be measured. The biggest drawback of the technique, as Fuller et al [7] point out, is that the strain history of all fluid particles is not the same, even though the extensional rate is constant. The closer particles are to the stagnation point, the higher is their residence time in the flow. This problem may be remedied by an analysis of the flow to determine the averaging which is effectively taking place in the tensile measurement.

The other drawback is that strain cannot be varied in this instrument. The streamline pattern is independent of tube diameter and distance between inlets, and therefore the amount of stretching is fixed. It is also not yet clear how to calculate the strain. This value will be needed for comparison with measurements from other instruments.

11. The M1 Project

The state of the art in measuring the extensional viscosity of mobile liquids was clearly exposed by a project organized by Dr. Tam Sridhar of Monash University. He prepared a large quantity of a stable dilute solution of polyisobutylene in a mixture of kerosene and polybutene. The near-constant-viscosity fluid was called M1 and samples were sent to most laboratories where extensional viscosity measurements are made. The results were presented at the International Conference on Extensional Flow, held in Combloux, France, March 20-23, 1989. Data of η_E vs. $\dot{\varepsilon}$ were provided by all the measurement techniques described in this paper, plus one or two others. Values of η_E at the same extensional rate differed by as much as three orders of magnitude. This variation is astounding, particularly because each technique provides a reasonably reliable measurement of η_E, i.e. the accuracy of each η_E measurement was no worse than $\pm 25\%$. The differences can be due only to different strain histories, but few at the meeting anticipated that strain history could be such a dominant factor. In fact, it is hard to rationalize the polymer chain physics which produces such variation. All of this creates additional pressure to understand the generation of stresses in extensional flow and to develop well-defined methods of measuring these stresses.

The results of the Combloux meeting are appearing in a special issue of the Journal of Non-Newtonian Fluid Mechanics, expected to appear in the spring or summer of 1990.

12. Concluding Remarks

It is evident from the variety of techniques to measure η_E and from the variety of results which they produce, that we are far from settling on a method to measure this elusive fluid property. The M1 project has brought home the fact that strain history must be incorporated in the interpretation of η_E data.

Experimentalists still face a steep grade. It is evident that all the current techniques are limited and there is ample room for new, imaginative methods. It is exceedingly difficult to create flows which are purely extensional and which have both a well-defined stress field and an equally well-defined strain history. Our own experience in rheometer design suggests that it is not necessary to seek a flow which is purely extensional and free of shear; it may be possible to account for shearing, if it is well defined. Perhaps this and other unconventional approaches will lead to a fully satisfactory way of measuring extensional viscosity for mobile liquids.

13. Acknowledgement

The support of the National Sciences and Engineering Research Council of Canada is gratefully acknowledged. Figures 1 and 4 are presented with permission of the Society of Rheology, and Figures 7 and 8 with the permission of Elsevier Science Publishers.

References

1. Binding, D.M. and Walters, K., J. Non-Newtonian Fluid Mechanics, **30**, 233-250 (1988).
2. Bird, R.B., Armstrong, R.C. and Hassager, O., "Dynamics of Polymeric Liquids", Vol. 1, John Wiley & Sons, New York (1987).
3. Chan, R.C., Gupta, R.K. and Sridhar, T., J. Non-Newtonian Fluid Mechanics, **30**, 267-283 (1988).
4. Chang, J.C. and Denn, M.M., J. Non-Newtonian Fluid Mechanics, **5**, 369-385 (1979).
5. Dealy, J.M., J. Rheology, **28**, 181-195 (1984).
6. Ferguson, J. and Missaghi, K., J. Non-Newtonian Fluid Mechanics, **11**, 269-281 (1982).
7. Fuller, G.G., Cathey, C.A., Hubbard, B., and Zebrowski, B., J. Rheology, **31**, 235-249 (1987).
8. Gupta, R.K. and Sridhar, T., Chapter 8 of "Rheological Measurement", Ed. A.A. Collyer and D.W. Clegg, Elsevier Applied Science, London (1988).
9. Hasegawa, T. and Iwaida, T., J. Non-Newtonian Fluid Mechanics, **15**, 279-307 (1984).

10. Hasegawa, T., and Fukutoni, K., Proceedings of Xth International Congress on Rheology, Sydney, page 1.395 (1988).

11. Hasegawa, T. and Nakamura, H., "Experimental Study of the Elongational Stress of Dilute Solutions in Orifice Flows", submitted to J. Non-Newtonian Fluid Mechanics.

12. James, D.F., McLean, B.D., and Saringer, J.H., J. Rheology, **31**, 453-481 (1987).

13. MacSporran, W.C., J. Non-Newtonian Fluid Mechanics, **8**, 119-138 (1981).

14. James, D.F., Chandler, G.M. and Armour, S.J., "A Converging Channel Rheometer for the Measurement of Extensional Viscosity", to appear in J. Non-Newtonian Fluid Mechanics.

15. Matta, J., "Liquid Stretching Using a Falling Cylinder" to appear in the J. Non-Newtonian Fluid Mechanics.

16. Matthys, E.F., J. Rheology, **32**, 773-788 (1988).

17. Metzner, A.B. and Metzner, A.P., Rheologica Acta, **9**, 174-181 (1970).

18. Mikkelsen, K.J., Macosko, C.W., and Fuller, G.G., Proceedings of Xth Int. Congress on Rheology, Sydney, page 2.125 (1988).

19. Moan, M. and Mageur, A., J. Non-Newtonian Fluid Mechanics, **30**, 343-354 (1988).

20. Munstedt, H., J. Rheology, **24**, 847-867 (1980).

21. Petrie, C.J.S., "Elongational Flows", Pitman, London (1979).

22. Secor, R.B., Schunk, P.R., Hunter, T.B., Stilt, T.F., Macosko, C.W., and Scriven, L.E., J. Rheology, **33**, 1329-1358 (1989).

23. Sridhar, T. and Gupta, R.K., Rheologica Acta, **24**, 207-209 (1985).

24. Tanner, R.I., "Engineering Rheology", Clarendon Press, Oxford (1985).

David F. James
Department of Mechanical Engineering
University of Toronto
Toronto, M5S 1A4, Canada

K.R. RAJAGOPAL, M. MASSOUDI AND J.M. EKMANN

Mathematical modeling of fluid-solid mixtures

1. INTRODUCTION

The flows of a mixture of solid particles and a fluid have relevance to several important technological applications. Pneumatic transport of solid particles, fluidized bed combustors, flow of dirty lubricants, flows in a hydrocyclone, and several flows of particulate suspensions are but some of the examples that abound. There are primarily two approaches which have been used for modeling such solid-fluid systems. In the first approach the model is obtained by some kind of averaging of the equations of motion for a single particle of the fluid or solid, which is modified to account for the presence of the second component, over a large enough characteristic unit which is yet small compared to the dimensions of the system as a whole. The averaging leads to terms for the interaction between the constituents for which one needs constitutive relations. The works of Anderson and Jackson [5] and Drew and Segel [10] provide prime examples for this approach.

The second method of modeling multi-component systems is mixture theory. This theory, which traces its origins to the work of Fick [12], was first put into a rigorous mathematical format by Truesdell [22]. The theory, is in a sense, a homogenization approach in which each component is regarded as a single continuum and at each instant of time, every point in space is considered to be occupied by a particle belonging to each component of the mixture. That is, each component of the mixture is homogenized over the whole space occupied by the mixture. Balance laws are then written for each component which takes into account interaction with the other constituents. A historical development of the theory can be found in the review articles by Atkin and Craine [6], Bowen [7], and the various review articles in the appendix of the book by Truesdell [23]. El-Kaissy [11], Ahmadi [3, 4], and Massoudi [15] have used such an approach for modeling solid-fluid systems.

In this short note, we discuss the modeling of solid-fluid mixtures using the second approach. We present a theory which in the absence of fluid in the mixture reduces to a theory appropriate for granular materials and in the absence of particles reduces to a linearly viscous fluid.

2. MODELING

In almost all the modeling of solid particles and a fluid, the solid particles are also assumed to behave like a linearly viscous fluid (cf. Murray [17], Anderson and Jackson [5], El-Kaissy [11]) with a viscosity μ^s and an associated pressure field p^s. However, the exact meaning of the term p^s is unclear, and moreover it leads to an indeterminacy in the system of equations (cf. Massoudi [15]), with one more unknown than the number of equations. This indeterminacy is overcome by assuming a priori a relationship between the solid pressure p^s and the fluid pressure p^f (cf. Drew [9], Homsy et al. [14], Anderson and Jackson [5]). A popular assumption is that $p^s = p^f$. This assumption seems to stem from ideas in two phase flows where such an assumption might indeed be appropriate when the two-phases are say water and steam. However, in the case of a mixture of a granular solid and fluid, not only is the motion of a solid pressure p^s not meaningful, but to assume that this quantity is the same as that of the pressure in the fluid seems even more inappropriate. To assume a priori that

$$p^s - p^f = f(\nu) ,$$

where ν is the volume distribution function, is based on even flimsier grounds. Another pitfall of many of the models is that they do not reduce to the model for a linearly viscous fluid or that for a granular solid in the two extreme limits.

Recently, Massoudi [15] advocated the modeling of the stress in the solid constituent of the mixture by a constitutive expression appropriate to flowing granular solids, and the stress in the fluid constituent of the mixture by a linearly viscous fluid. This modeling is to be done in such a manner that the model reduces to that for a linearly viscous fluid when the volume fraction of the solid goes to zero, and to a granular solid model when the volume fraction of the fluid goes to zero.

We shall assume that the fluid and solid phases are dense enough to be modeled as homogeneous continuous media so that we may exploit the theory of interacting continua (cf. Bowen [7], Atkin and Craine [6], Truesdell [23]). This theory assumes that each point in the mixture is occupied by a particle belonging to each of the homogenized constituents. Conservation laws are written for each constituent which take into account the possibility of mechanical and chemical interactions between the constituents.

Based on our knowledge of modeling in the theory of granular solids and a linearly viscous fluid, it would be natural to assume all the constitutive functions $A^{\alpha}, \mathbf{q}^{\alpha}, \mathbf{T}^{\alpha}$, $\mathbf{I}$ etc., depend on

$$\rho^s, \rho^f, \nabla\rho^s, \nabla\rho^f, \nabla\nabla\rho^s, \nabla\nabla\rho^f, \theta, \nabla\theta, \mathbf{u}^f - \mathbf{u}^s, \mathbf{D}^s, \mathbf{D}^f, \tag{2.1}$$

where A is the specific Helmholtz free energy, $\mathbf{q}$ is the heat flux vector, $\mathbf{T}$ is the stress, $\mathbf{I}$ is the interaction, and where ρ is the density, θ is the temperature, $\mathbf{u}$ is the velocity and $\mathbf{D}$ is the stretching tensor. The superscripts f and s denote the properties associated with the fluid and solid, respectively. In addition to the quantities in equation (2.1), the interaction term $\mathbf{I}$ could depend on a frame-indifferent quantity derivable from the accelerations $\mathbf{a}^s$ and $\mathbf{a}^f$ of the solid and the fluid acceleration, respectively, to account for the virtual mass effect. Then, using methods that are by now standard in continuum mechanics (cf. Shi, Rajagopal and Wineman [21], Atkin and Craine [6], Dai, Rajagopal and Wineman [8]), we can obtain restrictions and forms for such constitutive expressions. Here, we shall not discuss such an approach. Another approach to modeling the mixture is to postulate the constitutive expressions by simply generalizing the structure of the constitutive relations from a single constituent theory. This is what we shall do here. We shall also restrict ourselves to a purely mechanical theory and not worry about the thermal quantities. In this case

$$\mathbf{T}^s = \mathbf{T}^s\,(\rho^f, \rho^s, \nabla\rho^s, \nabla\rho^f, \nabla\nabla\rho^s, \nabla\nabla\rho^f, \mathbf{u}^f - \mathbf{u}^s, \mathbf{D}^s, \mathbf{D}^f)\,, \tag{2.2}$$

$$\mathbf{T}^f = \mathbf{T}^f\,(\rho^f, \rho^s, \nabla\rho^s, \nabla\rho^f, \nabla\nabla\rho^s, \nabla\nabla\rho^f, \mathbf{u}^f - \mathbf{u}^s, \mathbf{D}^s, \mathbf{D}^f)\,. \tag{2.3}$$

In the case of a single constituent granular solid, the Cauchy stress $\mathbf{T}^s$ would depend on the arguments which have the superscript s associated with them (cf. Massoudi and Boyle [16]). Such a model would be a generalization of the model derived by Goodman and Cowin [13] to describe the mechanics of granular solids. Due to the presence of the term $\nabla\nabla\rho^s$, such a theory would include the model proposed by Kortweg to describe the effects of capillarity (cf. Truesdell and Noll [24]). It is also worth observing that such models exhibit the phenomena of normal stress differences.

In general, the constitutive expressions for $\mathbf{T}^s$ and $\mathbf{T}^f$ in (2.2) and (2.3) would depend on the kinematical quantities associated with both the constituents. However, we shall assume that $\mathbf{T}^s$ and $\mathbf{T}^f$ depend only on the kinematical quantities associated with the solid and fluid, respectively. This assumption is sometimes called the "principle of phase separation" (cf. Drew [9]). This is not a principle but just an assumption (cf. Adkins [1, 2]). For the sake of simplicity we shall assume that $\mathbf{T}^s$ and $\mathbf{T}^f$ do not depend on $\nabla\nabla\rho^s$ and $\nabla\nabla\rho^f$, respectively. We shall assume that

$$\mathbf{T}^s = a_0\mathbf{1} + a_1\mathbf{D} + a_2\mathbf{N} + a_3\mathbf{D}^2 + a_4\mathbf{N}^2 + a_5(\mathbf{DN} + \mathbf{ND})$$
$$+ a_6(\mathbf{D}^2\mathbf{N} + \mathbf{ND}^2) + a_7(\mathbf{DN}^2 + \mathbf{N}^2\mathbf{D}) + a_8(\mathbf{D}^2\mathbf{N}^2 + \mathbf{N}^2\mathbf{D}^2) \quad (2.4)$$

where

$$\mathbf{N} \equiv \nabla\rho \otimes \nabla\rho\,, \quad (2.5)$$

and the coefficients a_i, $i = 0,1,\cdots,8$ depend on

ρ, tr$\mathbf{D}$, tr$\mathbf{D}^2$, tr$\mathbf{D}^3$, tr$\mathbf{N}$, tr$\mathbf{N}^2$, tr$\mathbf{N}^3$, tr$(\mathbf{DN})$, tr$(\mathbf{DN}^2)$, tr$(\mathbf{ND}^2)$, tr$(\mathbf{N}^2\mathbf{D}^2)$.

If we require that $\mathbf{T}^s$ depends linearly on $\mathbf{D}^s$ and $\mathbf{N}$, this model simplifies considerably:

$$\mathbf{T}^s = \left\{\beta_0(\rho^s) + \beta_1(\rho^s)\,\text{grad}\,\rho^s \cdot \text{grad}\,\rho^s + \beta_2(\rho^s)\,\text{tr}\mathbf{D}^s\right\}\mathbf{1}$$
$$+ \beta_3(\rho)\,\nabla\rho^s \otimes \nabla\rho^s + \beta_4(\rho^s)\mathbf{D}^s. \quad (2.6)$$

The spherical part of stress in (2.6) can be interpreted as the solid pressure p^s. In this sense, the material is like a compressible fluid which has an equation of state for the pressure. We shall assume that

$$\mathbf{T}^f = (1-\nu)\left\{\left[-p^f\mathbf{1} + \lambda^f(\rho^f)(\text{tr}\mathbf{D}^f)\right]\mathbf{1} + 2\mu^f(\rho^f)\mathbf{D}^f\right\}, \quad (2.7)$$

where ν is the volume fraction function defined through

$$\nu \equiv \frac{\rho^s}{\rho_R^s}, \tag{2.8}$$

in which ρ_R^s denotes the reference density for the granular solid, and λ^f and μ^f are coefficients of viscosity. Notice that in our modeling we have neglected the dependence of stresses on $\nabla\nabla\rho^s$ and $\nabla\nabla\rho^f$. Introducing higher gradients in the theory always introduces the attendant difficulty of specifying additional boundary conditions. Later, we shall see that even dependence on $\nabla\nu$ introduces difficulties with regard to boundary conditions.

Henceforth, we shall assume $\rho_R^s = \text{constant}$, i.e., incompressible grains. In the above models, we shall also assume that β_0, β_2, and $\beta_4 \to 0$ as $\nu \to 0$, and thus $\mathbf{T}^s \to 0$ as $\nu \to 0$. Also, notice that $\mathbf{T}^f \to 0$ as $\nu \to 1$. The total stress $\mathbf{T}$ is defined through

$$\mathbf{T} = \mathbf{T}^s + \mathbf{T}^f .$$

Thus $\mathbf{T} \to \mathbf{T}^f$ as $\nu \to 0$ and $\mathbf{T} \to \mathbf{T}^s$ as $\nu \to 1$.

The balance of linear momentum for the two constituents are

$$\text{div}\, \mathbf{T}^s - \mathbf{I} + \rho^s \mathbf{b} = \rho^s \frac{d\mathbf{u}^s}{dt}, \tag{2.9}$$

$$\text{div}\, \mathbf{T}^f + \mathbf{I} + \rho^f \mathbf{b} = \rho^f \frac{d\mathbf{u}^f}{dt}, \tag{2.10}$$

where $\mathbf{b}$ is the external body force field.

It would be appropriate at this juncture to point to a significant difference between the modeling proposed here and other approaches to modeling solid-fluid mixtures. Unlike previous approaches, there is no indeterminate solid pressure p^s. In this context, representation for $\mathbf{T}^s$ resembles that of a compressible material in which a specific constitutive expression is given for p^s, namely the spherical part of $\mathbf{T}^s$. This term is not picked in an ad hoc manner but comes from representation theory (cf. Savage [19]).

We have to pick a constitutive expression for $\mathbf{I}$. Early constitutive assumptions for $\mathbf{I}$ did not satisfy basic invariance requirements. Anderson and Jackson [5] modeled the virtual mass effects through the relative acceleration term, which is not frame indifferent. Similar errors were made in many of the early attempts at modeling the interaction term. For the purpose of illustration, we shall assume that

$$\mathbf{I} = \alpha_1 \text{ grad } \rho^s + \alpha_2 \text{ grad } \rho^f + \alpha_3(\mathbf{u}^s - \mathbf{u}^f) + \alpha_4 \frac{D}{Dt}(\mathbf{u}^s - \mathbf{u}^f), \quad (2.11)$$

where α_3 is related to the drag coefficient and α_4 is related to the virtual mass coefficient and α_i are functions of ρ, ν, and the term $\frac{D}{Dt}(\mathbf{u}^s - \mathbf{u}^f)$ is a frame-invariant form derived from the accelerations. The interaction term could have a more general representation, which could include terms that model the Magnus effect, Faxen forces and Basset forces.

Next, we shall write down the equations of motion for the constituents. It follows from (2.6) and (2.7) that

$$\rho^s \frac{d\mathbf{u}^s}{dt} = \text{div}\left\{\left[\beta_0(\nu) + \beta_1(\nu) \text{ grad } \nu \cdot \text{grad } \nu\right] \mathbf{1} + \beta_3(\nu) \text{ grad } \nu \otimes \text{grad} \nu\right.$$

$$+ \beta_2(\nu)(\mathrm{tr}\mathbf{D}^s)\,\mathbf{1} + \beta_4(\nu)\,\mathbf{D}^s\Big\} + \rho^s\mathbf{b} - \mathbf{I}\,, \tag{2.12}$$

and

$$\rho^f \frac{d\mathbf{u}^f}{dt} = \mathrm{div}\,\Big\{(1-\nu)\Big[-p^f\mathbf{1} + \lambda^f(\nu)(\mathrm{tr}\mathbf{D}^f)\,\mathbf{1} + 2\mu^f(\nu)\mathbf{D}^f\Big]\Big\} + \mathbf{I} + \rho^f\mathbf{b}\,. \tag{2.13}$$

3. APPLICATIONS

As we observed in the introduction, the mechanics of solid-fluid mixtures has relevance to fluidization and pneumatic transport of solids. We discuss briefly the equations governing the flow of such a mixture in a pipe as an example. We shall assume that

$$\mathbf{u}^s = \mathbf{u}^s(r)\,\mathbf{e}_r\,, \tag{3.1}$$

$$\mathbf{u}^f = \mathbf{u}^f(r)\,\mathbf{e}_r\,, \tag{3.2}$$

$$\nu = \nu(r)\,. \tag{3.3}$$

Substituting expressions (3.1)-(3.3) into (2.9) and (2.10), we obtain

$$\frac{d\beta_0}{dr} + \left(\frac{d\nu}{dr}\right)^2 \frac{d\beta_1}{dr} + 2\beta_1\left(\frac{d\nu}{dr}\right)\left(\frac{d^2\nu}{dr^2}\right) + \frac{d\beta_3}{dr}\left(\frac{d\nu}{dr}\right)^2$$

$$+ 2\beta_3\left(\frac{d\nu}{dr}\right)\left(\frac{d^2\nu}{dr^2}\right) + \frac{\beta_3}{r}\left(\frac{d\nu}{dr}\right)^2 - \alpha_1\,\rho_R^s\left(\frac{d\nu}{dr}\right) - \alpha_2\rho_R^f\,\frac{d(1-\nu)}{dr} = 0\,, \tag{3.4}$$

$$\frac{1}{2}\,\beta_4(\nu)\left\{\frac{1}{r}\frac{du^s}{dr} + \frac{d^2u^s}{dr^2}\right\} + \frac{1}{2}\frac{d\beta_4}{dr}\frac{du^s}{dr} - \alpha_3(u^s - u^f) = 0\,, \tag{3.5}$$

$$\frac{\partial}{\partial r}\big((1-\nu)\, p^f\big) = \alpha_1\, \rho_R^s \frac{d\nu}{dr} + \alpha_2\, \rho_R^f \frac{d(1-\nu)}{dr} \tag{3.6}$$

$$\frac{\partial}{\partial z}\big((1-\nu)\, p^f\big) = \alpha_3\left(u^s - u^f\right) + (1-\nu)\,\mu^f(\nu)\left[\frac{1}{r}\frac{du^f}{dr} + \frac{d^2u^f}{dr^2}\right] + \frac{d}{dr}\big((1-\nu)\,\mu^f(\nu)\big)\frac{du^f}{dr}\,. \tag{3.7}$$

In deriving the above equations, we have neglected the external body force field. Notice that (3.4) is a second order equation in ν. Thus, we have to prescribe two boundary conditions on ν. However, what is natural to the problem is the amount of the solid particles fed into the system, an integral condition like

$$\int_0^R \nu r\,dr = \widehat{m}\,. \tag{3.8}$$

This, however, provides us with only one condition. In order to make the problem determinate, we could either specify a symmetry condition (which would not be valid if, say, gravity were to be included), or the value of ν on the boundary. This condition has to be determined experimentally. For instance, the experimental results of Segre and Silberberg [20] suggest that

$$\nu = 0 \quad \text{at} \quad r = R\,. \tag{3.9}$$

Thus, if the theory included terms involving $\nabla\nabla\rho^s$, we would in fact need yet another boundary condition. Prescribing the boundary conditions on the velocity of the fluid is straightforward. We assume the fluid adheres to the boundary. However, in regard to the velocity of solid particles, it is not clear that the adherence condition is the obvious one. However, it seems quite reasonable to assume such a condition. Thus,

$$u^s \to 0 \quad \text{at} \quad r \to R\,, \tag{3.10}$$

$$u^f = 0 \quad \text{at} \quad r = R\,. \tag{3.11}$$

The symmetry conditions of the flow require that

$$\frac{du^s}{dr} = 0 \quad \text{at} \quad r = 0\,, \tag{3.12}$$

and

$$\frac{du^f}{dr} = 0 \quad \text{at} \quad r = 0\,. \tag{3.13}$$

If an external body force field like gravity were present and the particles tend to migrate downwards, we would be incorrect in assuming a volume distribution of the form (3.3). The system (3.4)-(3.7) subject to the boundary condition (3.8)-(3.13) is being studied by Rajagopal et al. [18].

BIBLIOGRAPHY

[1] J.E. Adkin, Phil. Trans. Royal Soc. London, Vol. 255A, 607 (1963).

[2] J.E. Adkins, Phil. Trans. Royal Soc. London, Vol. 255A, 635 (1963).

[3] G. Ahmadi, Intl. J. Non-Linear Mech., 15, 251 (1980).

[4] G. Ahmadi, Acta Mechanica, 44, 299 (1982).

[5] T.B. Anderson and R. Jackson, Ind. Eng. Chem. Fundamentals, Vol. 6, 527 (1967).

[6] R.J. Atkin and R.E. Craine, Q.J. Mech. Appl. Math., 29, 290 (1976).

[7] R.M. Bowen, in A.C. Eringen (Ed.), Continuum Physics, Vol. 3, Academic Press (1976).

[8] F. Dai, K.R. Rajagopal and A.S. Wineman, In Press, Acta Mechanica.

[9] D.A. Drew, Arch. Rational Mech. Anal., 62, 149 (1976).

[10] D.A. Drew and L.A. Segel, Studies in Applied Math, 50, 205 (1971).

[11] M.M. El-Kaissy, The Thermomechanics of Multi-phase Systems with Applications to Fluidized Continua, Ph.D. Thesis, Stanford University (1975).

[12] A. Fick, Ann-der Physik, 94, 56 (1855).

[13] M.A. Goodman and S.C. Cowin, Arch. Rational Mech. Anal., 44, 269 (1972).

[14] G.M. Homsy, M.M. El-Kaissy and A. Didwania, Intl. J. Multiphase Flow, 6, 305 (1980).

[15] M. Massoudi, Application of Mixture Theory to Fluidized Beds, Ph.D. Thesis, University of Pittsburgh (1986).

[16] M. M. Massoudi and J. Boyle, DOE/METC 88-4077, Technical Note (1987).

[17] J.D. Murray, J. Fluid Mech., 21, 465 (1965).

[18] K.R. Rajagopal, G. Johnson, M. Massoudi, In preparation.

[19] S.B. Savage, J. Fluid Mech., 92, 53 (1979).

[20] G. Segre and A. Silberberg, J. Fluid Mech., 14, 115 (1962).

[21] J.J. Shi, K.R. Rajagopal and A.S. Wineman, Intl. J. Engng. Sci., 19, 871 (1981).

[22] C. Truesdell, Lincei-Rend. Sc. fis. mat. e. nat., Series 8, 22: 33-38 and 158-166 (1957).

[23] C. Truesdell, Rational Thermodynamics, 2nd Ed., Springer-Verlag, New York (1984).

[24] C. Truesdell and W. Noll, The Non-Linear Field Theories of Mechanics, Handbuch der Physik, Vol. III/3, Springer-Verlag (1965).

K.R. RAJAGOPAL AND A.S. WINEMAN

Developments in the mechanics of interactions between a fluid and a highly elastic solid

INTRODUCTION

The general theory of interacting continua based on principles of modern continuum mechanics has been available for some time. The historical development and comprehensive surveys of the progress in this subject are presented in the review articles by Bowen [3], Atkin and Craine [1], Bedford and Drumheller [2], Passman et al. [19] and in several appendices in the book by Truesdell [29] on Rational Thermodynamics. A critical review of the literature makes it clear that there have been very few applications of the theory to solving boundary value problems of physical interest. This is due to several reasons, the main one being the difficulty in specifying the partial tractions, in traction or mixed boundary value problems.

Recently, Rajagopal, Wineman and Gandhi [22] proposed a method for generating appropriate boundary conditions for a class of problems involving solid-fluid mixtures in which the boundary of the mixture is saturated. This work, which is an outgrowth of previous research carried out by Shi, Rajagopal and Wineman [26], has provided a means for solving traction and mixed boundary value problems involving a mixture of a fluid and a non-linearly elastic solid. In this review article, we will present a number of results which have been obtained by solving boundary value problems within the context of the theory of interacting continua. The predictions of the theory agree well with experimental results, where available.

The outline of the paper is as follows. Part I is concerned with the theory. The underlying concepts and governing equations are summarized in Section 1. Constitutive equations are presented in Section 2. The notion of saturation and the boundary condition derived on the basis of such an assumption by Rajagopal, Wineman and Gandhi [22] are discussed in Section 3. The case in which a fluid is dispersed throughout a solid and is at rest is treated in Part II. Homogeneous deformations, especially those involving shear, are discussed in Section 4. The combined extension and twisting of a circularly cylindrical mixture is outlined in Section 5. An expression for the torsional stiffness valid for small values of twist is presented . This relation is independent of material properties and the degree of swelling. Several problems involving non-homogeneous deformations of a cylinder in contact with a fluid bath are treated in Section 6. Part III contains results for the steady state diffusion of a fluid through highly elastic isotropic and anistropic solids of differing geometries. The effects of shearing and stretching the solid, and the manner in which the anisotropy of the solid affects the diffusion process are discussed.

PART I

1. PRELIMINARIES

In this section, for the sake of continuity and completeness, we provide a brief discussion of the concepts and basic balance laws of the theory of interacting continua.

We shall consider a mixture of two constituents, a solid and a fluid. In the theory of interacting continua, it is assumed that at any instant of time t, at each point in the space occupied by the mixture, there exist two particles, one belonging to each constituent. Each constituent is then treated as a continuum. Thus, in effect each constituent is homogenized to co-exist in the region occupied by the mixture.

As in theories of a single continua, each constituent of the mixture has associated with it, its own kinematical quantities. Let $\mathbf{X}^s$ and $\mathbf{X}^f$ denote the position of a particle of the solid and fluid, in their reference configurations, respectively. The motions of the solid and fluid are denoted by

$$\mathbf{x}^s = \chi^s(\mathbf{X}^s, t), \quad \mathbf{x}^f = \chi^f(\mathbf{X}^f, t), \tag{1.1}$$

respectively. We assume that these motions are one-to-one, continuous and invertible. The velocity vectors associated with these motions are

$$\mathbf{u}^s = \frac{d\chi^s}{dt} \quad \text{and} \quad \mathbf{u}^f = \frac{d\chi^f}{dt}, \tag{1.2}$$

where $\frac{d}{dt}$ denotes the material time derivative, the deformation gradient associated with the solid S_1 is given by

$$F = \frac{\partial \chi^s}{\partial \mathbf{X}^s} . \qquad (1.3)$$

Let ρ_R^s and ρ_R^f denote the mass densities of the solid and fluid in their perspective reference configurations, measured per unit volume of the unmixed state. Let ρ^s and ρ^f denote the densities of the solid and fluid at time t, measured per unit volume of the mixture. The total density ρ and mean velocity $\mathbf{w}$ of the mixture are defined by

$$\rho = \rho^s + \rho^f , \qquad (1.4)$$

$$\rho \mathbf{w} = \rho^s \mathbf{u}^s + \rho^f \mathbf{u}^f . \qquad (1.5)$$

(i) Balance of mass

We shall assume that there is no interconversion of mass between the two constituents. However, the general theory allows for the possibility of interconversion of mass between the constituents. The local form for the conservation of mass for the solid is

$$\rho^s \, det\, \mathbf{F} = \rho_R^s . \qquad (1.6)$$

The local form for conservation of mass for the fluid is the usual

Eulerian form $$\frac{\partial \rho^f}{\partial t} + \operatorname{div}\left(\rho^f \mathbf{u}^f\right) = 0 , \qquad (1.7)$$

where $\frac{\partial}{\partial t}$ denotes the partial derivative with respect to time.

(ii) Balance of linear and angular momenta

It is assumed that a partial stress tensor is associated with each constituent. Let σ^s and σ^f denote the partial stress tensors associated with the solid and fluid, respectively. It is also assumed that each constituent exerts a force on the other, i.e., there is a momentum supply to each constituent by the other constituents. We shall denote this interactive body force by **b**. Let $\hat{l}$ denote the external body force per unit mass. Then, the local form of the balance of linear momentum for the solid and fluid are

$$\text{div}\,\sigma^s - \mathbf{b} + \rho^s \hat{l} = \rho^s \frac{d\mathbf{u}^s}{dt} \tag{1.8}$$

$$\text{div}\,\sigma^f + \mathbf{b} + \rho^f \hat{l} = \rho^f \frac{d\mathbf{u}^f}{dt} . \tag{1.9}$$

It follows from the balance of angular momentum that the total stress $\sigma^s + \sigma^f$ is symmetric,

$$\sigma^s + \sigma^f = (\sigma^s)^T + (\sigma^f)^T . \tag{1.10}$$

However, the partial stresses are not necessarily symmetric.

(iii) Surface conditions

Let **m** denote an outward unit normal vector on the boundary of the mixture. Let $\mathbf{p}^s$ and $\mathbf{p}^f$ denote the surface tractions on the boundary associated with σ^s and σ^f, respectively. The surface conditions are

$$\mathbf{p}^s = (\sigma^s)^T\mathbf{m} , \quad \mathbf{p}^f = (\sigma^f)^T\mathbf{m} . \tag{1.11)$_{1,2}$ $$

Thus,

$$\mathbf{p}^s + \mathbf{p}^f = \left[(\sigma^s)^T + (\sigma^f)^T\right]\mathbf{m} = \left[\sigma^s + \sigma^f\right]\mathbf{m}. \tag{1.12}$$

In general, in boundary value problems, the total traction $\mathbf{p} = \mathbf{p}^s + \mathbf{p}^f$

acting on the boundary of the mixture is known. However, the manner in which the total traction separates into the partial tractions is not known.

(iv) Thermodynamical considerations

For the purpose of brevity, we shall not present statements of the energy balance, and the entropy production inequality. We shall quote the relevant results based on thermodynamical considerations as needed. It will be assumed that both constituents have the same temperature θ. We also denote by A the Helmholtz free energy of the mixture per unit mass of the mixture.

2. CONSTITUTIVE EQUATIONS

Each constituent is assumed to be incompressible in its reference configuration. It is further assumed that the volume of the mixture is the sum of the volumes occupied by the solid and fluid before mixing. This volume additivity imposes the following constraint on the mixture (cf. Mills [18]):

$$\frac{\rho^s}{\rho_R^s} + \frac{\rho^f}{\rho_R^f} = 1 . \tag{2.1}$$

It is interesting to note that, according to (2.6) and (3.1), $det\mathbf{F} \neq 1$ unless $\rho^f = 0$, in which case there is no fluid and the mixture reduces to a single constituent solid.

Based on single constituent constitutive theories for non-linear elastic solids and viscous fluids, the response functions for the mixture are assumed to depend on $\mathbf{F}$, ρ^f, θ, $\mathbf{u}^s$, $\mathbf{u}^f$, grad ρ^f, grad $\mathbf{u}^s$, grad $\mathbf{u}^f$ and $\partial \mathbf{F}/\partial \mathbf{X}^s$. It follows from a standard and straightforward analysis based on the application of an interpretation of the second law of thermodynamics, that the specific Helmholtz free energy for the mixture has the form (cf. Shi, Rajagopal and Wineman [26]),

$$A = \tilde{A}\,(\mathbf{F}, \rho^f, \theta)\,. \tag{2.2}$$

Under the assumption of isotropy of the solid in its reference configuration, the Helmholtz free energy function can be expressed in terms of principal invariants of the strain tensor $\mathbf{B} = \mathbf{F}\mathbf{F}^T$ as

$$A = A(I_1, I_2, I_3, \rho^f, \theta)\,, \tag{2.3}$$

where

$$I_1 = \mathrm{tr}\mathbf{B}\,, \tag{2.4}$$

$$I_s = \frac{1}{2}\left[(\mathrm{tr}\mathbf{B})^2 - \mathrm{tr}\mathbf{B}^2\right], \tag{2.5}$$

$$I_3 = \mathit{det}\,\mathbf{B}\,. \tag{2.6}$$

In view of (2.6), (3.1), (3.6) and the definition of **B**,

$$I_3 = \left[1 - \frac{\rho^f}{\rho_R^f}\right]^{-1/2},$$

and A in (3.3) reduces to

$$A = A\,(I_1, I_2, \rho^f, \theta)\,. \tag{2.7}$$

In addition, the expressions for the partial stresses and the interactive body force consist of two groups of terms, those calculated from A and those which depend on the velocity fields of the solid and fluid. Dependence on the latter introduces several viscosity coefficients. The remainder of this article is concerned with situations in which viscosity effects are ignored. Part II is concerned with results for fluids at rest with respect to a deformed solid. In Part III, the fluid is assumed to be inviscid. With this in mind, the components of the partial stresses and interactive body force are given by

$$\sigma^s_{ki} = \phi\delta_{ki} - p\,\frac{\rho^2}{\rho^s_R}\,\delta_{ki} + 2\rho\left\{\left(\frac{\partial A}{\partial I_1} + I_1\,\frac{\partial A}{\partial I_2}\right)B_{ki} - \frac{\partial A}{\partial I_2}\,B_{km}\,B_{mi}\right\} \tag{2.8}$$

$$\sigma^f_{ki} = -\,\phi\delta_{ki} - p\,\frac{\rho^f}{\rho^f_R}\,\delta_{ki} - \rho\rho^f\,\frac{\partial A}{\partial \rho^f}\,\delta_{ki} \tag{2.9}$$

$$b_k = \frac{\partial \phi}{\partial x_k} - p\,\frac{\partial}{\partial x_k}\left(\frac{\rho^s}{\rho^s_R}\right) + \rho^s\,\frac{\partial A}{\partial \rho^f}\,\frac{\partial \rho^f}{\partial x_k} - \rho^f\left\{\left(\frac{\partial A}{\partial I_1} + I_1\,\frac{\partial A}{\partial I_2}\right)\delta_{il} - \frac{\partial A}{\partial I_2}\,B_{il}\right\} + \alpha\,\frac{\rho^s}{\rho^s_R}\,\frac{\rho^f}{\rho^f_R}\left(u^s_k - u^f_k\right). \tag{2.10}$$

The scalar p arises due to the constraint of volume additivity. The scalar ϕ was introduced into the theory of interacting continua by Green and Naghdi [14] from thermodynamic considerations, and is expressed in terms of the Helmholtz free energy for the mixture and for the constituents. ϕ is of interest only if partial stresses are to be calculated, since this will not be the case in the results to be presented here, we drop reference to it without loss of generality.

The constitutive parameter α accounts for a contribution to the interaction body force due to the relative motion between the solid and the fluid. The interaction between the solid and the fluid is evident from equations (2.8), (2.9), (2.10) where the partial stress of each constituent is affected by the properties of the other constituent.

The total stress is defined by

$$T_{ij} = \sigma^s_{ij} + \sigma^f_{ij} \quad . \tag{2.11}$$

By (3.1) and (3.9), the constitutive equation for the total stress is

$$T_{ij} = -\left(p + \rho\rho^f \frac{\partial A}{\partial \rho^f}\right)\delta_{ij}$$

$$+ \partial\rho \left\{\left(\frac{\partial A}{\partial I_1} + I_1 \frac{\partial A}{\partial I_2}\right)B_{ij} - \frac{\partial A}{\partial I_2} B_{ik} B_{kj}\right\} . \tag{2.12}$$

Finally, we confine attention to isothermal conditions and thus reference to the temperature θ will be omitted.

3. NOTION OF SATURATION AND THE ASSOCIATED BOUNDARY CONDITION

In a typical boundary value problem for a mixture, the governing equations involve unknowns associated with all the constituents. The solution of these equations requires the specification of a sufficient number of boundary conditions. While the total traction on the boundary of the mixture may be known, there is no clear way to specify the partial surface tractions, and in our case the partial tractions $\mathbf{p}^s$ and $\mathbf{p}^f$. Usually, only the total traction $(\mathbf{p}^s + \mathbf{p}^f)$ is known. Thus, we somehow need some additional information on the boundary, say some independent relationship between $\mathbf{p}^s$, $\mathbf{p}^f$, and possibly some of the properties and the values of the kinematics of the solid and fluid at the boundary. Such a relationship should of course stem from a physically meaningful and acceptable criterion. Rajagopal, Wineman and Gandhi [22] have discussed a method for augmenting the boundary conditions, for a particular class of problems. This class and the associated boundary condition are described below.

Consider the class of problems in which the boundary of the mixture is in contact with a fluid bath. The material elements of the mixture boundary are assumed to be in a saturated state. In a series of articles on the equilibrium swelling of rubber vulcanizates, Southern and Thomas [27] clearly indicate that the boundary is indeed saturated: "When a sheet of rubber vulcanizate is immersed in a solvent, the surface layer takes up its equilibrium amount of liquid, virtually instantaneously. The liquid subsequently diffuses into the bulk of the rubber". That is, the elements are in an equilibrium state in which the deformation, stress and fluid content are fixed. No further fluid can be accumulated at the boundary once it is saturated, and thus any fluid that enters it has to leave it.

Now consider a thought experiment where a cube of the mixture is subjected to the same state of homogeneous deformation as that of a material element on the saturated boundary of the original problem. The cube is assumed to be immersed in the fluid bath and in equilibrium. This equilibrium state[1] can be characterized by the condition that in an infinitesimal change of the state of the system, the variation in the Helmholtz free energy equals the infinitesimal work done on the system (cf. Denbigh [9]). As is shown below, this requirement leads to a relationship between the state of stress, the deformation of the solid and the density of the fluid in the homogeneously deformed cube. We now require that this condition be valid for the material elements on the boundary of the original solid-fluid mixture.

Let m denote the mass of the homogeneously swollen saturated cube, and recall that A denotes the Helmholtz free energy per unit mass of the mixture . The equilibrium condition is then

$$\delta W = \delta(mA) \tag{3.1}$$

where δW represents the infinitesimal work done. For an isotropic solid, the Helmholtz free energy per unit mass of the mixture is as in (2.7). Then, it can be shown (cf. Rajagopal, Wineman and Gandhi [22]), that the following relation holds:

[1]The equilibrium state, i.e., the saturated state is characterized by the variation of the Gibbs free energy of dilution being zero at that state. However, under certain further assumptions this statement is equivalent to the one stated above (cf. Rajagopal, Gandhi and Wineman [22], Treloar [28]).

$$T_{ij} = \left[\rho_R^f A + \rho \frac{\partial A}{\partial \rho^f}\left(\rho_R^f - \rho^f\right)\right] \delta_{ij}$$

$$+ 2\rho \left[\left(\frac{\partial A}{\partial I_1} + I_1 \frac{\partial A}{\partial I_2}\right) B_{ij} - \frac{\partial A}{\partial I_2} B_{ik} B_{kj}\right] . \qquad (3.2)$$

At saturation, (3.2) relates the total stress, the deformation of the solid as measured by **B** and the current density of the fluid ρ^f. More importantly, there is no indeterminate scalar as in the constitutive equation (2.12). On comparing (3.2) and (2.12), it is seen that the indeterminate scalar p assumes a specific form,

$$-p = \rho_R^f \left(A + \rho \frac{\partial A}{\partial \rho^f}\right) . \qquad (3.3)$$

This relation implies that the saturated mixture behaves as if it were a compressible material in equilibrium.

The following comments of Treloar [28], within the context of the phenomena of swelling of homogeneously deformed bodies in rubber elasticity, are very pertinent. "The physical significance of this result is that whereas in an incompressible rubber (i.e., whether unswollen or swollen to a specified or fixed extent) the volume, and hence the state of strain, is unaffected by the superposition of an arbitrary hydrostatic pressure; in a compressible rubber, or in a rubber considered to be in equilibrium with respect to the swelling liquid, this is no longer true, since any such superimposed hydrostatic pressure will reduce the volume or liquid content. In this respect, therefore, a swollen rubber in continuous equilibrium with a surrounding liquid, may be regarded from a purely formal standpoint, as having mechanical properties equivalent to those of a compressible material."

Most discussions of thermodynamic considerations and constitutive equations which have appeared in the literature introduce A, the Helmholtz free energy per unit mass of the mixture. Since the mass of the mixture is unknown, it may be useful to introduce the Helmholtz free energy per unit mass of the solid denoted by $\widehat{A}$. Since the volume occupied by the swollen solid coincides with the volume occupied by the mixture, A and $\widehat{A}$ are related by

$$\rho A = \rho^s \widehat{A} \,. \tag{3.4}$$

The saturation condition can be expressed in terms of $\widehat{A}$ by using (3.4). For an isotropic material, $\widehat{A} = \widehat{A}\left(I_1, I_2, \rho^f\right)$, and (3.2) becomes

$$T_{ij} = \frac{\rho_R^f (\rho^s)^2}{\rho_R^s} \frac{\partial \widehat{A}}{\partial \rho^f} \delta_{ij} + 2\rho^s \left[\left(\frac{\partial \widehat{A}}{\partial I_1} + \frac{\partial \widehat{A}}{\partial I_2} \right) B_{ij} - \frac{\partial \widehat{A}}{\partial I_2} B_{ik} B_{kj} \right] \tag{3.5}$$

Consider the general problem in which a fluid is at rest in a non-homogeneously deformed solid. The mixture is in contact with a fluid bath, and the system has reached an equilibrium (saturated) state. The field equations consist of (1.8) and (1.9), with $\mu^f = \mu^s = 0$, and constitutive equations (2.8) -(2.10), (in which ϕ is now omitted). As boundary conditions, one specifies surface displacements, total surface tractions and the condition that the surface material elements are in a saturated state, as expressed by (3.2) .

Motivated by the global variational statement pertaining to the whole mixture being saturated (cf. Rajagopal, Gandhi, Wineman

[22]) Gandhi and Usman [11] have shown that the boundary value problem implies the field equation:

$$div\, \mathbf{T}^{SAT} = 0\,, \tag{3.6}$$

the same displacement boundary conditions, and

$$\left(\mathbf{T}^{SAT}\right)^{T} \mathbf{m} = \mathbf{t}\,, \tag{3.7}$$

where $\mathbf{T}^{SAT}$ denotes the stress tensor in the saturation equation of state (3.2), and **t** is the total surfact traction. Thus, using the generalization of (3.1) namely (cf. Rajagopal, Gandhi and Wineman [22])

$$\delta \int_{V} \rho\, A dU = \int_{S} \mathbf{t} \cdot \delta U \; dA\,, \tag{3.8}$$

in which V is the current region occupied by the mixture, S the corresponding surface, and δU an arbitrary displacement field increment, Gandhi and Usman [12] arrive at (3.6) and (3.7) which is completely consistent with the assumed form for the boundary condition obtained by just assuming that the boundary of the mixture is saturated.

PART II

4. HOMOGENEOUS DEFORMATIONS

In this section, we discuss two problems involving a homogeneously deformed solid, in which a fluid is uniformly dispersed and at rest.

There can be a strong interaction between the amount of swelling of a block of solid and its deformation. In order to see this, consider a cube of solid which sits in a bath of liquid in a state of triaxial extension. We assume a deformation of the form

$$x_1^s = \lambda_1 X_1^s\,, \quad x_2^s = \lambda_2 X_2^s\,, \quad x_3^s = \lambda_3 X_3^s \tag{4.1}$$

where λ_α, $\alpha = 1, 2, 3$, is the stretch ratio measured from the unstrained, unswollen state. The saturation relation (3.5) becomes

$$T_{ii} = \frac{\rho_R^f (\rho^s)^2}{\rho_R^2} \frac{\partial \widehat{A}}{\partial \rho^f} + 2\rho^s \left[\left(\frac{\partial \widehat{A}}{\partial I_1} + I_1 \frac{\partial \widehat{A}}{\partial I_2} \right) \lambda_1 - \frac{\partial \widehat{A}}{\partial I_2} \lambda_i^4 \right], \text{ no sum on i} \tag{4.2}$$

where, by (2.4)

$$I_1 = \lambda_1^2 + \lambda_2^2 + \lambda_3^2\,. \tag{4.3}$$

We will consider the special case of butyl rubber, with benzene and heptane as the solvents. Experimental and theoretical results on these systems are discussed in Chapter 7 of the treatise by Treloar [28].

For these materials, the Helmholtz free energy per unit mass of the solid is given in Treloar [28] as

$$\rho_R^s \widehat{A} = \frac{\rho_R^s R\theta}{2Mc} (I_1 - 3) + f\left(\rho^f\right), \tag{4.4}$$

in which R is the universal gas constants,

($R = 8.317 \times 10^7$ dyne cm/mol°K), θ is the absolute temperature, and M_c is the molecular weight between cross lines. In order to specify $f\left(\rho^f\right)$, we first introduce the volume fraction of the solid ν, which is the ratio of the volume of the rubber block in its reference state to its swollen volume. By (1.3), (1.6), (2.1) and (4.1)

$$\nu = \frac{1}{\det \mathbf{F}} = \frac{1}{\lambda_1\lambda_2\lambda_3} = \frac{\rho^s}{\rho^s_R} = 1 - \frac{\rho^f}{\rho^f_R}\,. \tag{4.5}$$

Then $f(\rho^f) = \hat{f}(\nu)$, where

$$\frac{\rho^f_R(\rho^s)}{\rho^s_R}\frac{\partial\hat{A}}{\partial\rho^f} = -\nu^2\frac{\partial\hat{f}}{\partial\nu} = \frac{R\theta}{V_1}\left\{ln\,(1-\nu) + \nu + \chi\nu^2\right\}. \tag{4.6}$$

In (4.6), V_1 is the molar volume of the liquid. The expression within the brackets was derived by Flory-Huggins, where χ is the Flory-Huggins mixing parameter and depends on the particular solid-fluid combination.

Substitution of (4.4) and (4.6) into (4.2) gives

$$T_{ii} = \frac{R\theta}{V_1}\left\{ln\,(1-\nu) + \nu + \chi\nu^2\right\} + \rho^2\frac{R\theta}{M_c}\lambda_i^2, \quad \text{no sum on i}$$

$$= \frac{R\theta}{V_1}\left\{ln\,(1-\nu) + \nu + \chi\nu^2 + \frac{\rho^s_R V_1}{M_c}\nu\lambda_i^2\right\}. \tag{4.7}$$

For saturated states in which the pressure in the fluid bath is zero and the block is in uniaxial extension along the x_1-axis, $T_{22} = T_{33} = 0$, which implies that $\lambda_2 = \lambda_3$. Then $\nu = 1/\lambda_1\lambda_2^2$,

$$T_{11} = \frac{R\theta}{V_1}\left\{ln\,(1-\nu) + \nu + \chi\nu^2 + \frac{\rho^s_R V_1}{M_c}\nu\lambda_1^2\right\} \tag{4.8$_1$}$$

$$ln\,(1-\nu) + \nu + \chi\nu^2 + \frac{\rho^s_R V_1}{M_c}\frac{1}{\lambda_1} = 0\,. \tag{4.8$_2$}$$

When $T_{11} = 0$, $\lambda_1 = \lambda_2 = \lambda_3 = \nu^{-1/3}$ and the block has undergone free swelling. Let a value of λ_1 be specified so that $\lambda_1 > \nu^{-1/3}$. Then, ν can be found from $(4.8)_2$ and T_{11} from $(4.8)_1$. According to Treloar, T_{11} will be positive. Conversely, if $\lambda_1 > \nu^{-1/3}$, T_{11} will be compressive.

A comparison of the predictions of $(4.8)_2$ with experiments on butyl rubber in benzene and heptane is shown in Figure 1 (cf. Figure 7.10 of Treloar [28]).

For the case of heptane, the lower left point corresponds to $T_{11} = 0$, where $\lambda_1 = 1.6$ and the current volume of the block is about four times the original volume. Note the rapid change of volume with stretch ratio λ_1. Similar results for the compressive case are also presented in Treloar [28].

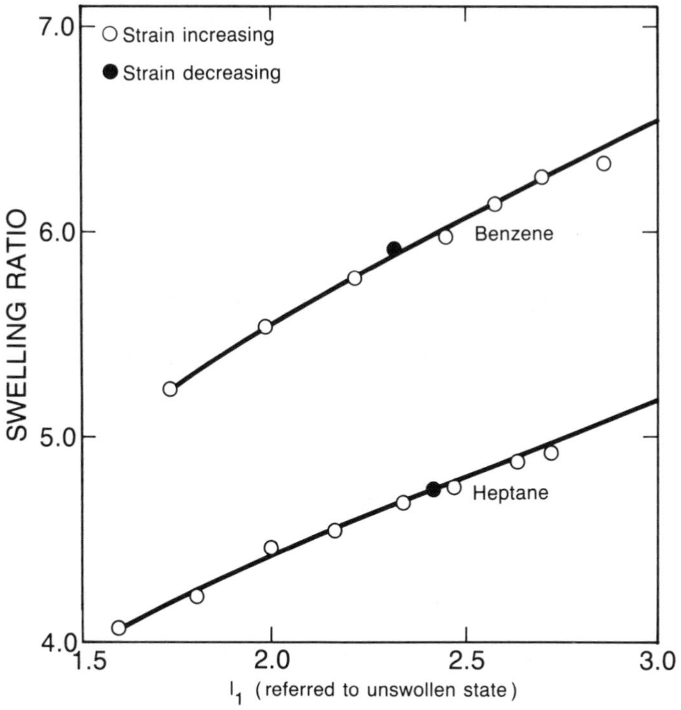

Figure 1. Swelling ratio.

Another interesting result arises in the case when fluid is uniformly dispersed through a block which is homogeneously deformed to a state of shear superposed on triaxial extension. Then (1.1) is given by

$$x_1^s = \lambda_1 X_1^s + \kappa\lambda_2 X_2^s ,$$

$$x_2^s = \lambda_2 X_2^s ,$$

$$x_3^s = \lambda_3 X_3^s . \qquad (4.9)$$

Let the block in its reference state be bounded by pairs of surfaces $X_i = \pm X_1^0$. Consider the surfaces $X_i = \pm X_1^0$. The total normal traction N, and shear traction L on say x_1^0, are expressed in terms of the relations

$$N_1 = \frac{T_{11} + \kappa^2 T_{22} - 2T_{12}\kappa}{1 + \kappa^2} \qquad (4.10)$$

$$L = \frac{\kappa\,(T_{11} - T_{22}) + T_{12}\,(1-\kappa^2)}{1 + \kappa^2} , \qquad (4.11)$$

where T_{ij} are the components of the total stress tensor.

Rajagopal and Wineman [23] established the following universal relation for blocks of mixtures which are deformed as described in (4.9):

$$T_{11} - T_{22} = \left(\frac{\lambda_1^2 + \lambda_2^2\,\kappa^2 - \lambda_2^3}{\kappa\,\lambda_2^2} \right) T_{12} . \qquad (4.12)$$

They showed that (4.12) holds for both the general constitutive equation (2.12) and for the saturation condition given by (3.2) and (3.5).

Suppose the block is maintained in this state of deformation by shear tractions and pressure from the bath,

$$N_1 = T_{22} = T_{33} = -p_0 . \qquad (4.13)$$

Then, (4.10) and (4.13) imply that

$$-p_0 = T_{11} - 2T_{12}\,\kappa \qquad (4.14)$$

(4.12) and (4.14), together with $T_{22} = -p_0$ then imply that

$$\lambda_1^2 = \lambda_2^2\,(1 + \kappa^2) . \qquad (4.15)$$

Rajagopal and Wineman [23] derived this result under the assumption that there were no normal tractions on the block, i.e, $p_0 = 0$. Equation(4.15) thus generalizes their result to blocks in baths under pressure.

It is to be noted that (4.12) and (4.15) are 'universal relations', that is they are valid for all mixtures of isotropic solids and fluids, independent of their specific properties. Note that the "universal relations" hold not only in saturation but also when the solid is infused with an arbitrary quantity of the fluid.

Suppose it is desired to determine the actual dimensions of a saturated sheared block on which the normal stress are given by (4.13). For the deformation (4.9), the saturation condition (4.5) becomes

$$T_{ij} = \frac{R\theta}{v_1}\left\{ln\,(1-\nu) + \nu + \chi\nu^2\right\}\delta_{ij} + \nu\,\frac{\rho_R^s R\theta}{M_c}\,B_{ij} , \qquad (4.16)$$

where

$$\nu = \frac{1}{\det \mathbf{F}} = \frac{1}{\lambda_1\lambda_2\lambda_3} , \qquad (4.17)$$

and

$$B_{11} = \lambda_1^2 + \kappa^2\lambda_2^2, \quad B_{22} = \lambda_2^2, \quad B_{33} = \lambda_3^2,$$

$$B_{12} = \kappa\lambda_2^2.$$

By virtue of (4.16), the condition $T_{22} = T_{33}$ implies that $\lambda^2 = \lambda^3$.

Equation (4.15) and (4.17) then give

$$v = \frac{1}{\lambda_2^3 (1 + \kappa^2)^{-1/2}} . \tag{4.18}$$

Now set i = 2 and j = 2 in (4.16) to get

$$-p_0 = \frac{R\theta}{v_1}\left\{ln\,(1-v) + v + \chi v^2\right\} + v\,\frac{\rho_R^s R\theta}{M_c}\lambda_2^2 . \tag{4.19}$$

Substitution of v from (4.18) gives an equation for λ_2 in terms of the amount of shear κ, the pressure in the bath p_0 and the properties χ, v, ρ_R^s and M_c. Once λ_2 has been determined, λ_1 follows from (4.15), the volume fraction v, or degree of swelling $1/v$ follows from (4.17), and the shear traction T_{12} can be evaluated using (4.16).

5. A UNIVERSAL RELATION IN TORSION

We consider a solid circular cylinder whose dimensions in its reference configuration are radius R_0 and length L_0. The cylinder is a fluid bath subjected to combined finite extension and torsion. The cylinder absorbs fluid, but does not necessarily become saturated. The fluid is dispersed throughout the cylinder and at rest.

We assume that, with respect to cylindrical coordinates, the deformation of the cylinder is described as

$$r = r(R)\ ,$$

$$\theta = \Theta + \psi\lambda Z\ ,$$

$$z = \lambda Z\ , \qquad (5.1)$$

where (r, θ, z) denote the coordinates in the deformed state of a particle which is at (R, Θ, Z), in the reference state. The radial deformation in $(5.1)_1$ accounts for non-uniform radial expansion due to swelling of the cylinder. In $(5.1)_3$, λ is the axial stretch ratio and ψ in $(5.1)_2$ is the rotation per unit current length. Let the resultant axial force and twisting moment on the ends of the cylinder be denoted, respectively, by F and M. These are calculated using the total stress T_{ij} and hence represent the total axial force and twisting moment applied to the cylinder.

Gandhi, Rajagopal and Wineman [10] considered the case of small twist superposed on finite extension of a mixture of an isotropic solid and fluid, i.e., $|\psi| << 1$. They showed that the problem decouples into (1) a problem of uniaxial tension and , (2) a torsion problem.

In problem (1) the deformation is homogeneous, and fluid is uniformly dispersed throughout the cylinder. The axial stretch ratio λ is specified and the stretch ratio in the cross section, denoted as $\hat{\lambda}$, depends on the amount of fluid in the mixture.

The main result is that the torsional stiffness M/ψ and the axial force satisfy the following relation

$$\lim_{\psi \to 0} \frac{M/\psi}{F} = \frac{R_0^2}{2} \frac{\lambda^2 \hat{\lambda}^2}{\lambda^2 - \hat{\lambda}^2} . \tag{5.2}$$

This relation is independent of the form of the Helmholtz free energy A and is hence a 'universal relation'. It is valid for all states in which the elastic solid is swollen with fluid, in the absence of fluid, the mixture becomes an incompressible isotropic solid and $\lambda\hat{\lambda}^2 = 1$, equation (5.2) then reduces to the classical expression established by Rivlin [25].

6. NON-HOMOGENEOUS DEFORMATION

We now describe an example in which a cylinder of isotropic solid is in a fluid bath and subjected to a non-homogeneous deformation. The system reaches a saturated state in which the fluid is at rest and dispersed non-uniformly throughout the mixture region. This distribution is maintained by the density and strain gradients which appear in the expression for the interaction body force. This example describes an experiment which can be compared with predictions of the theory.

The example is that of combined finite extension and torsion of $\lambda_r(0) = \lambda_\theta(0)$a cylindrical mixture, which was treated by Gandhi, Usman, Wineman and Rajagopal [12]. The deformation is given by (5.1), where ψ can now be large. The problem is formulated as a system of three coupled first order non-linear ordinary differential equations for the indeterminate scalar p, $\lambda_r = dr/dR$ and $\lambda_\theta = r/R$, two are obtained from the radial components of (1.8) and (1.9) with inertia neglected, the third equation is the compatibility relation

$$\frac{d\lambda_\theta}{dR} = \frac{\lambda_r - \lambda_\theta}{R} \ . \tag{6.1}$$

Boundary conditions were specified as follows:

(1) The solution must be bounded at R=0,

$$\lambda_r(0) = \lambda_\theta(0) \ . \tag{6.2}$$

(2) The total traction on the outer surface $R=R_0$ of the cylinder is specified. Here, it is assumed that this surface is traction free, which implies that

$$T_{rr}(R_0) = 0 \ . \tag{6.3}$$

(3) The outer surface of the cylinder is in a saturated state. This implies that the radial component of (4.2) vanishes,

$$\rho_R^f A + \rho\left(\rho_R^f - \rho^f\right)\frac{\partial A}{\partial \rho^f} + \partial\rho\left[\left(\frac{\partial A}{\partial I_1} + I_1\frac{\partial A}{\partial I_2}\right)B_{rr} - \frac{\partial A}{\partial I_2}B_{rr}^2\right] = 0 \ . \tag{6.4}$$

This boundary value problem was solved numerically for the Helmholtz free energy per unit mass of mixture defined by

$$A = \frac{v}{\rho}\left[\frac{R\theta\rho_0^s}{2M_c}(I_1 - 3) + \frac{R\theta}{v_1}\left[\frac{1-v}{v}\,ln\,(1-v) + \chi(1-v)\right]\right] \ , \tag{6.5}$$

which is obtained by combining (4.4), (5.4), (5.5) and (5.6). The particular values used in the calculation were selected to match conditions in the experimental work of Loke, Dickinson and Treloar [17] on natural rubber and toluene as the swelling liquid:

$\rho_R^s = 0.9016$ gm/cc, $\rho_R^f = 0.862$ gm/cc, $v_1 = 106.0$ cc/mole,

$M_c = 8891.0$ gm/mole, $\chi = 0.400$, $\hat{R} = 8.317 \times 10^7$ dym cm/mole

$\theta = 303.16°K$, $\lambda = 1.938$.

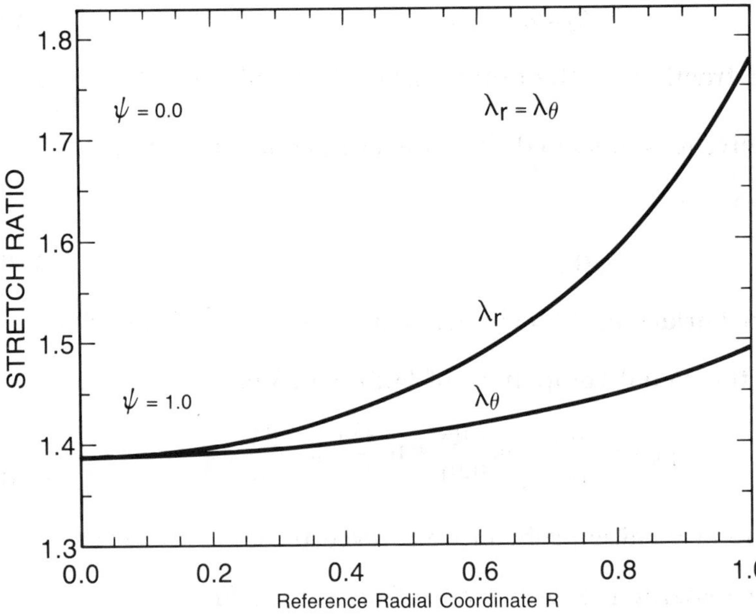

Figure 2. Variation of stretch ratios with the reference radial coordinate.

Figure 2 shows the radial variation of the radial stretch ratio λ_r and the circumferential stress ratio λ_θ for two different values of the angle of twist ψ. For the case of no twist, the cylinder is swollen homogeneously, and $\lambda_r = \lambda_\theta = \text{constant}$. When there is finite twist siginificant gradients in λ_r and λ_θ arise. Computational results also show that the total radial stress T_{rr} is compressive and increases in magnitude with Ψ for each fixed value of R. The total circumferential stress $T_{\theta\theta}$ is compressive in the core and tensile near the surface of the cylinder. Its magnitude also increases with the radius. The ratio of the volume V in the swollen twisted state to the volume in the unswollen reference state is shown in Figure 3. As Ψ increases, this ratio decreases indicating that fluid leaves the mixture region. Let V_0 denote the volume of the untwisted saturated cylinder. The ratio of the change in volume $V - V_0 = \Delta V$ to V_0 is compared with the experimental results of Loke, Dickinson and Treloar [28] in Figure 4. The computational results based on mixture theory predict the same qualitative and quantitative trends as are observed in the experimental results.

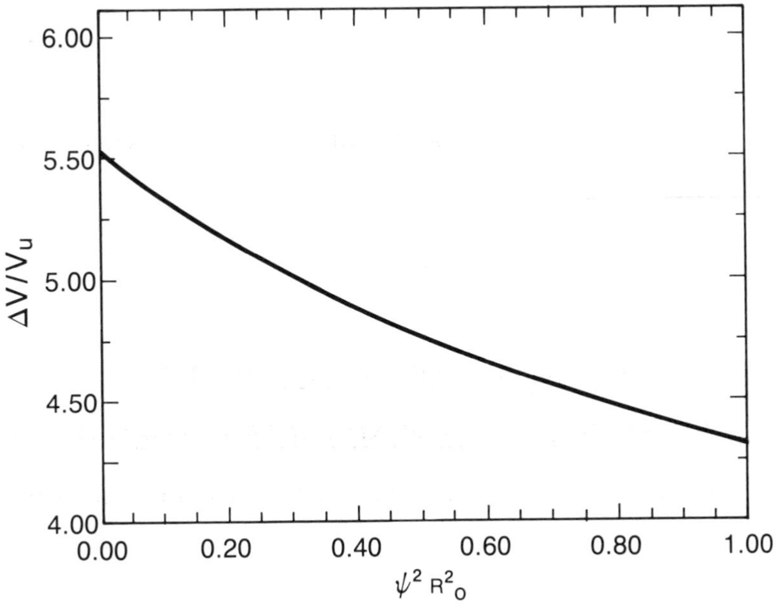

Figure 3. Variation of the current volume with twisting.

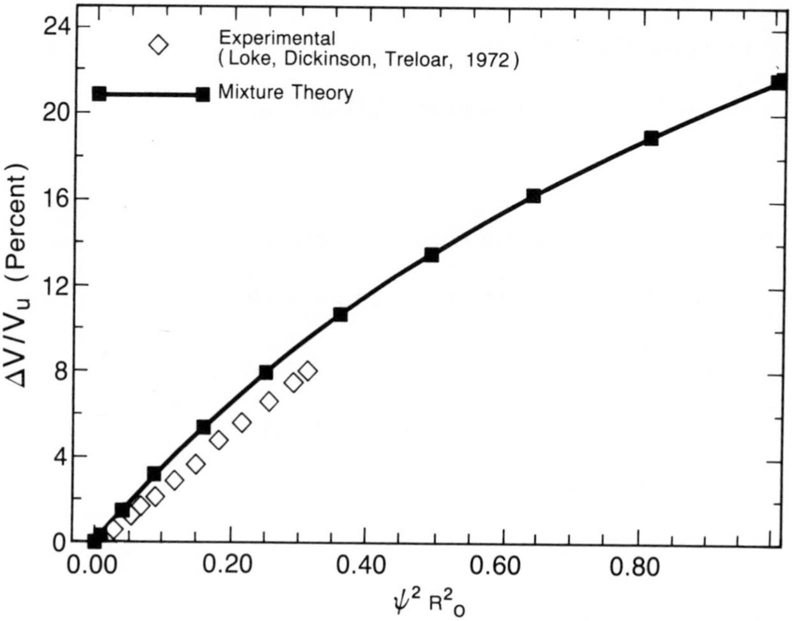

Figure 4. Change in volume with twisting.

The result clearly shows that significant effects arise because of solid and fluid interaction.

PART III

7. FLOW THROUGH ISOTROPIC AND ANISOTROPIC NON-LINEARLY ELASTIC SOLIDS UNDERGOING LARGE DEFORMATIONS

In the previous sections we were concerned with the statics of solids infused with a fluid. We shall now discuss the diffusion of fluids through isotropic and anisotropic solids undergoing finite deformations.

Diffusion of Fluids Through Isotropic Non-Linearly Elastic Solids

Several boundary value problems involving the steady state diffusion of a fluid through an isotropic solid in simple geometries like layers, cylindrical shells and spherical shells have been carried out. We provide a brief account of the same.

7a. DIFFUSION OF A FLUID THROUGH AN ISOTROPIC LAYER

Consider the steady-state diffusion of an ideal fluid normal to the surface of an isotropic layer (cf. Figure 5). We shall assume a deformation of the form

$$x = \lambda X, \quad y = \lambda Y, \quad z = z(Z)\,, \tag{7.1}$$

where (X, Y, Z) and (x, y, z) denote the initial and current position of a solid particle, respectively.

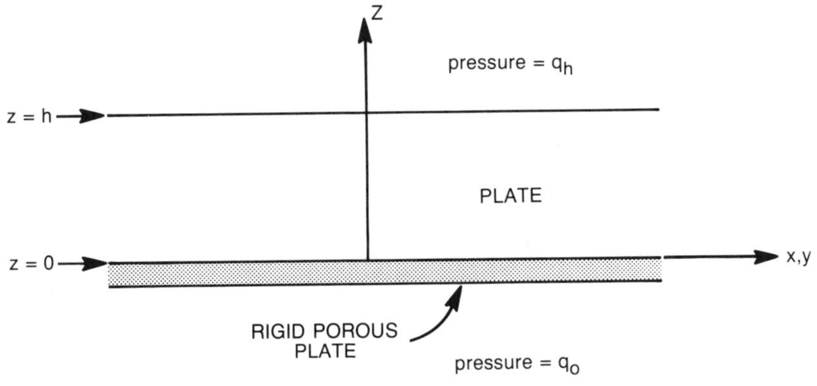

Figure 5. Flow through an elastic layer.

We shall assume that the flow is purely normal to the layer, and hence the velocity field for the fluid is given by

$$u_x = u_y = 0\,, \quad u_z = u_z(Z)\,, \tag{7.2}$$

where u_x, u_y, and u_z are the x, y, and z components of the velocity, respectively.

The conservation of mass for the solid reduces to

$$\lambda^2 z' \rho^s = \rho_R^s\,, \tag{7.3}$$

andthe conservation of mass for the fluid reduces to

$$\rho^f u_z = F = \text{constant}\,, \tag{7.4}$$

where F denotes the flux across any surface Z=constant.

The problem reduces to solving the following highly non-linear second order ordinary differential equation[2] (cf. Shi, Rajagopal and Wineman [26])

$$\bar{z}'' = -\frac{\alpha F h_0}{\left(\rho_R^f\right)^2 K \lambda^2}\left\{ -\frac{2\rho^*}{\lambda^2} + 2\bar{z}'\left[2\bar{\rho} + (1-2\rho^*)\frac{\left(\lambda^2\bar{z}'-1\right)}{\lambda^2\bar{z}'}\right]\right.$$

$$-\frac{1}{\lambda^2\bar{z}'}\left[\left(4\bar{z}'+\lambda^2\right)\left(\bar{\rho}-\rho^*\frac{\left(\lambda^2\bar{z}'-1\right)}{\lambda^2\bar{z}'}\right)-\left(2(\bar{z}')^2+\lambda^2\bar{z}'-1\right)\frac{\rho^*}{\lambda^2\left(\bar{z}'\right)^2}\right]$$

$$\left. + \bar{\rho}\frac{1}{\lambda^2\left(\bar{z}'\right)^2}\right\}^{-1}, \tag{7.5}$$

subject to the boundary conditions

$$\bar{z}(0) = 0\,, \tag{7.6}$$

---

[2]There is a typographical error in [26], on the right hand side of the equation which corresponds to (7.5) above.

and

$$T^s_{33} + T^f_{33} = -q_h , \tag{7.7}$$

where q_h is the pressure at $z = h$.

Shi, Rajagopal and Wineman [26] augment the boundary conditions by requiring that the layer is saturated, at $z = h$ and $z = 0$.

Thus,

$$ln[1-\nu(0)] + \nu(0) + \chi\nu^2(0) + \frac{\rho_{10}V_1}{M_c}\left(\frac{1}{\lambda^4\nu(0)} - \frac{\nu(0)}{2}\right) = (-q_h + q_0)\frac{V_1}{RT} . \tag{7.8}$$

Of course, the specific structure of the additional boundary condition depends on the form of the specific Helmholtz free energy function A for the mixture. We shall not go into a detailed discussion of the solution here. While predictions for the mass flux using the above approach and Fick's law agree with experimental results (cf. Paul and Ebra-Lima [21]) for small values of the pressure difference, the results based on a Fick's law approach is even qualitatively different for large values of the pressure difference (cf. Figure 6). The predictions based on the theory of interacting continua agrees both qualitatively and quantitatively even at large pressure differences. Furthermore, the approach based on the theory of interacting continua can determine the extent of swelling, i.e., the stretch ratios, and also the density variation through the layer.

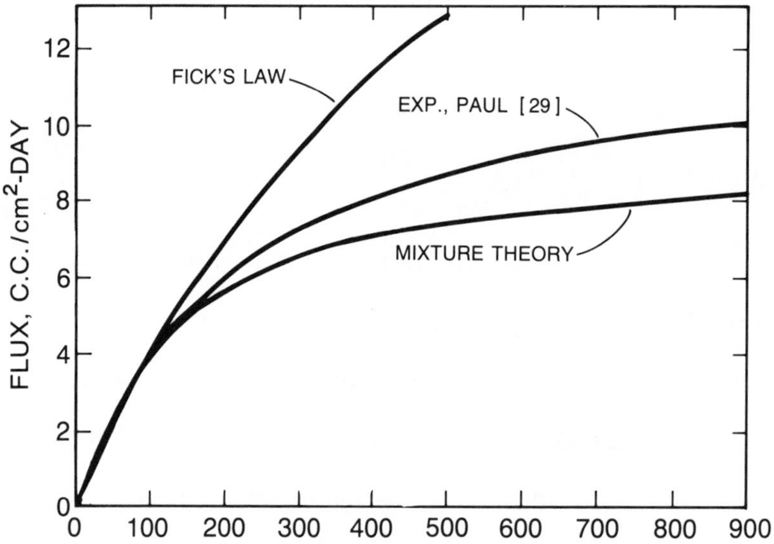

Figure 6. Comparison between theory and experiment.

Gandhi, Rajagopal and Wineman [13] studied the steady-state diffusion of a fluid through an isotropic elastic layered which has been sheared and stretched (cf. Figure 7). They assume a motion of the form

$$x = \lambda X + g(Z), \quad y = \lambda Y, \quad z = f(Z)\,, \tag{7.10}$$

where λ is a constant. Once again, we assume the fluid velocity of the form

$$u_x = 0, \quad u_y = 0, \quad u_z = w(z) . \tag{7.11}$$

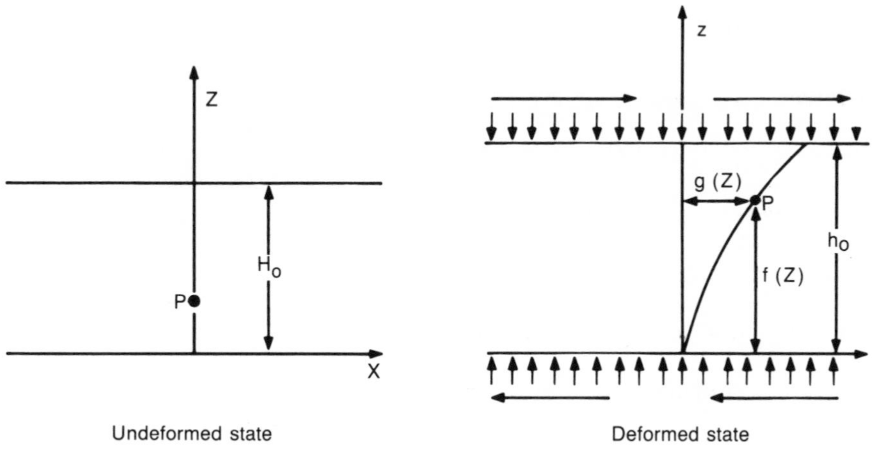

Figure 7. Diffusion through a sheared plate.

The boundary value problem reduces to solving a set of coupled non-linear ordinary differential equations of the form

$$\phi_1 \left(f', g', f'', g'', \lambda, F\right) = 0 , \tag{7.12}$$

$$\phi_2 \left(f', g', f'', g'', \lambda, F\right) = 0 , \tag{7.13}$$

subject to non-linear algebraic boundary conditions of the form

$$\overline{Z}\,(0) = 0 , \tag{7.14}$$

and

$$T^s_{33} + T^f_{33} = q_h . \tag{7.15}$$

The boundary conditions which stem from the assumption of saturation are given by

$$\psi_1 (f', g', \lambda) = -q_h , \tag{7.16}$$

$$\psi_2 (f', g', \lambda) = -q_0 . \tag{7.17}$$

Here F is the mass flux, and q_h and q_0 are the pressures on the upper and lower surfaces, respectively. The coupled system (7.12), (7.13) subject to (7.14)-(7.17) were solved numerically. We discuss the results below:

(i) In the case when there is no lateral stretching or shear, the computed results compare very favorably with the results in [13] and the experimental results of Ebra-Lima [21].

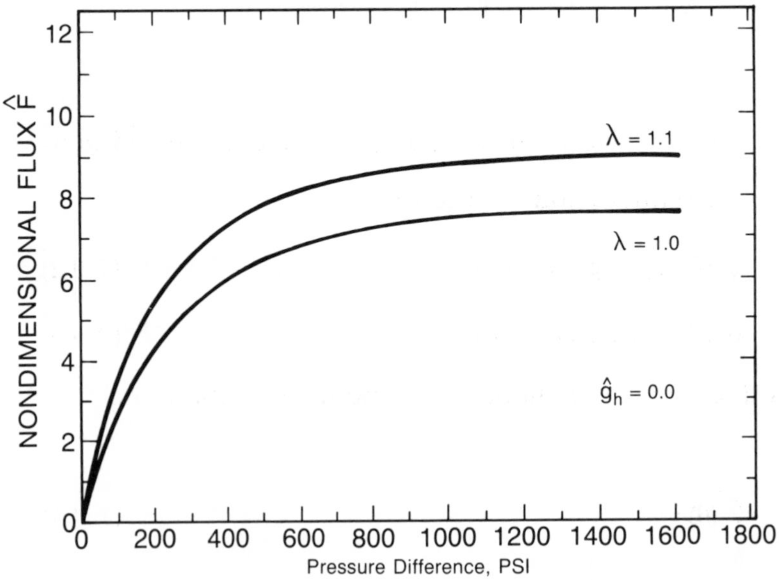

Figure 8. Effect of lateral stretching on the flux for plate diffusion.

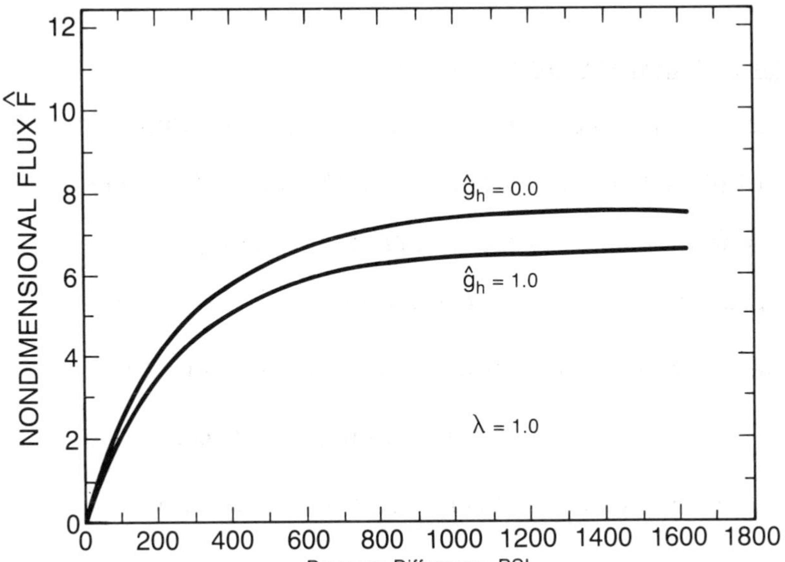

Figure 9. Effect of shearing on the flux for plate diffusion.

(ii) The mass flux increases with increased lateral stretching (cf. Figure 8). This is to be expected as increasing the lateral stretching increases the pore size.

(iii) Shearing tends to decrease the mass flux (cf. Figure 9). This is possibly due to the misalignment of the pore structure due to the shearing.

(iv) It was found that the stretch ratios and the density vary non-linearly through the thickness of the layer. The layer swells non-homogeneously and the swelling is quite significant.

<u>Diffusion Through a Hollow Circular Cylinder</u>

Gandhi, Wineman and Rajagopal [13] studied the radial diffusion of fluid through a hollow isotropic circular cylinder, which has been bonded to a rigid porous cylinder on the outside, due to the difference in the pressures on the inner and outer surfaces. Let R_i and R_0 denote the inner and outer radii of the cylinder, and let q_i and q_0 denote the pressures on the inner and outer surfaces. Gandhi, Rajagopal and Wineman [13] consider a deformation of the form:

$$r = r(R), \quad \theta = \Theta, \quad z = \lambda Z .$$

We shall document neither the governing equations nor the boundary conditions here. We shall go directly into a discussion of the results. We shall call the diffusion from inside the cylinder towards the outside, "outward diffusion", and the diffusion from the outside toward the center of the cylinder "inward diffusion". In the case of "inward diffussion":

(i) The mass flux increases with an increase in the stretch ratio λ, the pressure difference remaining fixed.

(ii) The thinner the cylinder, the greater the mass flux due to the pressure difference, for a fixed axial stretch ratio.

(iii) It is found that for a reasonably thick cylinders, the gradients of the stretch ratio are quite significant.

(iv) As the pressure difference increases, that is when we try to push more fluid through, the tangential stretch ratio at the inner surface continues to decrease substantially. The inner surface swells inward.

Diffusion of a Fluid Through a Thick Spherical Shell

Rajagopal, Shi and Wineman [24] studied the pressure induced radial diffusion of a fluid through a thick walled spherical shell. Unlike the problem for the cylindrical shell, the spherical shell was not constrained by a rigid porous support, either inside or outside. The problem once again reduces to solving a set of non-linear equations subject to non-linear boundary conditions. In the case of "outward diffusion", the inner and outer radii of the spherical shell increases with an increase in flux. The variation in the radial stretch ratio, the circumferential stretch ratio and the density of the fluid do not vary very significantly for the range of pressure differences considered. While the variations of the stretch ratios and the density were significant through a cylindrical shell, the variations were quite small in this problem. However, it should be borne in mind that the spherical shell was unconstrained by a rigid prorous support, and thus allowed to swell and stretch freely due to the applied pressure while the cylindrical shell was constrained by the pressure of a rigid support. It seems reasonable that increasing the mass flux while restricting the size causes a more rapid variation in the stretch ratio and the density which could have been alleviated by an expansion. This explanation is indeed borne out by recent work on diffusion of a fluid through a transversely isotropic spherical shell which is bonded to a rigid porous spherical shell on the outside. Details of the problem can be found in Dai and Rajagopal [7].

7b. DIFFUSION OF FLUIDS THROUGH ANISOTROPIC SOLIDS

Here we shall briefly discuss some recent studies on the diffusion of fluids through transversely isotropic and orthotropic solids undergoing finite deformation. Diffusion through such solids has relevance to problems in biomechanics (diffusion through the arterial wall, swelling of biological tissues and ligaments, cardiac edema, etc., (cf. Patel and Vaishnav [20], Vaishnav, Young, Janicki and Patel [30])), absorption of moisture by wood products (cf. Johnson [15], Johnson and Urbanik [16]), geotechnology (cf. Carroll [4], Carroll and Katsube [5]), the swelling of polymer composites and rubber, and the problem of ultrafiltration.

As in the case of diffusion of fluids through an isotropic solid, we have to augment the boundary conditions, and we do this once again by extending the technique developed by Rajagopal, Wineman and Gandhi [22]. We shall not get into a detailed discussion of the boundary value problems that were considered but outline the main results.

Diffusion of Fluids Through a Stretched and Sheared Anisotropic Non-Linear Elastic Cylinder

In the case of diffusion of the fluid through a transversely isotropic cylinder, the specific Helmholtz free energy for the mixture has the form (cf. Dai and Rajagopal [6]):

$$A = A(I_1, I_2, I_3, E_{33}, E_{13}^2 + E_{23}^2, \rho_2) .$$

The expression for A was expanded as a Taylor series in the variables and terms up to second order were retained. Thus, essentially the specific Helmholtz free energy is quadratic in the variables.

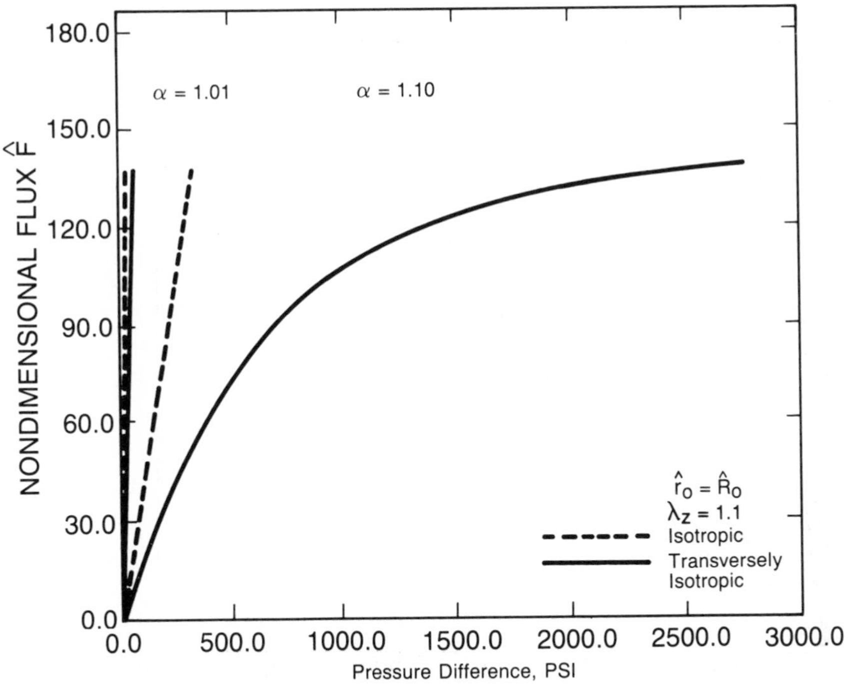

Figure 10. Effect of the cylinder thickness on the flux for the outward diffusion.

While the anisotropy of the material has a pronounced effect on the quantitative values of the flux and swelling, the results are qualitatively similar to the diffusion through an anisotropic cylinder. once again, in the case of outward diffusion, a ceiling flux is achieved (cf. Figure 10), while in the case of inward diffusion, for the range of pressure differences considered, there is no ceiling flux (cf. Figure 11)). Once again, shearing the cylinder tends to decrease the mass flux due to pore misalignment, while stretching the cylinder enhances the mass flux (cf. Figures 12 and 13), due to increase in pore size.

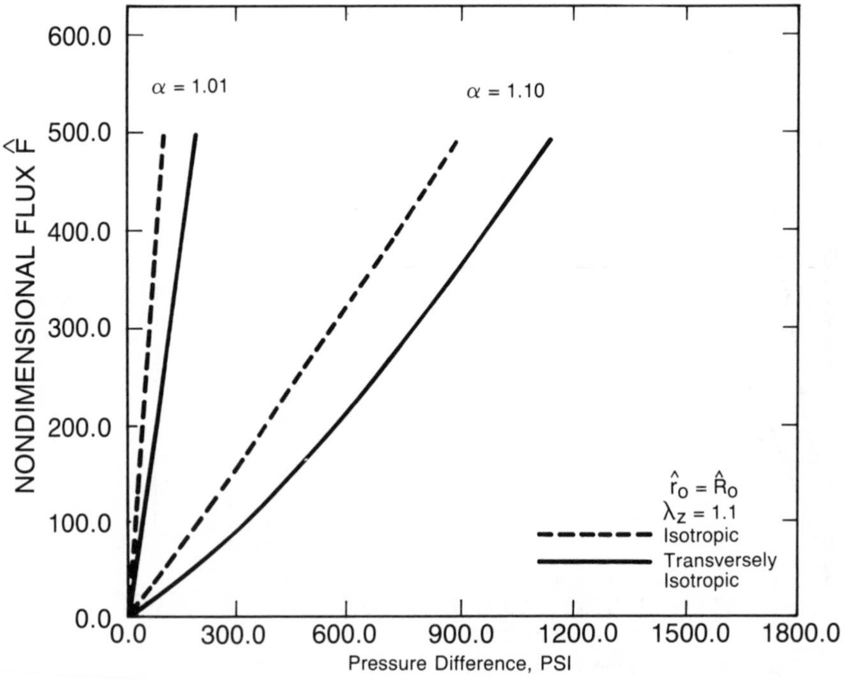

Figure 11. Effect of the cylinder thickness on the flux for inward diffusion.

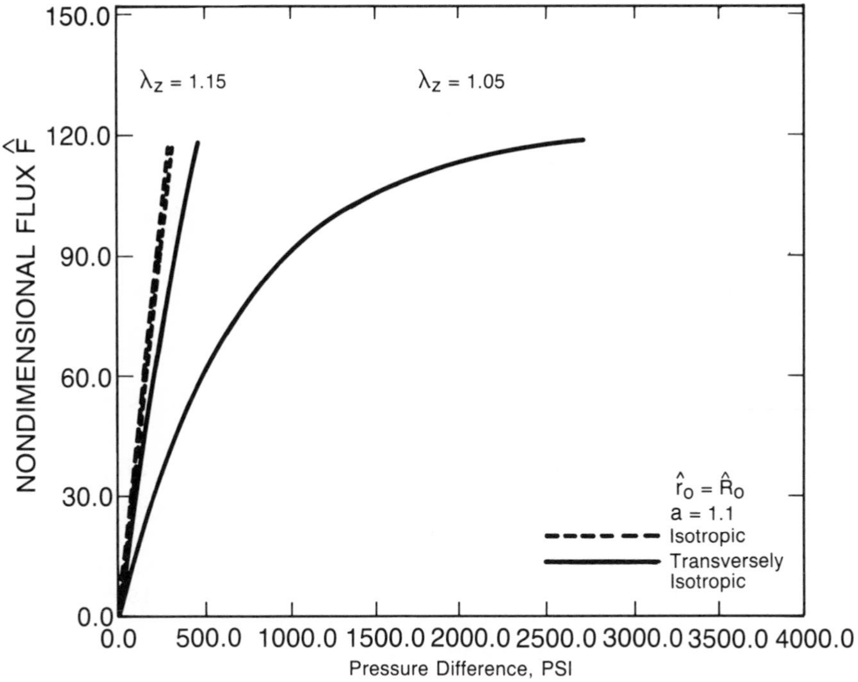

Figure 12. Effect of the cylinder thickness on the flux for inward diffusion.

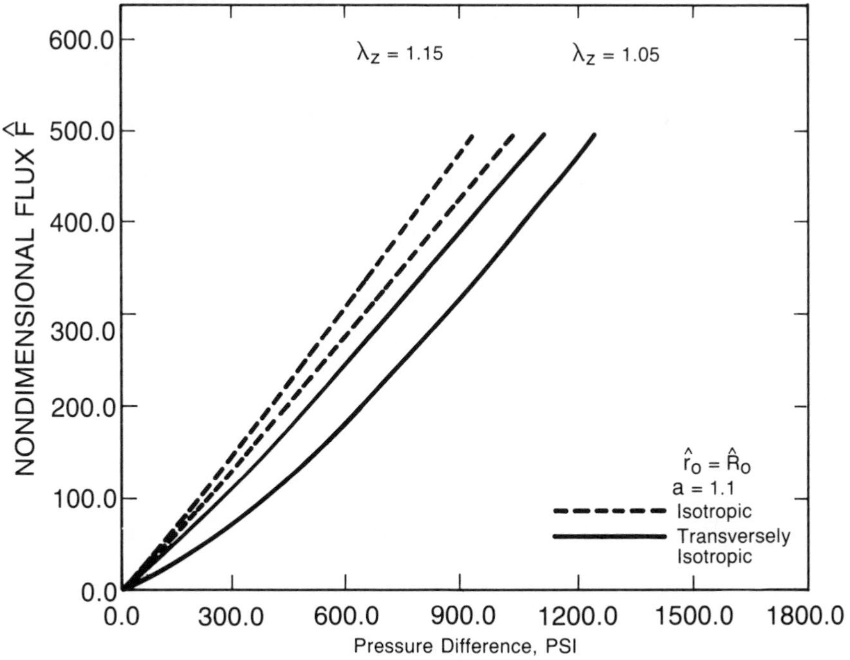

Figure 13. Effect of the axial deformation on the flux for inward diffusion.

The specific Helmholtz free energy function for a mixture of an ideal fluid and an orthotropic non-linearly elastic solid has the form (cf. Dai, Rajagopal and Wineman [8]):

$$A = A(E_{11}, E_{22}, E_{33}, E_{12}^2 + E_{13}^2, E_{23}^2, \rho_2) .$$

As before, the form for A was approximated by using a Taylor's series expansion. It was found that the extent of anisotropy in the radial direction. By this we mean the values of the material coefficeint multiplying terms invovling E_{11} have a more pronounced effect on the diffusion process than the other material coefficients. In general, anisotropy tends to inhibit the diffusion process. Effects of shearing and stretching the cylinder share the same qualitative structure as the diffusion through a transversely isotropic cylinder.

Diffusion of Fluids Through Transversely Isotropic Spherical Shells

The diffusion of a fluid through a thick walled spherical shell which is bonded to a rigid porous shell on the outside was studied with a view to delineate the effect of anisotropy and radial prestretching of the shell. While the anisotropy of the shell in the radial direction inhibits the mass flux through the shell, radially prestretching the shell has the exact opposite effect (cf. Dai and Rajagopal [7]). The fluid density was found to vary in a highly non-linear fashion through the thickness of the shell (cf. Figures 14 and 15).

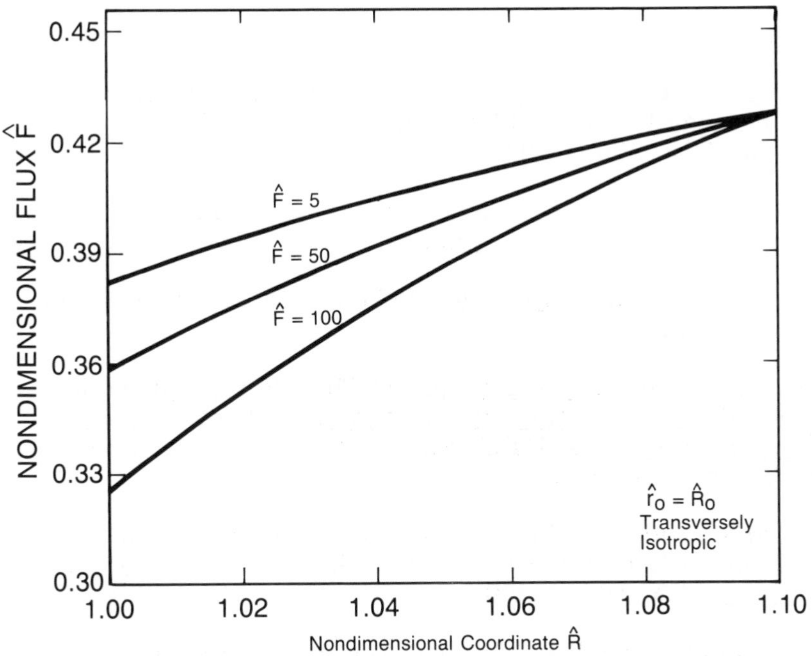

Figure 14. Variation of the fluid density along the shell thickness for outward diffusion.

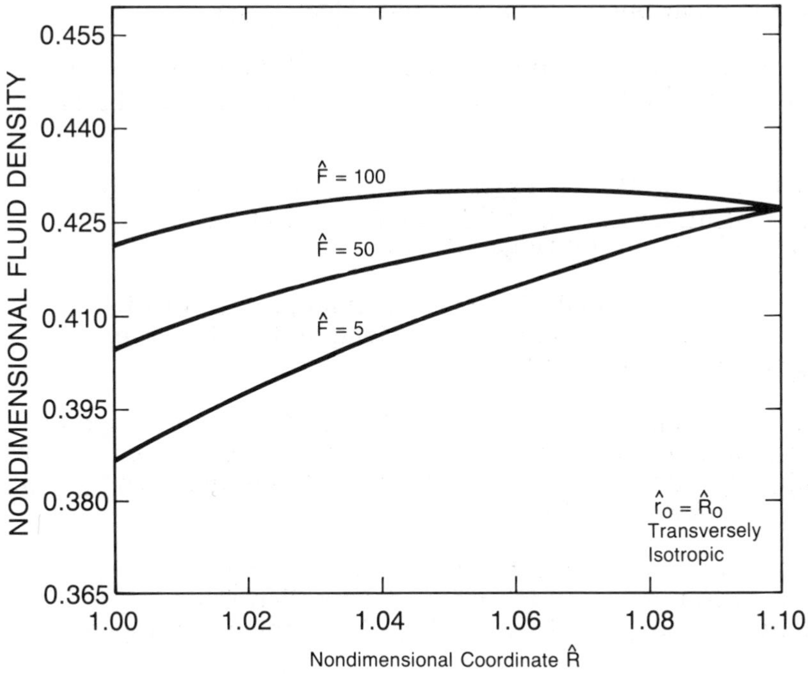

Figure 15. Variation of the fluid density along the shell thickness for inward diffusion.

REFERENCES

[1] Atkin, R.J. and Craine, R.E., "Continuum Theories of Mixtures: Basic Theory and Historical Development," Q. J. Mech. Appl. Math., **XXIX** (2), 209 (1976).

[2] Bedford, A. and Drumheller, D.S. "Theory of Immiscible and Structured Mixtures," Int. J. Engng. Sci., **21**, 863 (1983).

[3] Bowen, R.M., Continuum Physics, (edited by A.C. Eringen), Vol. III, Academic Press (1975).

[4] Carroll, M.M., Mechanical Response of Fluid-Saturated Porous Materials, Proc. of the 15th Intl. Congress on Theoretical and Applied Mechanics, eds. F.R.J. Rimrott and B. Taborrok, North-Holland, New York (1980).

[5] Carroll, M.M. and Katsube, N., "The Role of Terzaghi Effective Stress in Linearly Elastic Deformation, J. of Energy Resources Technology, **105**, 109 (1983).

[6] Dai, F. and Rajagopal, K.R., "Diffusion of Fluids Through Transversely Isotropic Solids", In press, Acta Mechanica.

[7] Dai, F. and Rajagopal, K.R., "Diffusion Through an Anisotropic Thick Spherical Shell," In press, Acta Mechanica.

[8] Dai, F., Rajagopal, K.R. and Wineman, A.S., "Diffusion Through Non-Linear Orthotropic Cylinders," In press, Int. J. of Engng. Sci.

[9] Denbigh, K., The Principles of Chemical Equilibrium with Application in Chemistry and Chemical Engineering, Cambridge University Press, Cambridge (1971).

[10] Gandhi, M.V., Rajagopal, K.R. and Wineman, A.S., "A Universal Relation in Torsion for a Mixture of Solid and Fluid," J. of Elasticity, **15**, 155 (1985).

[11] Gandhi, M.V. and Usman, M., "Equilibrium Characterization of Fluid-Saturated Continua and an Interpretation of the Saturation Boundary Condition Assumption for Solid-Fluid Mixtures," Int. J. Engng. Sci., **27**, 539 (1989).

[12] Gandhi, M.V., Usman, M., Wineman, A.S. and Rajagopal, K.R., "Combined Extension and Torsion of a Swollen Cylinder Within the Context of Mixture Theory," Acta Mechanica, (to appear).

[13] Gandhi, M.V., Wineman, A.S., and Rajagopal, K.R., "Some Non-Linear Diffusion Problems Within the Context of the Theory of Interacting Continua," Int. J. Engng. Sci., **25**, 1441 (1987).

[14] Green, A.E. and Naghdi, P.M., "On Basic Equations for Mixtures," Q. J. Mech. Appl. Math., **427** (1969).

[15] Johnson, J.A., Review of the Interaction of the Mechanical Behavior with Moisture Movement in Wood, in General Constitutive Relation for Wood and Wood Based Materials, report of a workshop in 1978, Syracuse University, Syracuse, NY.

[16] Johnson, M.J. and Urbanik, T.J., "A Non-Linear Theory for Elastic Plates with Application to Characterizing Paper Properties," J. of Appl. Mech., **51**, 146 (1984).

[17] Loke, K.M., Dickinson, M. and Treloar, L.R.G., "Swelling of a Rubber Cylinder in Torsion: Part 2. Experimental," Polymer, **13**, 203 (1972).

[18] Mills, N., "Incompressible Mixtures of Newtonian Fluids," Int. J. Engng. Sci., **4**, 97 (1966).

[19] Passman, S., Nunziato, J. and Walsh, E.K., "A Theory of Multiphase Mixtures," Sandia Report, SAND 82-2261 (1983).

[20] Patel, D.J. and Vaishnav, R.N., Basic Hemodynamics and Its Role in Disease Processes, University Park Press, Baltimore (1980).

[21] Paul, D.R. and Ebra-Lima, O.M., "Pressure Induced Diffusion of an Organic Liquid Through a Highly Swollen Polymer Membrane," J. Appl. Polymer Sci., **14**, 220 (1970).

[22] Rajagopal, K.R., Wineman, A.S. and Gandhi, M., "On Boundary Conditions for a Certain Class of Problems in Mixture Theory," Int. J. Engng. Sci., **24**, 1453 (1986).

[23] Rajagopal, K.R. and Wineman, A.S., "New Universal Relations for Non-Linear Isotropic Elastic Material," J. of Elasticity, **17**, 75 (1987).

[24] Rajagopal, K.R., Shi, J.J. and Wineman, A.S., "The Diffusion of a Fluid Through a Highly Elastic Spherical Membrane," Int. J.

Fluid Through a Highly Elastic Spherical Membrane," Int. J. Engng. Sci., **21**, 1171 (1983).

[25] Rivlin, R.S., "Large Elastic Deformations of Isotropic Materials, VI, Further Results in the Theory of Torsion, Shear and Flexure," Phil. Trans. Roy. Soc. London A242, 173 (1949).

[26] Shi, J.J., Rajagopal, K.R. and Wineman, A.S., "Applications of the Theory of Interacting Continua to the Diffusion of a Fluid Through a Non-Linear Elastic Medium," Int. J. Engng. Sci., **9**, 871 (1981).

[27] Southern, E. and Thomas, A.G., "Effects of Constraints of Equilibrium Swelling of Rubber Vulcanazites," J. Polymer Sci., **3**, 641 (1965).

[28] Treloar, L.R.G., The Physics of Rubber Elasticity, Oxford University Press, Oxford (1975).

[29] Truesdell, C., Rational Thermodynamics, Springer, New York (1984).

[30] Vaishnav, R.N., Young, J.T., Janicki, J.S. and Patel, D.J., Non-Linear Isotropic Properties of Aorta, Bio. Physics J., **12**, 1008 (1982).

ACKNOWLEDGEMENT

K.R. Rajagopal would like to thank the National Science Foundation for its support.

S.L. PASSMAN AND D.A. DREW

A simple multicomponent fluid theory with accurate physics

1 Introduction

In a *fluid solution* the constituents are so intimately intermixed that each can be thought of as being present in all of a body of such a solution for all times. The classical continuum theory of mixtures of fluids is a model for such solutions. A *multicomponent mixture* is different from a fluid solution in that its constituents occupy distinctly different portions of the body. These parts never overlap. Modeling such a mixture as a continuum is a difficult problem, and is the subject of intense debate in the current literature. We focus our attention on a saturated suspension of approximately spherical solid particles in a liquid. We discuss a simple model for such a suspension. The model is theoretically correct in a sense we make precise. The suspension has definite microstructure, but the model is so idealized that the only vestige of that microstructure is the volume fraction of the solid constituent. This is quite adequate to yield a mathematically interesting theory that is capable of describing a range of physical phenomena.

2 Continuum Theories, Corpuscular Theories, and Averaging Theories

For a fluid solution, the assumption that the material is described by a generalization of ordinary fluid mechanics is equivalent to the assumption that an ordinary fluid is a continuum. That argument is, if the scale of the set of boundary value problems of interest is very large compared to dimensions of the molecules described in works on modern particle physics, then the continuum equations are expected to hold. It is extremely difficult to render that argument into a form acceptable in classical mathematical analysis. Nonetheless it has wide acceptance because of its practical utility.[1] Here, we call any theory in which molecules have a role a *corpuscular model* or a *corpuscular theory*. Yet another, and radically different, argument is possible. The classical kinetic theory of Maxwell [6] takes as its basic entities very small particles. These comprise a set of measure zero in physical space. An exact and elegant set of arguments then leads to precisely the same equations as those assumed in the continuum theory of compressible fluids. Furthermore, these arguments may be generalized to fluid solutions. These arguments are extremely pleasing in that formal manipulations in the context of a corpuscular theory leads to equations of the same form as assumed in a continuum theory. In a sense, however,

[1]See Truesdell and Toupin [12] for extensive discussion of this matter.

the agreement is illusory. The theories appeal to entirely different models of the same physical reality. The agreement is partly a result of ingenious formal manipulation of symbols. The symbols used in a corpuscular theory have different meanings from those in a continuum theory. The proofs of agreement between the theories depend on identifying these symbols with one another. Even the corpuscular model itself may be questioned, for it is but a gross physical approximation of the corpuscular model of matter presently accepted by physicists and described in works on modern particle physics.

For multicomponent mixtures, yet another difficulty arises. For example in a suspension of solid particles in a fluid, there is yet a second "molecular" scale, that of the particles. Most often, materials of this type have particles that are very large compared to the molecules described in works on modern particle physics. Usually it is assumed that classical continuum mechanics holds for both the particles and the enveloping fluid. Then the same argument as before, that is, that a necessary condition for the theory to be valid is that the scale of the boundary value problems of interest be large compared to the scale of the particles, is asserted yet a second time on a different level. Continuum mixture theories have been constructed for such materials, and some have been successful in predicting physical phenomena. Of course molecular models can be constructed on this level also, and in fact they are quite common and quite highly developed.[2] Yet another set of arguments is commonly used. Again it is assumed that classical continuum mechanics holds for both the particles and the enveloping fluid. Now, however, continuum equations are derived for the mixture by performing averaging operations over regions of space large compared to the scale of the particles but small compared to the scale of the boundary value problems. Alternatively, averaging may be done over suitable intervals of time, or over ensembles of experiments with similar initial and boundary conditions. For almost all averaging methods, suitable definitions and algebraic manipulations suffice to render the resulting equations of balance into the same forms, despite the fact that their domains of definition are different, possibly disjoint sets on the same spaces, or even sets defined on different spaces. Moreover, these forms may be shown to be equivalent to those derived from continuum mixture theories. The difficulty occurs, as it does in the case of comparison of Maxwell's theory with classical fluid mechanics, that in the continuum theory the fields are smooth almost everywhere, while in the corpuscular theory the fields are very unsmooth. The complex operations and definitions intrinsic to averaging seldom do anything to alleviate the inconsistency concerning smoothness. If this point is ignored, as it usually is, it is fairly clear that for multicomponent mixtures the *forms* of the equations are so robust that they can withstand much, but not absolute, mathematical abuse, and still turn out the same. Then it is plausible to argue that though the mathematical basis of the theories are not yet fully understood, there is good evidence as to the forms of the equations.

---

[2]Sources can be traced from the recent and interesting work of Jenkins and McTigue [4].

3 Notation

We restrict our arguments to three-dimensional physical space. Time is denoted by t and spatial co-ordinates by $\mathbf{x}$. We use the tensor notation of Gibbs[3] [3]. In that notation, scalars are denoted by lightface minuscules, and vectors by boldface minuscules. Tensors (of second order) are denoted by boldface majuscules. The spatial gradient of a vector $\mathbf{v}$ is denoted by $\nabla \mathbf{v}$ and the divergence of a tensor $\mathbf{A}$ is $\nabla \cdot \mathbf{A}$. The identity tensor is $\mathbf{I}$.

We require kinematics sufficiently general so that each component has its own velocity. Furthermore, we assume that the velocity $\mathbf{v}$ of the composite continuum may be described in some reasonable way. This gives rise to three distinct material derivatives, one with respect to each constituent, and one with respect to the total material. The former two are denoted by backwards primes. Thus accelerations for the constituents are $\grave{\mathbf{v}}_a$ defined thus:

$$\grave{\mathbf{v}}_a = \frac{\partial \mathbf{v}_a}{\partial t} + \mathbf{v}_a \cdot \nabla \mathbf{v}_a. \tag{1}$$

The latter is given by the conventional equation, *e.g.*,

$$\dot{\mathbf{v}} = \frac{\partial \mathbf{v}}{\partial t} + \mathbf{v} \cdot \nabla \mathbf{v}. \tag{2}$$

4 Mathematical Structure

In this paper, we restrict our attention to continuum theory, so that corpuscular theories and averaging theories have no formal role. As is normal in continuum theories of mixtures, we retain both corpuscular theories and averaging theories as motivational tools for some of our arguments. The particular range of physical materials we wish to model should be described equally well by any of the three types of theory, thus any of the three types should be equivalent to the other two, if each is properly formulated.

We discuss the following matters:

1. Balance principles for constituents
2. Balance principles for the mixture
3. Constitutive principles
4. Restrictions on constitutive principles
5. Determinism of the system of equations

We note that these matters have been discussed numerous times in the literature[4]. Here we describe a theory constructed on the basis of clear axioms, though of course we do so on an informal basis. The axiomatic structure is important because a careful perusal of the literature shows that every single mathematical structure for multicomponent mixtures has been questioned, often with reasonable mathematical or physical basis, somewhere or other. Without a clear structure, construction of a theory simply reduces to postulating field equations. Though this is sometimes done in mathematical modeling of physical phenomena, it is not appropriate to a mature field of physical science. We do not think

---

[3]This is closely related to the tensor notation set forth by Truesdell and Noll [11].

[4]See Truesdell [9] for a discussion, but not a definitive discussion, of some of the discussions.

it is appropriate to the study of multicomponent mixtures, because such materials are governed by mechanics, probably the most mature field of physical science.

We do not consider the thermodynamics of multicomponent mixtures. This is a vital and interesting subject, and bears substantial discussion. However, that discussion would add considerably to the length of this paper, while doing little to enhance the points we wish to make. All of the equations we write are in accordance with the principal ideas of thermodynamics as we understand them.

4.1 Balance Principles for Constituents

We assume each of the constituents has a mass per unit volume ρ_a, a volume fraction α_a, a velocity $\mathbf{v}_a$, a body force per unit mass $\mathbf{b}_a$, and a stress $\mathbf{T}_a$. Each of these quantities is a time-dependent field over a region of three-dimensional physical space. The quantities that are subject to balance equations are mass and momentum. The primitive concept of the balance equation for mass is that a body of one of the constituents is chosen, and then it is postulated that the mass of that body is balanced by the loss or gain of mass of that body to other bodies in the mixture. Likewise the time rate of change of momentum of the body is balanced by the integral of forces over the body and the loss or gain of momentum to other bodies in the mixture. The local forms of these equations are

$$\grave{\alpha}_a + \mathbf{v}_a \cdot \nabla \alpha_a = c_a, \tag{3}$$

$$\alpha_a \rho_a \grave{\mathbf{v}}_a = \alpha_a \rho_a \mathbf{b}_a + \nabla \cdot \mathbf{T}_a + \mathbf{m}_a, \tag{4}$$

where in order to simplify the equations somewhat, we have restricted our attention to incompressible constituents, and where c_a is the mass exchange and $\mathbf{m}_a$ the momentum exchange of the constituent.

4.2 Balance Principles for the Mixture

In general, a multicomponent mixture will exhibit rheological behavior somewhat more complex than a classical continuum. This fact is well-known, and is cited in many publications on applications [2]. However, mass for the mixture is conserved and the total time rate of change of momentum of the mixture is balanced by the total force on the mixture. For these statements to have meaning, the mass and momentum of the mixture, as well as the various forces on the mixture, must be interpreted in terms of the constituents. A set of interpretations that *suffice* are

$$\begin{aligned} \rho &= \textstyle\sum_a \rho_a, \\ \rho \mathbf{v} &= \textstyle\sum_a \rho_a \mathbf{v}_a, \\ \rho \mathbf{b} &= \textstyle\sum_a \rho_a \mathbf{b}_a, \\ \mathbf{T} + \rho \mathbf{v}\mathbf{v} &= \textstyle\sum_a (\mathbf{T}_a + \rho_a \mathbf{v}_a \mathbf{v}_a). \end{aligned} \tag{5}$$

Some of these definitions are easy to motivate in the sense of continuum mechanics, while others are not. All are motivated naturally in classical kinetic theory. It follows that the conservation laws

$$\begin{aligned} \textstyle\sum_a c_a &= 0, \\ \textstyle\sum_a (\mathbf{m}_a + c_a \mathbf{v}_a) &= 0, \end{aligned} \tag{6}$$

are *equivalent* to

$$\dot{\rho} + \mathbf{v} \cdot \nabla \rho = 0, \tag{7}$$

$$\rho \dot{\mathbf{v}} = \rho \mathbf{b} + \nabla \cdot \mathbf{T}. \tag{8}$$

The definitions (5) certainly are not unique, and in fact other definitions suffice to yield the equivalence of (6) with the set (7) and (8). Generally, though, such definitions require that other quantities such as the total body force be defined in ways which are not easily motivated either in the context of accepted corpuscular theories or in the context of continuum mechanics. For example, it is reasonably common to see assertions that $(5)_4$ is not motivated physically, and should be replaced by

$$\mathbf{T} = \textstyle\sum_a \mathbf{T}_a. \tag{9}$$

In that case the only alternatives are either to replace the definition of the total body force $(5)_3$ by a definition that is extremely difficult to motivate, or to reject Newton's momentum equation as expressed by (8). It is a bit surprising that the second alternative ever has been suggested, but it has.

4.3 Constitutive Equations

We focus our attention on the balance equations for mass and momentum of the constituents, (3) and (4). Generally the body forces are taken as externally specified, and the stress tensors as symmetric. The resulting system has more unknowns than equations, and it is intuitive that more equations are needed [8]. This squares with the physical expectation that different expressions for the stress, and for the mass and momentum exchanges, are needed to model different materials. Such expressions are called *constitutive equations*, and are the focus of much study in the classical mechanics of continua. For multicomponent mixtures the choice of these expressions is quite difficult, and is often done with the aid of so-called "micromechanical models". The study of such models is an interesting subsidiary field in classical continuum mechanics. For multicomponent mixtures it dominates the literature, at least in terms of bulk and subject of debate. In the context of continuum theory, numerous "principles" have been propounded either as axioms or as guidelines for formulating constitutive equations.[5] These "principles" often are useful aids in formulation of theories. If they are taken so seriously that they are accepted as axioms, the result often is that either they render the set of putative constitutive equations so large as to be almost useless, or so small as to be almost empty, depending on which subset of the nearly endless list is chosen. The "principles" may be guidelines for mathematical modeling of constitutive phenomena, but they are not substitutes for modeling physical phenomena. The exceptions are frame-indifference and material isotropy. We discuss those separately.

[5] Eringen [1] gives a long but not definitive list.

An example of constitutive equations is the following:

$$\begin{aligned} \mathbf{T}_s = & \; -p_s(\alpha_s)\mathbf{I} & +\mu_{ss}(\alpha_s)\mathbf{D}_s + \mu_{sf}(\alpha_s)\mathbf{D}_f, \\ \mathbf{T}_f = & \; -p_f\mathbf{I} & +\mu_{fs}(\alpha_s)\mathbf{D}_s + \mu_{ff}(\alpha_s)\mathbf{D}_f, \end{aligned} \tag{10}$$

$$c_a = 0, \tag{11}$$

$$\mathbf{m}_s = \lambda_1(\alpha_s)(\mathbf{v}_s - \mathbf{v}_f) + \lambda_2(\alpha_s)\nabla\alpha_s, \tag{12}$$

where $\mathbf{D}_a$ is the symmetric part of $\nabla\mathbf{v}_a$. Here and subsequently, we consider a mixture of two constituents, a fluid denoted by the subscript "$_f$", and a suspended agglomeration of solid particles denoted by the subscript "$_s$". The mixture is assumed to be saturated, so that

$$\alpha_s + \alpha_f = 1. \tag{13}$$

4.4 Restrictions on Constitutive Equations

Two distinct ideas provide useful restrictions on the constitutive equations. The first usually is labeled a "principle", and thus often is lumped with the "principles" described above. Nonetheless, it is a genuine axiom, and is to be treated as such. This *principle of material frame-indifference* is: *The response of the material is independent of the frame of the observer.* This is purely a statement about constitutive equations. It is almost tautological to point out that the balances of momentum (8) contain accelerations, and thus depend upon frame. Therefore when the constitutive equations are substituted into the balances of momentum, the resulting equations *will* depend on the frame of the observer. Sometimes the principle of material frame-indifference alone *forces* a constitutive equation to be described by an isotropic function, but most often the resulting restriction is considerably more general.

Material constitutive equations also may be restricted by the *principle of material isotropy* [11]. This principle is sufficiently complex so that we find it inconvenient to describe it fully in this paper. The particular materials we consider here are fluid-like. Thus for every reference configuration, their isotropy group is the full orthogonal group. This result, taken with the principle of material frame-indifference, implies that *for this type of material*, all constitutive functions are isotropic. Representation theorems for isotropic functions are well-known and usually highly restrictive. This fact has been exploited for multicomponent fluids [7]. The constitutive equations (10), (11), and (12) all are frame-indifferent and have as their isotropy group the full orthogonal group.

4.5 Determinism of the System of Equations

In theories for multicomponent fluids, the "closure problem" often occurs. This results from the fact that, if the field equations for a multicomponent fluid are formulated in what might be considered to be a "natural way", the number of equations is less than the number of unknowns. The usual method for solving this problem is to adduce hydrodynamic arguments external to the theory, but linking variables already appearing in the theory in a logical way. Another solution has been suggested, independently, by Massoudi and Boyle [5], Jenkins and McTigue [4], and Passman [8], all at approximately the same time. Here, we have used a version of the theory formulated by Passman [8]. In it, the agglomeration of particles is considered to be a compressible fluid, so that $p_s = \hat{p}_s(\alpha)$, and the suspending liquid is considered to be an incompressible fluid, so that p_f is a quantity to be determined by the solution to the boundary value problem.

5 Boundary Value Problems

A theory, no matter how well formulated, is but empty formalism until it is tested against physical experience. This is a particularly difficult task in the case of multicomponent mixtures because of the lack of definitive experimental results. Still, an essential step in performing the comparison is solution of boundary value problems associated with particular experiments. For the theory considered here, we have found a number of exact solutions to boundary value problems corresponding to flows produced in viscometers. These solutions appear in [10].

6 Conclusion

Multicomponent fluids are physically complex. It is not surprising then that they are mathematically and experimentally complex, so that modeling them is a difficult task. Over the last few years, the path for constructing continuum models has become clear. We have reviewed it here. As an example, we have given a particularly simple theory. This theory is amenable to exact solutions.

References

[1] A. C. Eringen, Nonlinear Theory of Continuous Media. McGraw-Hill, New York (1962).

[2] F. Gadala-Maria and A. Acrivos, Shear-Induced Structure in a Concentrated Suspension of Solid Spheres. Journal of Rheology 24, 799–814 (1980).

[3] J. W. Gibbs and E. B. Wilson, Vector Analysis, Yale University Press, New Haven (1909).

[4] J. T. Jenkins and D. F. McTigue, Transport Processes in Concentrated Suspensions: The Role of Particle Fluctuations. Two-Phase Waves in Fluidized Beds, Sedimentation, and Granular Flows, edited by D. D. Joseph, Springer-Verlag, New York (1990).

[5] M. Massoudi and E. Boyle, Continuum Theories of Granular Materials with Applications to Fluidized Beds. Morgantown Energy Technology Center Technical Note DOE/METC-88-4077(DE88001051), May, 1987.

[6] J. C. Maxwell, Illustrations of the Dynamical Theory of Gases. The Philosophical Magazine (4) 19, 19–32 and 20, 21–37 (1860).

[7] S. L. Passman, Forces on the Solid Constituent in a Multiphase Flow, Journal of Rheology 30, 1077–1083 (1986).

[8] S. L. Passman, Stress in Dilute Suspensions. Two-Phase Waves in Fluidized Beds, Sedimentation, and Granular Flows, edited by D. D. Joseph, Springer-Verlag, New York (1990).

[9] C. Truesdell, Rational Thermodynamics. First edition, McGraw-Hill, New York (1979); Second edition, Springer-Verlag, New York (1984).

[10] S. L. Passman and D. A. Drew, Exact Solutions in a Theory of Multicomponent Mixtures. To appear.

[11] C. Truesdell and W. Noll, The Non-Linear Field Theories of Mechanics. Handbuch der Physik, Volume III, Part 3, Springer-Verlag, Berlin (1966).

[12] C. Truesdell and R. Toupin, The Classical Field Theories. Handbuch der Physik, Volume III, Part 1, Springer-Verlag, Berlin (1960).

Acknowledgments—Our thanks to N. Bixler and D. F. McTigue their comments on this manuscript. The work was performed while Passman was at the Pittsburgh Energy Technology Center on temporary assignment from Sandia National Laboratories. Drew's work was supported by the United States Army Research Office. Sandia National Laboratories is operated under contract DE-AC04-DP00789 from the United States Department of Energy.

STEPHEN L. PASSMAN
Pittsburgh Energy Technology Center
Pittsburgh, Pennsylvania 15236
and
DONALD A. DREW
Department of Mathematical Sciences
Rensselaer Polytechnic Institute
Troy, New York 12181

M. GRMELA

An interplay of thermodynamics and rheology

I. INTRODUCTION

Mathematical models are needed to organize, understand and predict results of observations. The choice of the type of observations is determined by the nature of the physical systems that are observed, by our needs and by our intended applications. A simplicity of mathematical models is essential for satisfying the expectations placed on them. In order that a mathematical model be simple, the state space chosen for the model has to be kept as small as possible. In other words, the state variables (elements of the state space) that are chosen to characterize states of the system under consideration have to be related as closely as possible to results of the chosen type of observations. Consequently, the requirement of simplicity leads to many different models each formulated to organize, understand and predict results of a specific type of observations. A family of descriptions and models is thus obtained in this way. We shall refer to this family as a multilevel description and multilevel family of models. We say that a level of description (level 1) is more microscopic (or less macroscopic) than another (level 2) if more microscopic details of the system can be expressed through the state variables chosen on the level 1 than through the state variables chosen on the level 2. If we would like to avoid the multilevel nature of modeling, we would have to either choose a microscopic model that, in principle, should be able to organize, understand and predict all the types of observations in our disposition or limit the types of observations so that only one level of description would be sufficient. The former alternative leads to a model that does not satisfy the requirement of simplicity, the latter alternative is practically realizable when dealing with simple fluids like for example water but it cannot be used in the context of complex fluids like for example polymeric fluids.

Following the route of multilevel modeling (the only route leading to useful models of polymeric liquids) the following questions arise: Do the models in the family share some common structure? How can we combine the models? How can we derive a more macroscopic model from a more microscopic model? (see Refs. [14], [25]). My aim in this lecture is to show that an analysis of the above questions is a natural extension of thermodynamics. The lecture is organized as follows. The levels of description that are most frequently used in the context of polymeric fluids are reviewed in Section II. Relationship between equilibrium thermodynamic description and a more microscopic description in both equilibrium and dynamical context is discussed in Section III and Section IV. In Section V, the question of how equilibrium properties like e.g. specific heat, solubility depend on the flow to which the fluid under consideration is submitted is considered. This question will be shown to be closely related to the problem of relating two general levels of description. A complementary question of how rheological properties like e.g. various viscosity coefficients depend on temperature and pressure is discussed in Section VI. The focus of the lecture is on the overall structure. Some selected details are however mentioned in illustrations and problems accompanying every section. I shall use the symbol ● to denote the end of illustrations examples and problems. Finally, I would like to emphasize that this Lecture is not meant to be a systematic review of the existing literature about the interaction of thermodynamics and rheology but rather a presentation of the subject that reflects strongly my point of view and my interests.

II. MULTILEVEL DESCRIPTION OF POLYMERIC FLUIDS

The choice of state variables (the choice of the level of description) depends on: (i) our needs and intended applications, and (ii) intrinsic complexity of the fluids. For example, we shall choose quantum mechanics to interpret results of slow neutron scattering experiments performed on water and hydrodynamic description to interpret observations of the flow of water in the Gulf Stream. One of the main special features of polymeric fluids is that there is no single level of description that would suffice to satisfy the needs arising for example in industrial polymer processing. In addition to hydrodynamic description (the state variables in hydrodynamic description are: $\rho(\underline{r},t)$, $e(\underline{r},t)$, $\underline{u}(\underline{r},t)$ denoting the fields of mass, energy and velocity densities respectively; $\underline{r}$ is the position vector; in order to shorten the notation, we shall refer to hydrodynamic state variables as f_{hydro}) we need a state variable that keeps track of internal (molecular) structure of the polymeric fluids. We shall denote this state variable by the symbol φ. The complete state variable, denoted by the symbol f, is thus $f \equiv (f_{hydro}, \varphi)$. The state space, i.e. the space composed of all physically admissible state variables f, will be denoted by the symbol M (i.e. $f \in M$).

Now, we review some frequently used internal state variables φ. First, we shall distinguish between the state variables, denoted φ_{intra}, characterizing internal structure of macromolecules and the state variables, denoted φ_{inter}, characterizing the intermolecular arrangement. We have thus $\varphi \equiv (\varphi_{intra}, \varphi_{inter})$. The most general state variable φ_{inter} or φ_{intra} is a n-particle distribution function. The larger n, the more detailed (microscopic) is the description. Following the experience collected in kinetic theory of polymeric fluids [41], [4], [7], we suggest the following three types of the state variables φ_{inter} or φ_{intra}: (i) $\psi(\underline{r}, \underline{R}, t)$ called configuration space distribution function, $\underline{R}$ is the end-to-end vector of a macromolecule, $\int d\underline{r}^3 \int d^3\underline{R}\, \psi(\underline{r}, \underline{R}, t) = 1$; (ii) $\underline{\underline{c}}(\underline{r}, t)$ (see also [30], [12], [44], [49], [31], called a conformation tensor, is a symmetric tensor characterizing deformations of macromolecules if it stands for φ_{intra} and an arrangement (e.g. the network) of macromolecules if it stands for φ_{inter}; the conformation tensor $\underline{\underline{c}}(\underline{r}, t)$ can be related to $\psi(\underline{r}, \underline{R}, t)$ by $c_{\alpha\beta}(\underline{r}, t) = \int d^3\underline{R}\, R_\alpha R_\beta\, \psi(\underline{r}, \underline{R}, t)$; $\alpha, \beta = 1, 2, 3$; (iii) $\underline{n}(\underline{r}, t)$, called a director vector field, characterizing the orientation of the macromolecules, $|\underline{n}(\underline{r}, t)| = 1$, $\underline{n}$ and $-\underline{n}$ are considered to be physically undistinguishable, the director vector can be related to the conformation tensor $\underline{\underline{c}}(\underline{r}, t)$ by $\underline{n} = \underline{e}_1 / |\underline{e}_1|$, where $\underline{e}_1$ is the eigenvector of $\underline{\underline{c}}$ corresponding to the largest eigenvalue. A most obvious extension of the above state variables can be achieved by replacing $\psi(\underline{r}, \underline{R}, t)$ by $\psi_i(\underline{r}, \underline{R}, t)$, i = 1,....,n (n is an integer greater than one) and similarly for $\underline{\underline{c}}$ and $\underline{n}$. This extension expresses a more detailed view of polymeric fluids in which n different types of macromolecules or macromolecular arrangements are distinguished.

An important question that arises is the direct measurability of the chosen internal state variables φ. This question is then also related to the problem of the specification of boundary conditions for φ that are needed if the model involving φ is used for example to calculate flows arising in polymeric processing operations. An analysis of these questions is out the scope of this Lecture (for a remark concerning the boundary conditions see Section V). We mention only that in order to apply successfully a model involving an internal state variable φ, we do not need to be able to measure it directly. Of course, it is preferable if φ can be directly measured. There is more possibility then to compare predictions of the model with results of observations. For example, the conformation tensor $\underline{\underline{c}}$ can be directly measured by optical methods [60].

III. EQUILIBRIUM THERMODYNAMICS

The question that I want to discuss in this section is the following. How do the properties of macromolecules determine equilibrium properties, like for example the P-V-T relation, specific heats and phase transitions, of polymeric fluids?

First, we recall the setting of classical equilibrium thermodynamics. For the sake of brevity, we shall consider a one component system confined to a space region $\Omega \in \Re^3$. The volume V of Ω is kept constant. The state variables chosen in classical equilibrium thermodynamics are: $f_{th} \equiv (e,n) \in \Re^2$, e is the energy and n is the number of moles. The individual features of the fluid under consideration is expressed in classical equilibrium thermodynamics by a function $s = s(e,n)$, s is called an entropy. The function $s(e,n)$ can be interpreted geometrically as a two dimensional surface (called hereafter Gibbs surface) in the five dimensional surface with coordinates $(e, n, \frac{1}{T}, -\frac{\mu}{T}, s)$, where T denotes the temperature and μ the chemical potential. The Gibbs surface is the image of the mapping $(e,n) \to (e,n,\partial s/\partial e\,(e,n), \partial s/\partial n\,(e,n), s(e,n))$. Next, we introduce another function Φ, called thermodynamic potential, by

$\Phi(e,n;\frac{1}{T},\frac{\mu}{T}) = -s + \frac{1}{T}e - \frac{\mu}{T}n$. Solutions of $\partial\Phi/\partial e = 0$, $\partial\Phi/\partial n = 0$ will be denoted by the symbol (e_{eq}, n_{eq}) and called equilibrium states. We introduce $-PV/T = \Phi(e_{eq}, n_{eq}; \frac{1}{T}, \frac{\mu}{T})$, where P is the pressure. In this way we arrive at the relation $P = P(\mu,T)$ that is the dual (Gibbs-Duhem) form of the thermodynamic relation $s = s(e,n)$. We note that it is the thermodynamic potential Φ that leads us from the state variables (s,e,n) to the conjugate variables (T,P,μ).

If instead of the state variables (e,n) we choose more microscopic state variables f (for example those introduced in the previous section) we formulate equilibrium thermodynamics as follows. We introduce again the thermodynamic potential $\Phi(f;\frac{1}{T},\frac{\mu}{T})$ by

$\Phi(f;\frac{1}{T},\frac{\mu}{T}) = -S(f) + \frac{1}{T}E(f) - \frac{\mu}{T}N(f)$, where S,E,N are entropy, energy and number of moles respectively. Solutions of $\partial\Phi/\partial f = 0$ (if f is a function then $\partial/\partial f$ means the Volterra functional derivative) are denoted f_{eq} and called equilibrium states. The thermodynamic relation $P = P(\mu,T)$ implied by the thermodynamic potential $\Phi(f;\frac{1}{T},\frac{\mu}{T})$ is

$-PV/T = \Phi(f_{eq};\frac{1}{T},\frac{\mu}{T})$. It is thus again the thermodynamical potential that leads us from the state variables f to P,μ,T and to the thermodynamic relation $P = P(\mu,T)$. In order to specify the thermodynamic potential $\Phi(f;\frac{1}{T},\frac{\mu}{T})$ we need to specify three functions $S(f)$, $E(f)$ and $N(f)$. These functions can be specified by using the insight offered by Gibbs' equilibrium statistical mechanics or by an analysis of the equations that govern the time evolution of the state variable f (see Section IV and Section V).

Example III.1

In the Simha-Somcynsky theory [67], [68] a polymeric fluid is regarded as in the cell model of liquids [32]. The state variables are chosen to be $f \equiv (n,y)$, where n is the number of moles (i.e. $N(f) = n$) and y is the fraction of occupied cells. We note that Simha and Somcynsky then suggest, by using the insight offered by the cell model of liquids, the functions S(f) and N(f) (see Ref. [32]). The P-V-T relation implied by the Simha-Somcynsky theory has been found to be in a good agreement with results of observations (see Ref. [68]). ●

Example III.2

The equilibrium theory in which φ are distribution functions has recently been introduced by Curro and his collaborators (see Ref. [66] for a recent review and references to the previous papers). In Curro's formulation only equations determining equilibrium distribution functions (corresponding to equations $\partial\Phi/\partial f = 0$ in our formulation) are introduced. The thermodynamic potential $\Phi(f;\frac{1}{T},\frac{\mu}{T})$ itself is not introduced. This observation leads to the following problem. ●

Problem III.1

(i) Can Curro's equations determining the distribution functions be cast into the form $\partial\Phi/\partial f = 0$? Identify the thermodynamic potential $\Phi(f;\frac{1}{T},\frac{\mu}{T})$.

(ii) Compare the Curro theory with the Simha-Somcynsky theory.

The first part of the problem is interesting because the requirement that equations determining equilibrium distribution functions f can be written in the form $\partial\Phi/\partial f = 0$ is in fact a restrictive conditions put on the choice of the closure approximations that are needed to close the infinite hierarchy of equations implied by the Gibbs theory[64], [42]. To answer the second part of the problem, I suggest the following first step. Let only intermolecular pair correlation function $g_{inter}(\underline{r})$ ($(\underline{r})$ is the intermolecular distance) be chosen to serve as a state variable in Curro's theory. We introduce $(C_{inter})_{\alpha\beta} = \int d^3\underline{r}\, r_\alpha r_\beta g_{inter}(\underline{r})$. I suggest that the state variable y introduced in the Simha-Somcynsky theory is related to $\underline{\underline{c}}_{inter}$ by $y^2 \sim \det(\underline{\underline{c}}_{inter})$ (recall that $\det\underline{\underline{c}}$ is proportional to the volume of the geometrical object defined by the tensor $\underline{\underline{c}}$). ●

Illustration III.1

Some details of calculations will be presented in this illustration. The fluid under consideration is a macromolecular solution. The problem is to model the transition between isotropic and anisotropic equilibrium states. The only state variable that will be considered explicitly is $\varphi_{intra} = \underline{\underline{c}}$. We shall assume moreover that $\underline{\underline{c}}$ is independent of the position vector $\underline{r}$. The part Φ_{intra} of the thermodynamic potential Φ that depends on $\underline{\underline{c}}$ is given by $\Phi_{intra}(\underline{\underline{c}};\frac{1}{T}) = -S_{intra}(\underline{\underline{c}}) + \frac{1}{T}E_{intra}(\underline{\underline{c}})$ (note that $N_{intra}(\underline{\underline{c}}) = 0$). The macromolecules in the solution are assumed to be inextensible. This property is expressed as the constraint $tr\underline{\underline{c}} = R_0^2$ (recall that $tr\underline{\underline{c}}$ is proportional to the square of the length, denoted by R_0 of the macromolecule). We shall normalize the conformation tensor $\underline{\underline{c}}$ in such a way that $R_0 = 1$. Since the macromolecules are inextensible, the intramolecular energy $E_{intra}(\underline{\underline{c}}) = 0$. Following Onsager [58] (who however used the distribution function $\psi(\underline{R})$ as the state variable) we suggest $S_{intra}(\underline{\underline{c}}) = \frac{1}{2}\ln\det\underline{\underline{c}} - \frac{K}{2}(1 - tr\underline{\underline{c}}\,\underline{\underline{c}})$, where K is a parameter proportional to the concentration of macromolecules. The first term is the entropy of rigid macromolecules [65], [20], [24] (see Illustration IV.1), in the second part we take into account the topological interaction among macromolecules [48], [20], [24]. The constraint $tr\underline{\underline{c}} = 1$ will be taken into account by the method of Lagrange multipliers. The equation $\partial\Phi/\partial\underline{\underline{c}} = 0$ is thus

$$\frac{\partial}{\partial c_{\alpha\beta}}\left(-\frac{1}{2}\ln\operatorname{dex}\underline{\underline{c}} + \frac{K}{2}(1 - tr\underline{\underline{c}}\,\underline{\underline{c}}) + \lambda\,(tr\underline{\underline{c}} - 1)\right) = 0,$$

where λ is the Lagrange multiplier. Without the loss of generality, we can assume that $\underline{\underline{c}}$ is diagonal. Let c_1, c_2, $1-c_1-c_2$ be the entries on the diagonal. Equation $\partial\Phi/\partial\underline{\underline{c}} = 0$ implies that $c_1=c_2=c$ and c is determined by $(c - \frac{1}{3})(4Kc^2 - 2Kc + 1) = 0$ We see that $c = \frac{1}{3}$ (describing isotropic state) is always a solution. Moreover, for $K \geq 4$ there are two other solutions (real roots of the equation $4Kc^2 - 2Kc + 1 = 0$) describing anisotropic states. The phase transition from isotropic to anisotropic states appears thus as a pitchfork bifurcation. Onsager [58] (see also Ref. [39]) has arrived at the same result but on different level of description (Onsager used the distribution function $\psi(\underline{R})$ as the state variable). The mathematical simplicity of our derivation is of advantage in particular if the fluids under consideration become more complex (e.g. the macromolecules are not stiff but flexible - see Refs. [40], [20], [24]). ●

<u>Problem III.2</u>

If the stiff macromolecules are replaced by semiflexible macromolecules then $S_{intra}(\underline{\underline{c}}) = -\frac{1}{2} a\, tr\, \underline{\underline{c}}^{-1} - \frac{K}{2}(1 - tr\, \underline{\underline{c}}\,\underline{\underline{c}})$, where a is proportional to L/l, L is the total length of the macromolecule and l is the so called Kuhn length (the length of the part of the macromolecule that can be regarded as stiff - see Refs. [40]). Show that the calculations made in Illustration III.1 lead now to

$c_1 = c_2 = c$, $\left(c - \frac{1}{3}\right)\frac{3}{c(1-2c)}\left(4\frac{K}{a}c^4 - 4\frac{K}{a}c^3 + \frac{K}{a}c^2 + c - 1\right)$ that determines the diagonal entry of $\underline{\underline{c}}$. The onset of the anisotropic state is thus $K/a = 47{,}321315$ (the smallest value of K for which equation $4\frac{K}{a}c^4 - 4\frac{K}{a}c^3 + \frac{K}{a}c^2 + c - 1 = 0$ has a real root) [20], [24]. Qualitatively, a similar result has been derived, but on the level of description that uses the distribution function as the state variable) by Khokhlov and Semenov [40]. ●

IV. <u>EQUILIBRIUM THERMODYNAMICS VERSUS DYNAMICS ON A MORE MICROSCOPIC LEVEL OF DESCRIPTION (ISOLATED SYSTEMS)</u>

The question considered in this section is the following. What are the equations that govern the time evolution of polymeric fluids that could be used to organize, understand and predict results of both rheological and thermodynamical measurements? The problem is thus the following.

Let a state variable f be chosen, we look for an equation $\frac{\partial f}{\partial t} = \ldots$ whose solutions can be used to: (i) recover the equilibrium theory presented in Section III from an analysis of asymptotic $(t \to \infty)$ properties of solutions,

(ii) organize, understand and predict results of rheological observations (observations of responses of fluids to externally controlled flows), and (iii) calculate the flows and the structure of polymeric fluids arising in polymeric processing operations. Still in other words, we look for equations governing the time evolution of the chosen state variable f that manifestly reveals the passage from the level of description that uses f as the state variable to the equilibrium thermodynamics level of description.

Results of Onsager [59], Casimir [6] and experience with many particular models of macroscopic physical systems (e.g. the Euler-Navier-Stokes-Fourier hydrodynamics, the Boltzmann kinetic theory, ...) indicate that the time evolution equations describing isolated systems on all levels of description have the following common structure:

$$\frac{\partial f}{\partial t} = L^{c}(f)\,\partial\Phi/\partial f - L^{o}(\partial\Phi/\partial f)\;\partial\Phi/\partial f \qquad (1)$$

By Φ we denote the thermodynamic potential, the same as the one introduced in Section III. The quantity $L^{c}(f)$ is a twice contravariant skewsymmetric tensor. In addition to being skewsymmetric, the tensor $L^{c}(f)$ is also a Poisson tensor. This means that $\{A,B\} = \langle \partial A/\partial f, L^{c}(f)\,\partial B/\partial f\rangle$, where A,B are sufficiently regular functions of f and <,> is the inner product chosen in the state space M ($f \in M$) is a Poisson bracket $\{A,B\}$ (i.e. $\{A,B\}$ satisfies the Jacobi identity $\{A,\{B,C\}\} + \{B\{C,A\}\} + \{C\{A,B\}\} = 0$). The tensor $L^{o}(p) = \partial^2\Psi/\partial p\partial p$, where $p = \partial\Phi/\partial f$ and Ψ, called dissipative potential, is a sufficiently regular function of p satisfying the following properties: (i) Ψ reaches its minimum at $p = 0$, (ii) $\Psi(0) = 0$, and (iii) $\Psi(p)$ is a convex function in a neighborhood of $p = 0$. Furthermore, we require that Φ, L^{c} and Ψ are chosen in such a way that eq. (1) implies conservation of mass

$$\frac{dN(f)}{dt} = 0$$

and conservation of energy (2)

$$\frac{dE(f)}{dt} = 0$$

Equation (1) has been first introduced by Onsager and Casimir to describe the time evolution in a small neighborhood of equilibrium states (we recall that equilibrium states are solutions of $\partial\Phi/\partial f = 0$). Onsager [59] and Casimir [6] have assumed that the tensor L^{c} (called hereafter a

Casimir tensor) and L^{o} (called hereafter an Onsager tensor) are constant tensors independent of f and $\partial\Phi/\partial f$. The nonlinear generalization of L^{o} has been introduced in Refs. [52], [9], [71] and the nonlinear generalization of L^{c} in Refs. [8], [13], [37], [53], [15], [16]. We note that if the equation governing the time evolution of f is known then by recasting it into the form of eq. (1) we discover the thermodynamic potential Φ that, as it has been shown in Section III, realizes the passage from the level of description that uses the state variable f to the thermodynamic level of description. We shall illustrate this in Illustration IV.1. Note that Φ plays in eq. (1) the role of the Lyapunov function for the approach to equilibrium states. If, on the other hand, we know the thermodynamic potential Φ from other considerations (e.g. from those based on Gibbs equilibrium statistical mechanics) we can use it to formulate the time evolution equations (this will be illustrated in Illustration IV.3).

Illustration IV 1

In this illustration we assume that the polymeric fluid under consideration is incompressible and isothermal (its temperature equals T). This assumption allows us to keep in f_{hydro} only the velocity field $\underline{u}(\underline{r},t)$. As the internal state variable we chose $\varphi_{intra} = \underline{\underline{c}}(\underline{r},t)$, where $\underline{\underline{c}}$ is the conformation tensor (see Section II). We assume moreover that the time evolution of $\underline{\underline{c}}$ is governed by

$$\frac{\partial c_{\alpha\beta}}{\partial t} = -u_{\gamma}\frac{\partial c_{\alpha\beta}}{\partial r_{\gamma}} + c_{\alpha\gamma}\frac{\partial u_{\beta}}{\partial r_{\gamma}} + c_{\beta\gamma}\frac{\partial u_{\alpha}}{\partial r_{\gamma}} + \Lambda\tau_{\alpha\beta}$$

$$\tau_{\alpha\beta} = k_B T\delta_{\alpha\beta} - 2c_{\alpha\beta} \tag{3}$$

representing the classical rheological model known as the upper convected Maxwell model [51]. The tensor $\underline{\underline{\tau}}$ denotes the extra stress tensor, Λ is the mobility that is assumed here to be a scalar and a constant. The summation convention is used throughout this lecture. The equations governing the time evolution of the velocity field is

$$\frac{\partial u_{\alpha}}{\partial t} = -u_{\gamma}\frac{\partial u_{\alpha}}{\partial r_{\gamma}} - \frac{\partial p}{\partial r_{\alpha}} - \frac{\partial \tau_{\alpha\gamma}}{\partial r_{\gamma}} \tag{4}$$

$$\frac{\partial u_\gamma}{\partial r_\gamma} = 0$$

Our problem now is to cast eqs. (3) and (4) into the form of eq. (1). It is easy to verify that eqs. (3) and (4) indeed become a particular realization of eq. (1) corresponding to: (see Ref. [22])

the Poisson bracket

$$\{A,B\} = -\int d^3\underline{r}\, u_\alpha \left[\frac{\partial A}{\partial u_\gamma}\frac{\partial}{\partial r_\gamma}\frac{\partial B}{\partial r_\alpha} - \frac{\partial B}{\partial u_\gamma}\frac{\partial}{\partial r_\gamma}\frac{\partial A}{\partial u_\alpha}\right]$$
$$-\int d^3\underline{r}\, c_{\alpha\beta}\left[\frac{\partial A}{\partial u_\gamma}\frac{\partial}{\partial r_\gamma}\frac{\partial B}{\partial c_{\alpha\beta}} - \frac{\partial B}{\partial u_\gamma}\frac{\partial}{\partial r_\gamma}\frac{\partial A}{\partial c_{\alpha\beta}}\right] \tag{5}$$
$$-\int d^3\underline{r}\left[\frac{\partial}{\partial r_\alpha}\left(\frac{\partial A}{\partial u_\gamma}\right)\frac{\partial B}{\partial c_{\gamma\beta}} + \frac{\partial}{\partial r_\beta}\left(\frac{\partial A}{\partial u_\gamma}\right)\frac{\partial B}{\partial c_{\gamma\alpha}} - \frac{\partial}{\partial r_\alpha}\left(\frac{\partial B}{\partial u_\gamma}\right)\frac{\partial A}{\partial c_{\gamma\beta}} - \frac{\partial}{\partial r_\beta}\left(\frac{\partial B}{\partial u_\gamma}\right)\frac{\partial A}{\partial c_{\gamma\alpha}}\right]$$

the entropy

$$S(\underline{u},\underline{\underline{c}}) = \frac{1}{2}\int d^3\underline{r}\, \ln\det\underline{\underline{c}}$$

the energy (6)

$$E(\underline{u},\underline{\underline{c}}) = \int d^3\underline{r}\,\frac{1}{2}u^2 + \int d^3\underline{r}\,\mathrm{tr}\,\underline{\underline{c}}$$

and the dissipative potential given by the equation

$$L^o_{\alpha\gamma} = \Lambda c_{\alpha\gamma} = \frac{\partial^2\Psi}{\partial c^{-1}_{\alpha\beta}\,\partial c^{-1}_{\gamma\beta}} \tag{7}$$

●

<u>Illustration IV.2</u>

The Cahn-Hilliard equation [5] extending the equilibrium van der Waals theory to a dynamical theory is another famous particular realization of eq. (1) corresponding to f ≡ field of mass density $\rho(\underline{r})$

$L^c \equiv 0,\quad \Psi = D\frac{\partial}{\partial r_\alpha}(\partial\Phi/\partial\rho(\underline{r}))\frac{\partial}{\partial r_\alpha}(\partial\Phi/\partial\rho(\underline{r}'))$, where $D > 0$ and Φ that arises in the van der Walls equilibrium theory. ●

<u>Illustration IV.3</u>

We take now a different point of view. In the two previous illustration the time evolution equation has been assumed to be known. We have cast then the equation into the form of eq. (1). Now, the formulation of the time evolution equation modeling a given physical system will be our objective. We shall proceed as follows. First, we choose the state variables. Second, we assume that the time evolution equation that we search is a particular realization of eq. (1). This assumption guarantees immediately that solutions of the time evolution equation agree with results of the observation of the approach to equilibrium and the behaviour at equilibrium. In order to specify a time evolution equation that has the structure of eq. (1) we have to specify: (i) the thermodynamic potential Φ (we can use for example the insight associated with Gibbs equilibrium statistical mechanics to suggest Φ), (ii) the Casimir tensor L^{c}, or equivalently the Poisson bracket $\{,\}$ (we can use for example the Lie-Poisson theory of the symplectic structure associated with the Lie algebra structure of the chosen state space - see e.g. Refs. [46], [33]), (iii) the dissipative potential Ψ. This strategy has been applied for example in the context of polymeric liquid crystals (see Refs. [22], [10], [3], [26], [24], [27]. We note in particular that both the time evolution equations for the chosen internal state variable and formulas for the scalar pressure, extra stress tensor and heat flux are obtained if this strategy is followed. ●

Problem IV.1

Cast the rheological models introduced in Refs. [1], [2], [19], [20] into the form of eq. (1). Note the advantages of this formulation. ●

Problem IV.2

So far in this section, we have considered only $\varphi = \varphi_{intra}$. Formulate rheological models that use as internal state variables

$\varphi \equiv \left(\underline{\underline{c}}_{intra}, \underline{\underline{c}}_{inter}\right)$. The conformation tensor $\underline{\underline{c}}_{inter}$ can be interpreted for example as characterizing a network of macromolecules [2] or characterizing tubes restricting the space that is available to macromolecules [7], [34]. ●

Problem IV.3

Suggest the time evolution equations extending the Simha-Somcynsky equilibrium theory and the Curro equilibrium theory (see Ref. [17]). ●

Finally, we shall reformulate eq. (1) into another form [28], [29] that reveals its geometrical meaning. Starting with the state space M, we introduce a larger space $T^{*}M\times\Re$ (by $T^{*}M$ we denote the cotangent bundle of M). The local coordinates in $T^{*}M\times\Re$ will be denoted by

$(f,p,z), f \in M, p \in T_f^* M$ ($T_f^* M$ is the cotangent space attached to f), $z \in \Re$. We introduce the following time evolution equation in the space $T^*M x \Re$:

$$\frac{\partial f}{\partial t} = -\frac{\partial K}{\partial p}$$
$$\frac{\partial p}{\partial t} = \frac{\partial K}{\partial f} - p\frac{\partial K}{\partial z} \qquad (8)$$
$$\frac{\partial z}{\partial t} = K - \left\langle p, \frac{\partial K}{\partial p} \right\rangle$$

where

$$K = \langle p, \partial\Psi/\partial p \rangle - \left\langle p, [\partial\Psi/\partial p]_{p = \partial\Phi/\partial f} \right\rangle - \left\langle p, L^c(f)\partial\Phi/\partial f \right\rangle \qquad (9)$$

We observe that the surface $\Im$ in $T^*M x \Re$ that is defined as the image of the mapping $f \rightarrow (f, \partial\Phi/\partial f(f), \Phi(f))$ is an invariant surface (i.e. the vector field given by the right hand side of eq. (8) is tangent to $\Im$) and eq. (8) restricted to $\Im$ is exactly eq. (1). Moreover, the flow generated by eq. (8) preserves the natural contact structure of the space $T^*M x \Re$. We see that in this formulation the time evolution is generated by a potential K (involving both L^c and L^o) and the thermodynamic potential Φ characterizes the geometrical structure (i.e. the structure of the surface $\Im$) on which the time evolution takes place. The formulation (8), (9) will also allow us to use powerful geometrical methods to analyze solutions of eq. (1).

V. DYNAMICS VERSUS DYNAMICS ON A MORE MICROSCOPIC LEVEL OF DESCRIPTION (EXTERNALLY FORCED SYSTEMS)

The questions discussed in this section are the following: How do the specific heat, solubility and other equilibrium properties depend on the flow to which the fluid under consideration is submitted? What are the appropriate boundary conditions for the internal state variables φ in calculations of processing flows? We consider thus in this section externally forced systems in contrast to Section IV where we have considered isolated systems. The main difference between isolated and externally forced systems is that isolated systems approach equilibrium states but externally forced systems do not. It was precisely the approach to equilibrium states that provided us in Section IV the tool allowing to formulate and analyze the dynamics. If we want to follow the same route in this section (we regard externally forced systems as a generalization of isolated systems - see Application V.1 below) we have to identify some type of approach (as $t \to \infty$) also in externally forced systems. I have suggested [18], [23], [25] that in the context of externally forced systems the approach that replaces the approach to equilibrium states in the context of isolated systems is the approach to a more macroscopic level of description on which the time evolution still takes place. This type of the approach is indeed observed in many (possibly all) externally forced systems. We recall for example that the system considered in the Bénard problem (a layer of a fluid subjected to the gravitational force and heated from below). It is a well known experience that this externally forced system is well described by hydrodynamic equations. This means that if more microscopic quantities (like for example the one particle distribution function) are observed then the approach to the hydrodynamic description is observed.

We shall now adapt the formulation (1), (2) to this more general case. First, we extend the concept of the thermodynamic potential. Let N denote the state space used on the more macroscopic level of description, elements of N will be denoted by the symbol g. In particular case when N is the state space of equilibrium thermodynamics then $N \in \Re^2$, $g \equiv (e, n)$ (see Section III). Now we have to specify how g is expressed in terms of f. We recall that this passage from f to g has been realized in Section III by introducing two functions $E(f)$, $N(f)$. Now, we shall write formally $g = d\aleph(f)$. For example if $f \equiv f(\underline{r}, \underline{v})$ (one particle distribution function) and g are hydrodynamic fields $(\rho(\underline{r}), e(\underline{r}), \underline{u}(\underline{r}))$ then

$$\rho(\underline{r}) = \int d^3\underline{v}\, f(\underline{r}, \underline{v}),\ e(\underline{r}) = \int d^3\underline{v}\, \frac{1}{2} m v^2 f(\underline{r}, \underline{v}),\ u_\alpha(\underline{r}) = \int d^3\underline{v}\, v_\alpha f(\underline{r}, \underline{v}).$$

The thermodynamic potential generalizing the thermodynamic potential introduced in Section III is introduced by $\Phi(f; g^*) = -S(f) + \langle g^*, \aleph(f) \rangle$, where <,> denotes the scalar product in the state space N, N^* is the space dual to N and $g^* \in N^*$. The above definition of the thermodynamic potential reduces clearly to the definition introduced in Section III if $N \in \Re^2$, $g \equiv (e, n)$. The entropy $S(f)$ depends on the choice of the state space N. It will not be therefore the same as the entropy $S(f)$ introduced in Section III and used in Section IV. Besides, the entropy $S(f)$ depends also on the external forces. Some arguments that could be used to introduce the function $S(f)$ will be reviewed later in this section. We point out that $M(f)$ depends, in general also on the external forces. For example the external forces modify often the energy $E(f)$

Motivated by the results of Section IV, we suggest that the approach of the time evolution formulated in the state space M to the time evolution in a more macroscopic state space N is governed by eq. (1). Equation (2) is however replace by

$$\frac{d\,\aleph(f)}{dt} = 0 \tag{10}$$

Note that in the context of the approach to equilibrium states (Section IV) $\aleph(f) = (E(f), N(f))$ and eq. (10) reduces thus to eq. (2). In addition to eqs. (1) and (10) we have to now specify also the time evolution in the state space N. We shall write it formally as

$$\frac{\partial g}{\partial t} = R(g) \tag{11}$$

For example, in the case of the Bénard problem eq. (11) is the hydrodynamic (Boussinesq) equation describing the system. In the case considered in Section IV, there is no time evolution in the equilibrium thermodynamic state space N and eq. (11) is thus replaced simply by $\partial g/\partial t = 0$. We note that the thermodynamic potential Φ plays again the role of the Lyapunov function that is now associated with the approach to a more macroscopic dynamics.

At this stage, I do not have any illustration of eqs. (1), (10), (11) analogous to Illustration IV.1. Some incomplete illustrations concerning the Bénard problem and the kinetic theory versus hydrodynamics can be found in Refs. [23], [25]. We can however also take the point of view introduced in Illustration IV.3. We shall assume that the time evolution in the state space M is governed by eqs. (1), (10), (11), equation (11) is assumed to be known. Then, we try by using our insight into the physical system under consideration, to identify the elements of the structure involved in eqs. (1) and (10), i.e. the two tensors L^c, L^o and the thermodynamic potential Φ. Now, we shall list some arguments that can be used to introduce the entropy function $S(f)$:

(i) Zubarev [74] and MacLennan [47] have attempted to extend Gibbs' arguments on which equilibrium statistical mechanics is based to externally forced systems. These arguments then can be used to introduced $S(f)$.

(ii) Maximum entropy formalism reviewed in [45] can also be used for this purpose.

(iii) According to Muschik [54] (see also [62], $S(f) = S_{isol}(f) + \tilde{\sigma}(f)$, where $S_{isol}(f)$ is the entropy of the isolated system obtained by switching off the external forces and $\tilde{\sigma}(f)$ is the entropy produced in the process of reaching the equilibrium state of the isolated system after the external forces have been switched off.

(iv) Extended irreversible thermodynamics [55], [43], [35], [56], [36] can be regarded as being closely related to the analysis of dynamics and thermodynamics presented in this lecture. The particularity of extended irreversible thermodynamics lies in the choice of the state variables φ. In extended irreversible thermodynamics $\varphi \equiv$ irreversible fluxes of energy and momentum or in another version [11], [57], $\varphi \equiv$ the time derivatives of the hydrodynamic state variables. The particular nature of the chosen state variables and the physical insight associated with these state variables is then used to add some structure to the basic structure expressed in eqs. (1), (10), (11). The additional structure can be then used to suggest the entropy function $S(f)$. It turns out that $S(f)$ suggested in extended irreversible thermodynamics is essentially the same $S(f)$ obtained by following Muschik's argument.

(v) Let the starting time evolution equation be eq. (1) that is modified by adding to its right hand side terms expressing external forces. We look for a new thermodynamic potential $\tilde{\Phi}$ such that the additional terms expressing the external forces can be absorbed in it. The modified eq. (1) consisting of eq. (1) with the additional terms becomes eq. (1) with no additional terms but with $\tilde{\Phi}$ replacing Φ. I believe that this can be achieved only in particular cases [69], [50]. An illustration in which the dependence of specific heat on the shear rate is studied in this way can be found in Ref. [21].

(vi) Arguments that have its origin in stochastic dynamics have been introduced by Keizer [38].
(vii) In Refs. [72], [73], [63], [61] the influence of the external forces is taken into account in $\aleph(f)$.

Each of the seven types of arguments introduced above can be used to introduced the entropy function and consequently also the thermodynamic potential. I believe however that a function can be called thermodynamic potential only after it has been put into the context of dynamics in which it plays the role of the Lyapunov function associated with the approach of a one level of description to a more macroscopic level of description.

This section is much less complete than Sections III and IV. It presents more a program than results. We end this section by bringing the discussion closer to the specific context of rheological measurements. In these measurements polymeric fluids are subjected to flows controlled from outside. The flows are thus considered as the external forcing. The fluids are described by the state variable $\tilde{f}$ that is the state variable f from which the field of velocity $\underline{u}(\underline{r},t)$ has been removed. We note that if the fluids under consideration is considered to be incompressible and isothermal then $\tilde{f} \equiv \varphi$. The equation governing the externally forced polymeric fluids is eq. (1) restricted to $\tilde{f}$. We shall write it formally as

$$\frac{\partial \tilde{f}}{\partial t} = \tilde{F}(\tilde{f}; \underline{u}, \underline{\underline{\nabla u}}) \tag{12}$$

The velocity and the gradient of velocity fields $\underline{u}$ and $\underline{\underline{\nabla u}}$ are considered in eq. (11) as parameters specifying external forces. From the time evolution equation for the velocity field $\underline{u}(\underline{r},t)$, that is missing in eq. (11), we recall only the expression

$$\underline{\underline{\tau}} = \underline{\underline{\tau}}(\tilde{f}, \underline{\underline{\nabla u}}) \tag{13}$$

for the extra stress tensor $\underline{\underline{\tau}}$. We also introduce a state variable g that is more macroscopic than the state variable $\tilde{f}$ by $g = \aleph(\tilde{f})$. For example, let $\tilde{f} = \varphi \equiv$ the configuration distribution function φ and g is the conformation tensor $\underline{\underline{c}}$ (i.e. $\aleph(\psi) = \int d^3\underline{R} R_\alpha R_\beta \psi(\underline{r}, \underline{R}, t)$). The problem now is to reformulate eq. (12) in the form of eqs. (1), (10), (11). By using the Muschik's argument, we suggest that

$S(\tilde{f}) = [S(f)]_{f=\tilde{f}} - \frac{1}{\Lambda} \tau_{\alpha\beta} \partial u_\alpha / \partial r_\beta$, where $1/\Lambda$ is the relaxation time

and $[S(f)]_{f=\tilde{f}}$ is the entropy introduced in Section IV. Consequently,

$\Phi(\tilde{f};q^*) = -S(\tilde{f}) + \langle g^*, \aleph(\tilde{f})\rangle$. Note that this thermodynamic potential depends explicitly on the external forces. It means that any thermodynamic quantity obtained from the thermodynamic potential depends now on the external forces.

Finally, we indicate how this thermodynamic potential can be used to solve the problem of boundary conditions. Let for example the boundary conditions be known for the extra stress tensor $\underline{\underline{\tau}}$. The rheological model that we want to use for calculating processing flows uses $\tilde{f}$ as the state variable. We thus need boundary conditions for $\tilde{f}$. If these is a one-to-one relation between the extra stress tensor $\underline{\underline{\tau}}$ and the chosen $\tilde{f}$ (this can arise for example if $\tilde{f} = \varphi \equiv$ one conformation tensor) then the boundary conditions for $\underline{\underline{\tau}}$ imply the boundary conditions for $\tilde{f}$. If however $\tilde{f}$ is a more microscopic state variable (for example if $\tilde{f} = \varphi \equiv$ a pair of conformation tensors) then we need to express $\tilde{f}$ in terms of the extra stress tensor $\underline{\underline{\tau}}$. This can be achieved by using the thermodynamic potential $\Phi(\tilde{f};g^*)$ with $\aleph(\tilde{f})$ chosen as $\aleph(\tilde{f}) = \underline{\underline{\tau}}$. We recall that solutions of $\partial\Phi/\partial\tilde{f} = 0$ (i.e. equilibrium states on the level of description on which g serves as the state variable) are the states $\tilde{f}_{eq}$, approached as $t \to \infty$, that are expressed in terms of g. We shall then use this relation to transform the boundary conditions for $\underline{\underline{\tau}}$ into the boundary conditions for g.

VI. DEPENDENCE OF RHEOLOGICAL PROPERTIES ON TEMPERATURE AND PRESSURE

In this section we ask the following question. How do the viscosity coefficients and other quantities characterizing rheological behaviour depend on temperature, pressure and in the case of more component fluids on concentrations of the components? The setting for discussing this question is provided by eqs. (12) and (13). Temperature T and pressure P appear implicitly in eq. (12) by requiring that if the external forces (i.e. the flow in our case) are switched off then eq. (12) is solved by an equilibrium state $\tilde{f}_{eq}$ corresponding to T and P (i.e. $\tilde{f}_{eq}$ is a solution of $\partial\Phi(\tilde{f};\frac{1}{T};\frac{u}{T})/\partial\tilde{f}=0$). By inserting $\tilde{f}_{eq}$ into $\underline{\underline{\tau}}$ given by eq. (13) we get $\underline{\underline{\tau}}=0$. If we solve now eq. (12) with the external forces switched on and insert the solution into eq. (13) we obtain $\underline{\underline{\tau}}(T,P)\neq 0$.

Let us consider a particular case. Let $\tilde{f}=\varphi\equiv(\varphi_{inter},\varphi_{intra})$ and let φ_{inter} be much less influenced by the flow than φ_{intra} so that we can assume that eq. (12) is solved by $\varphi=\left((\varphi_{inter})_{eq}^{(T,P)},\varphi_{intra}\right)$, where $(\varphi_{inter})_{eq}^{(T,P)}$ is the solution of eq. (12) with no flow. If we insert now this solution into eq. (13), the extra stress tensor $\underline{\underline{\tau}}$ will become dependent on P,T through its dependence on $(\varphi_{inter})_{eq}^{(T,P)}$. In the setting of the Simha-Somcynsky theory (see Section III), $\varphi_{inter}=y$ represents the fraction of occupied cells. Utracki [70] has observed that the experimental data about $\eta(T,P)$ (we use the symbol η to denote the shear viscosity coefficient) for a large class of polymeric fluids can be well organized if the dependence of η on T and P is graphically represented as $\eta(y_{eq}(T,P))$, where $y_{eq}(T,P)$ is the equilibrium state arising in the

Simha-Somcynsky theory (i.e. $y_{eq}(T,P)$ is a solution of

$\partial\Phi(n,y;\frac{1}{T};\frac{u}{T})/\partial y = 0$, where $\Phi(n,y;\frac{1}{T};\frac{u}{T})$ is the thermodynamic potential introduced in the Simha-Somcynsky theory).

How does $\underline{\underline{\tau}}$ depend on the state variables φ? We recall that the formula (13) for the extra stress tensor $\underline{\underline{\tau}}$ arises as a consequence of eq. (1). It therefore reflects the dependence of L^c, L^o on φ. It is rare however that we have the physical insight that allows us to specify the dependence of L^c on φ, and in particular, L^o on $\partial\Phi/\partial\varphi$. In the case that we do not know this dependence we can proceed as follows. We regard Φ, L^c, L^o as unspecified functions of f and use eqs. (12) and (13) to derive some relations among results of rheological and thermodynamical measurements that will allow us, if compared withe experimental results, to specify $\Phi(f), L^c(f), L^o(\partial\Phi/\partial f)$. We shall now introduce an example that is not directly related to the T,P-dependence of $\underline{\underline{\tau}}$ but that illustrates the combinations of rheological measurements.

Illustration VI.1

Let eq. (12) be

$$\frac{\partial c_{\alpha\beta}}{\partial t} = \frac{\partial u_\alpha}{\partial r_\gamma} c_{\gamma\beta} + \frac{\partial u_\beta}{\partial r_\gamma} c_{\alpha\gamma} + \Lambda(tr\underline{\underline{c}})\, \tau_{\alpha\beta} \tag{14}$$

an eq. (13)

$$\tau_{\alpha\beta} = \rho k_\beta T\, \delta_{\alpha\beta} - 2\rho H(tr\underline{\underline{c}})\, c_{\alpha\beta} \tag{15}$$

ρ denotes the mass density, k_B is the Boltzmann constant. Both of these equations arise from eq. (1) introduced in Illustration IV.1; the mobility coefficient Λ is not assumed to be a constant now but it is an unspecified function of $tr\underline{\underline{c}}$, E(f) in eq. (6) is related to $H(tr\underline{\underline{c}})$ by

$H(tr\underline{\underline{c}}) = \frac{1}{2}\frac{dE(tr\underline{\underline{c}})}{dtr\underline{\underline{c}}}$. We note that if both Λ and H are constants then eqs. (14) and (15) represent the upper convected Maxwell model. We assume that the conformation tensor $\underline{\underline{c}}$ as well as the velocity gradient $\underline{\underline{\nabla u}}$ are independent of the position vector $\underline{r}$. An elementary analysis of eqs. (14) and (15) (see Refs. [17], [19]) shows that the following rheological predictions are independent of the choice of the functions H and Λ.

(1) The second normal stress difference equals zero. (2) $\rho k_B T N_1 = 2\tau_{12}^2$, where N_1 is the first normal stress difference. (3) The shear and the elongational viscosities are related so that the knowledge of one can be used to specify the other. The relation is expressed by the commutativity of the following diagram

$$\begin{array}{ccccc} (\tau_{12})_{sh} & \xrightarrow{\Pi_1} & \dot{\gamma} & \xrightarrow{\Pi_3} & \dot{\varepsilon} \\ \Big\downarrow \Pi_2 & & & \nearrow \Pi_4 & \\ (\tau_{11}-\tau_{22})_{el} & & & & \end{array}$$

where gradient of velocity corresponding to the shear flow is $\partial u_\alpha/\partial r_\beta = 0$ except $\partial u_1/\partial r_2 = \dot{\gamma}$ and the gradient of velocity corresponding to the elongational flow is $\partial u_\alpha/\partial r_\beta = 0, \alpha \neq \beta$ and $\partial u_1/\partial r_2 = \dot{\varepsilon}, \partial u_2/\partial r_2 = \partial u_3/\partial r_3 = -\frac{\dot{\varepsilon}}{2}$. The mapping Π_1 represents the shear viscosity (we recall that $(\tau_{12})_{sh} = -\eta(\dot{\gamma})\dot{\gamma}$ where $\eta(\dot{\gamma})$ is the shear viscosity). The mapping Π_4 represents the elongational viscosity (we recall that $(\tau_{11}-\tau_{22})_{el} = -\eta_E(\dot{\varepsilon})\dot{\varepsilon}$). The mapping Π_2 is given by

$$(\tau_{11}-\tau_{22})_{el} = (\tau_{12})_{sh}/2\rho T + (\tau_{12})_{sh}\left(\left(3(\tau_{12})_{sh}\right)^2/(2\rho T)^2 + 3\right)^{1/2}$$ and

the mapping Π_3 by $\dot{\varepsilon} = \dot{\gamma}\,(\tau_{12})_{sh}/(\tau_{11}-\tau_{22})_{el}$.

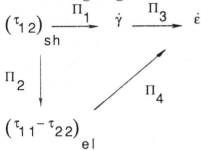

These three invariants can be then used to identify the polymer fluids that cannot be modeled by any of the models included in the family (14), (15). These will be the fluids for which the experimental results do not show at least one of the above three invariants. A more detailed analysis is then needed to actually specify the functions $H(\mathrm{tr}\,\underline{\underline{c}})$ and $\Lambda(\mathrm{tr}\,\underline{\underline{c}})$. ●

VII. CONCLUSION

What are the aspects of our methods of investigation to which we should be particularly grateful for our ability to comprehend the natural phenomena observed around us? I suggest that one of these aspects is the multilevel nature of our perception. We observe both details and overall phenomena and confront them continuously. Trying to formulate this mathematically, we note that thermodynamics is a mathematical formulation of the relation between the macroscopic description used in equilibrium thermodynamics and a more microscopic description. The thermodynamic formulation is then naturally extended to the formulation of the relationship between two general levels of description. We thus perceive thermodynamics as a mathematical formulation of the multilevel nature of our observations.

Carrying this view of thermodynamics to the investigation of polymeric fluids, we regard thermodynamics as an essential part of rheological modeling. In particular, we are led, by requiring that rheological models formulated on different levels of description are compatible, to a systematic and simple method of constructing rheological models (the method is illustrated in Illustration IV.3). We recall the principal steps of the method.

(i) Physical insight, based in particular on the information in our disposition about the molecular structure, concerning the polymeric fluid under consideration is collected.

(ii) Depending on the nature of the fluid and on our needs and intended applications, we select the state variables for the model.

(iii) The physical insight is expressed in the elements of the structure of eq. (1) (i.e. in the thermodynamic potential Φ and the two tensors L^c, L^o).

(iv) The governing equations of the rheological model together with the formula for the extra stress tensor are then obtained as the consequence of eq. (1).

(v) The process of solving the governing equations of the model (requiring often a discretization) is essentially a process of passing to a more macroscopic level of description. It is thus again a problem that can be analyzed by thermodynamical methods.

In the last two sections we have indicated the role that thermodynamics plays in the analysis of the following two questions. Let the polymeric fluid under consideration be subjected to a flow. How do its equilibrium properties like specific heat and solubility depend on the flow? How do the rheological properties like coefficients of viscosities depend on temperature and pressure?

ACKNOWLEDGEMENT

This work was supported by the National Science and Engineering Council of Canada and by the Province of Quebec.

REFERENCES

1. A. Ait Kadi, P.J. Carreau and M. Grmela, Rheologica Acta, 27, 241 (1988).
2. A. Ajji, P.J. Carreau, M. Grmela and H.P. Schreiber, J. Rheol. 33, 401 (1989).
3. A.N. Beris and B.J. Edwards, J. Rheol. (1989).
4. R.B. Bird, O. Hassager, R.C. Armstrong and C.F. Curtiss, Dynamics of Polymeric Fluids, Vol. 2, Wiley, N.Y. (1987).
5. J.W. Cahn, Acta Metal. 9, 795 (1961).
6. H.G. Casimir, Rev. Mod. Phys., 17, 343 (1945).
7. M. Doi and S.F. Edwards, The Theory of Polymer Dynamics, Oxford, Colderon (1986).
8. I.E. Dzyaloshinskii and G.E. Volovick, Ann. Phys., 125, 67 (1980).
9. D.G.B. Edelen, Int. J. Eng. Sci., 10, 481 (1972).
10. B.J. Edwards, A. Beris and M. Grmela, Generalized Constitutive Relations for Polymeric Liquid Crystals, Part I to appear in J. Non-Newtonian Fluid Mech.
11. L.S. Garcia Colin and M. Lopez de Haro, J. Noneq. Thermod., 7, 19, 95 (1982).
12. H. Giesekus, J. Non-Newtonian Fluid Mech., 11, 69 (1982).
13. M. Grmela, Talk presented at AMS Workshop, Boulder, July 1983, Contemp. Math., 28, 125 (1984).
14. M. Grmela, Polym. Eng. Science, 24, 673 (1984).
15. M. Grmela, Phys. Lett., 111, 36 (1985).
16. M. Grmela, Physica, D21, 179 (1986).
17. M. Grmela, J. Rheol., 30, 707 (1986).
18. M. Grmela, J. Chem. Phys., 85, 5689 (1986).
19. M. Grmela and P.J. Carreau, J. Non-Newtonian Fluid Mech., 23, 271 (1987).
20. M. Grmela and Chhon Ly, Phys. Lett., A120, 281 (1987).
21. M. Grmela, Phys. Lett, A120, 276 (1987).

22. M. Grmela, Phys. Lett., A130, 81 (1988).
23. M. Grmela, in Proceedings of 7th Symp. on Trends in Applications of Mathematics to Mechanics, eds. W. Eckhaus and J.F. Besseling, Springer Verlag, p. 329 (1988).
24. M. Grmela and Chhon Ly, Spatial Nonuniformities in Lyotropic Liquid Crystals, Preprint, Ecole Polytechnique de Montréal, January (1988).
25. M. Grmela, Common Structure in the Hierarchy of Models of Physical Systems, Proceedings of 6th Symp. on Continuum Models and Discrete Systems, ed. G.A. Maugin (1989).
26. M. Grmela, Phys. Lett., A137, 342 (1989).
27. M. Grmela, J. Phys., A Math. and Gen., 22, 4375 (1989).
28. M. Grmela, Contact Geometry, Thermodynamics and Nonlinear Onsager-Casimir Dynamics, submitted to Phys. Lett. A.
29. M. Grmela, Thermodynamical Lift of the Nonlinear Onsager-Casimir Vector Field, Proceedings of the Workshop on Hamiltonian Systems, Transformation Groups and Spectral Transform Methods, October (1989), CRM Université de Montréal.
30. C.L. Hand, J. Fluid Mech., 13, 33 (1962).
31. S. Hess, J. Noneq. Thermod., 11, 175 (1986).
32. D. Henderson, J. Chem. Phys., 37, 631 (1962).
33. D.D. Holm, J.E. Marsden, T. Ratiu and A. Weinstein, Phys. Reports, 123, 1 (1985).
34. R.J.J. Jongschaap, Elements of Microrheological Modeling, Dept. of Physics, Twente Univ., Enschede (1988).
35. D. Jou and J. Casas-Vazquez, J. Noneq. Therm., vol. 5, 91 (1980), 8, 127 (1983).
36. D. Jou, J. Casas-Vazquez and G. Lebon, Rep. Prog. Phys., 1105 (1988).
37. A.N. Kaufman, Phys. Lett. A., 100, 419 (1984).
38. J. Keizer, J. Chem. Phys., 82, 2751 (1985).
39. R.F. Keyser Jr. and H.P. Raveché, Phys. Rev. A., 17, 2067 (1978).
40. A.R. Khokhlov and A.N. Semenov, J. Stat. Phys., 38, 161 (1985).
41. J.G. Kirkwood in Documents in Modern Physics, edited by P.L. Auer, Gordon and Breach, N.Y. (1967).
42. F. Lado, J. Chem. Phys., 82, 4829 (1984).
43. G. Lebon, Bull. R. Soc., Belg. Clas. Sci., 64, 456 (1978).
44. A.I. Leonov, Rheol. Acta, 15, 85 (1976), 21, 683 (1982).
45. R.D. Levin and M. Tribus, editors, Maximum Entropy Formalism, MIT, Cambridge.
46. P. Liberman and C.-M. Marle, Symplectic Geometry and Analytical Mechanics, D. Reidel, Publ. Comp. (1987).
47. J.A. MacLennan, Phys. Fluids, 4, 1319 (1961).
48. W. Maier and A. Saupe, Z. Naturforschung A, 14, 882 (1959), 15, 287 (1960).
49. G.A. Maugin and R. Drouot, Int. J. Eng. Sci., 21, 705 (1983).
50. G.A. Maugin, R. Drouet and A. Morro, J. Non Newtonian Fluid Mech., 23, 201 (1987).
51. J.C. Maxwell, Philos. Mag., 35, 129, 185 (1968).
52. J.J. Moreau, C.R. Acad. Sci., Paris, 271, 608 (1970).

53. P.J. Morrison, Phys. Letters A., 100, 423 (1984), Physica D., 18, 410 (1986).
54. W. Muschik, J. Noneq. Thermod., 4, 277 (1979).
55. I. Müller, Z. Phys., 198, 329 (1967).
56. I. Müller, Thermodynamics, London Pitman (1985).
57. R.E. Nettleton and E.S. Freidkin, Physica A., 158, 672 (1989).
58. L. Onsager, Ann. N.Y. Acad. Sci., 51, 627 (1949).
59. L. Onsager, Phys. Rev., 37, 405 (1931), 38, 2265 (1931).
60. A. Peterlin, Ann. Rev. Fluid Mech., 8, 35 (1976).
61. N. Pistor and K. Binder, Colloids Polym. Sci., 200, 132 (1988).
62. I. Prigogine, Etude thermodynamique des phénomènes irréversibles, Desoer, France (1974).
63. C. Rangel-Nafaile, A.B. Metzner and K.F. Wissburn, Macromol., 17, 1187 (1984).
64. J.M. Richardson, J. Chem. Phys., 23, 230 (1985).
65. G.S. Sarti and G. Marrucci, Chem. Eng. Sci., 28, 1053 (1973).
66. K.S. Schweitzer and J.G. Curro, J. Chem. Phys., 91, 5059 (1989).
67. R. Simha and T. Somcynsky, Macromolecules, 2, 342 (1969).
68. R. Simha, J. Rheol., 30, 700 (1986).
69. T.J. Sluckin, Macromol., 14, 1676 (1981).
70. L.A. Utracki, in Advances in Rheology, Vol. 1, p. 353, editors: B. Mena, A. Garcia-Rejon and C. Rangel-Nafaile, UNAM, Mexico (1984).
71. N.G. van Kampen, Physica A., 67, 1 (1973).
72. G. Ver Strate and W. Philipoff, J. Polym. Eng. Sci., 14, 401 (1974).
73. B.A. Wolf, Macromol., 17, 615 (1984).
74. D.N. Zurabev, Nonequilibrium Statistical Mechanics, Consultant Bureau, N.Y. (1974).

Author Index

Subject Index